T0235197

Insect Physiological Ecology

Insect Physiological Ecology

Mechanisms and Patterns

Steven L. Chown
University of Stellenbosch

Sue W. Nicolson
University of Pretoria

OXFORD

UNIVERSITY PRESS

Great Clarendon Street, Oxford OX2 6DP

Oxford University Press is a department of the University of Oxford.
It furthers the University's objective of excellence in research, scholarship,
and education by publishing worldwide in

Oxford New York

Auckland Cape Town Dar es Salaam Hong Kong Karachi
Kuala Lumpur Madrid Melbourne Mexico City Nairobi
New Delhi Shanghai Taipei Toronto

With offices in

Argentina Austria Brazil Chile Czech Republic France Greece
Guatemala Hungary Italy Japan Poland Portugal Singapore
South Korea Switzerland Thailand Turkey Ukraine Vietnam

Oxford is a registered trade mark of Oxford University Press
in the UK and in certain other countries

Published in the United States
by Oxford University Press Inc., New York

© Oxford University Press 2004

The moral rights of the author have been asserted
Database right Oxford University Press (maker)

First published 2004

All rights reserved. No part of this publication may be reproduced,
stored in a retrieval system, or transmitted, in any form or by any means,
without the prior permission in writing of Oxford University Press,
or as expressly permitted by law, or under terms agreed with the appropriate
reprographics rights organization. Enquiries concerning reproduction
outside the scope of the above should be sent to the Rights Department,
Oxford University Press, at the address above

You must not circulate this book in any other binding or cover
and you must impose the same condition on any acquirer

British Library Cataloguing in Publication Data
Data available

Library of Congress Cataloging in Publication Data
Data available

Typeset by Newgen Imaging Systems (P) Ltd., Chennai, India
Printed in Great Britain
on acid-free paper by
Antony Rowe Ltd, Chippenham, Wilts.

ISBN 0 19 851548 0 (Hbk)
ISBN 0 19 851549 9 (Pbk)

10 9 8 7 6 5 4 3 2 1

Preface

Living in the Southern Hemisphere has several advantages. It means ready access to relatively wild, undisturbed, and animal-rich areas, a climate where the seasonal variation is usually not vast but where every year brings something of a surprise, and an extensive body of water, the Southern Ocean, that is alive with riches should one choose to venture out onto it. However, living in the south also has several disadvantages. Like-minded scientists often live a day or more's travel away, editors have until recently tended to view the equator as something of an asymmetric barrier, and local biological meetings almost inevitably leave certain areas, such as insect physiology, scratching about for an audience, if not for speakers. It is the latter disadvantage that provided the initial impetus for this book. The field of insect physiological ecology, at least in southern Africa, is small. In consequence, we regularly found ourselves in the same sessions at the biennial zoological or entomological meetings, each presenting a rather different perspective on physiological ecology: one more mechanistic, the other more comparative. Our interactions over at least a decade of meetings, both locally and abroad, have shown that these perspectives have much to offer each other. Moreover, our discussions have mirrored a growing realization in the field as a whole that an integration of mechanistic and broader-scale comparative physiology can result in novel and unexpected insights. One of the main points of this book is to explore these insights. Along the way, we also hint at the idea that the Southern Hemisphere might differ from the north in more ways than those alluded to above. These differences are not only biologically interesting, but might also have profound consequences for the way in which humans attempt to manage the global experiment they have set in motion.

Like most other authors we have an intellectual debt. We owe much of the way we think about the world to discussions with Andy Clarke, Peter Convey, Trish Fleming, Kevin Gaston, Allen Gibbs, Sue Jackson, Jaco Klok, John Lighton, Gideon Louw, Lloyd Peck, Brent Sinclair, Ken Storey, Ben-Erik Van Wyk, and Karl Erik Zachariassen. We are grateful to them for the insights they have readily contributed. SLC is particularly indebted to Kevin Gaston for sharing many ideas and for the thousands of messages that have crossed the equator. Brent Sinclair and Ken Storey gave us permission to use figures they had drawn. Kevin Gaston, Allen Gibbs, Ary Hoffmann, Sue Jackson, Melodie McGeoch, Brent Sinclair, Graham Stone, and Art Woods read one or more chapters or the entire manuscript. We are grateful for their comments, which were helpful, challenging and insightful.

We thank the following for permission to reproduce figures from works they have published: Academic Press, the American Association for the Advancement of Science, the American Institute of Biological Sciences, the American Physiological Society, Blackwell Publishing, Cambridge University Press, the Company of Biologists, the Ecological Society of America, Elsevier, Kluwer, the National Academy of Sciences of the USA, Oxford University Press, the Royal Society, the Society for Integrative and Comparative Biology, the Society for the Study of Evolution, Springer-Verlag, the University of Chicago Press and Wiley Interscience; also Tim Bradley for permission to reproduce Fig. 4.8. We are grateful to Sheena McGeoch and Anel Garthwaite for assistance with the figures.

Ian Sherman and Anita Petrie looked after publication. Ian provided much encouragement and advice throughout the project, and was unperturbed by our recent moves around South Africa, and one of us being appointed to departmental chair.

Our families were uniformly supportive, and endured busy weekends and few holidays, for which we are grateful.

Stellenbosch and Pretoria
December 2003

Contents

CHAPTER 1

Introduction

The goal of science is a consensus of rational opinion over the widest possible field.

Ziman (1978)

Few, if any, insect species occur everywhere. At global scales, geographic ranges are generally small, and only a handful of species is distributed across several continents or oceans. Among the many reasons for this preponderance of narrow geographic distributions are the major barriers presented by continental and oceanic margins and the limited dispersal powers of many species. However, even within continents, range size frequency distributions are right-skewed. Most species have ranges much smaller than the total size of the continent (Fig. 1.1), which indicates that geographic ranges are limited.

The factors responsible for range size limitation have been much debated. For example, MacArthur (1972) suggested that for Northern Hemisphere species, biotic factors, such as predation, disease, or competition, limit ranges to the south, whereas the

northern limits of species are set by abiotic factors such as temperature extremes. The ecological reasons for species' borders and the hypotheses proposed to explain borders are diverse (Hoffmann and Blows 1994). Horizontal (e.g. competition) and vertical (e.g. parasitism) interactions between species are thought to be significant factors limiting distributional ranges, as is the influence of the abiotic environment (e.g. Quinn *et al.* 1997; Davis *et al.* 1998; Hochberg and Ives 1999; Gaston 2003). Moreover, in many instances species interactions and abiotic conditions probably combine to produce range margins (Case and Taper 2000). This complexity means that range margins are unlikely to be set for the same reasons in any two species. Nonetheless, abiotic conditions probably play a role in setting at least a part of many species geographic range limits (Gaston 2003). That is, species are unable to

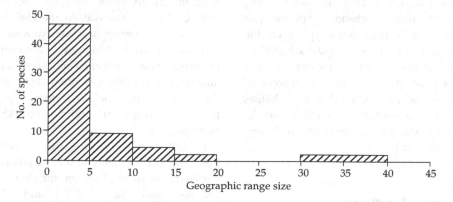

Figure 1.1 Species range size frequency distribution for keratin beetles (Trogidae) in Africa. The units are numbers of grid cells of 615 000 km².

Source: Gaston and Chown (1999*b*).

survive and reproduce under the full range of abiotic conditions that might be found on a continent. Quite why this view is held is largely a consequence of clear instances of species borders being correlated with a given climatic variable (e.g. Robinson *et al.* 1997), species range shifts being associated with changes in climate (usually temperature), both currently (Parmesan *et al.* 1999) and in the past (Coope 1979) (reviewed in Chown and Gaston 1999), and experimental work involving caging studies ($\approx$ reciprocal transplants) showing that individuals often struggle to survive a short distance beyond their natural ranges (Jenkins and Hoffmann 1999).

These findings, in turn, raise the question of why species have been unable to alter their tolerances of abiotic conditions and in so doing expand their ranges. From an evolutionary perspective there are several reasons why this might be the case, including genetic trade-offs, low heritability, low levels of genetic variation, and the swamping of marginal by central populations (reviewed in Hoffmann and Blows 1994; Gaston 2003). Indeed, there appear to be many grounds for assuming that species might never alter their physiologies in response to environmental change. However, this has clearly not happened, or else large parts of the Earth would be uninhabited. Thus, although there are limits to the conditions that insects can tolerate, they are not helpless in the face of changing abiotic conditions. Apart from moving, which many species clearly do, insects can respond to changing conditions in two ways. Over the short term, they can do so by means of phenotypic plasticity or short-term changes in their phenotype. Over the longer term their responses include adaptation, either of basal responses to the environment or of the extent of plasticity, or both. The outcome of these responses and the limited dispersal abilities of most organisms are what we see today—spatial and temporal variation in diversity, including physiological diversity.

1.1 Physiological variation

There are many ways in which physiological diversity (or variation) is manifested (Spicer and

Gaston 1999). Individuals vary through time for several reasons. Some individuals appear to be intrinsically variable, such as those of the cockroach *Perisphaeria* sp. (Blaberidae), where single individuals can show as many as four gas exchange patterns at rest (Fig. 1.2) (Marais and Chown 2003). Individual-level variation also arises as a consequence of ontogeny. For example, cold hardiness varies substantially between larval instars and between adults and larvae in sub-Antarctic flies (Vernon and Vannier 1996; Klok and Chown 2001). A variety of physiological traits also show changes associated with responses to the immediate environment. Indeed, such individual-level variation has been demonstrated in many species over a variety of time-scales, ranging from a few hours to an entire year (Lee *et al.* 1987; Hoffmann and Watson 1993; van der Merwe *et al.* 1997; Klok and Chown 2003). This phenotypic flexibility can be reversible, in which case it is considered acclimation or acclimatization, or fixed, and is then referred to as a developmental switch or polyphenism (Huey and Berrigan 1996; Spicer and Gaston 1999).

Cross-generational effects are also a form of phenotypic flexibility. Although the extent and significance of cross-generational effects, or the influence of parental or grandparental environmental history on an individual, has been widely examined for insect life history traits, the same is not true for physiological characters. Nonetheless, there have been some investigations of this kind, mostly involving *Drosophila* flies (Huey *et al.* 1995; Crill *et al.* 1996; Watson and Hoffmann 1996). Parental and grandparental exposure to stress can have substantial influences not only on early offspring fitness, but also on development time, and these responses can often differ in size and direction between male and female parents and between grandparents and parents (Magiafoglou and Hoffmann 2003).

Physiological variation is also a consequence of genotypic diversity. While this variation can differ considerably depending on the trait of interest (Falconer and Mackay 1996), and whether the species is clonal (parthenogenetic) or not, it is nonetheless of considerable importance. Consistent among-individual variation (which can be assessed as repeatability of a trait) is a prerequisite for natural selection (Endler 1986), which in turn is one of the major reasons why there is such a range of physiological diversity today. Curiously, the relationship between within- and among-individual variation in insect physiological traits has not been widely assessed despite its importance in determining whether the conditions for natural selection are met (Bech *et al.* 1999; Dohm 2002). It might be argued that laboratory selection (see Gibbs 1999) has adequately demonstrated that the conditions for selection have been fulfilled, but this is true only of a restricted set of traits and a relatively small number of taxa which are at home in the laboratory. For many other taxa, adaptation is simply assumed. To date, investigations of repeatability in insect physiological traits have largely been restricted to gas exchange and metabolic characteristics (Buck and Keister 1955; Chappell and Rogowitz 2000; Marais and Chown 2003), which have shown that repeatability tends to be both significant and high.

Individuals also vary through space as a consequence either of their membership of different populations of a given species or as consequence of their species identity. Indeed, inter-population and interspecific differences in physiological traits have been widely examined in insects over many years, and this comparative approach forms a cornerstone of physiology (Sømme 1995; Chapman 1998; Nation 2002). While the criteria by which this variation is judged to be adaptive have become more stringent (Kingsolver and Huey 1998; Davis *et al.* 2000), and the tools on which these decisions can be based more sophisticated (Garland *et al.* 1992; Freckleton *et al.* 2002), spatial variation in physiological characteristics, over large (Fig. 1.3), and small (Fig. 1.4) scales, continues to form much of the foundation for modern understanding of the evolution of physiological diversity (Spicer and Gaston 1999; Feder *et al.* 2000a).

1.2 How much variation?

Given that physiological traits can vary with time of day (Sinclair *et al.* 2003a), season (Davis *et al.* 2000), instar (Klok and Chown 1999), stage (Vernon

and Vannier 1996), and through space, it might be tempting to conclude that physiological traits are highly subtle in their variation. In other words, that insect physiological traits show such a bewildering complexity of variation (Hodkinson 2003), that they defy anything other than a species by species, stage by stage, season by season investigation, if their evolution and variation are to be comprehended.

If this were the case, physiological ecology could claim to know very little about insects. The reasoning is simple. Despite the many studies that have been done on a broad array of physiological traits, comprehensive knowledge is available for just a few taxa. These are the vinegar fly (*Drosophila melanogaster*), honeybee (*Apis mellifera*), Colorado potato beetle (*Leptinotarsa decemlineata*), migratory locust (*Locusta migratoria*), American cockroach (*Periplaneta americana*), tobacco hornworm (*Manduca sexta*), South American assassin bug (*Rhodnius prolixus*), and the mealworm (*Tenebrio molitor*) (see Chown *et al.* 2002a). Much of modern physiological understanding, which forms the foundation for physiological ecology, is based on investigations of these and a few other species (e.g. Grueber and Bradley 1994). However, even conservative estimates place insect species richness at roughly two million (Gaston 1991). Do these estimates mean that little is known about insect physiological responses to their environments, and that generalizations are not possible? To paraphrase Lawton (1992), the answer plainly lies in whether there are 10 million kinds of insect physiological responses. In our view there are not (see also Feder 1987). Rather, just as there is only limited variation about several population dynamic themes (Lawton 1992), there is limited variation about several major physiological themes for the traits in question (Chown *et al.* 2003). Several lines of evidence support this view:

1. Recent work has demonstrated that a considerable proportion of the variation in several physiological traits is partitioned at higher taxonomic levels (Chown *et al.* 2002a) (Table 1.1). Indeed, it is clear that a majority of the variation is often partitioned above the species level, and often at the family and order levels. This would not be

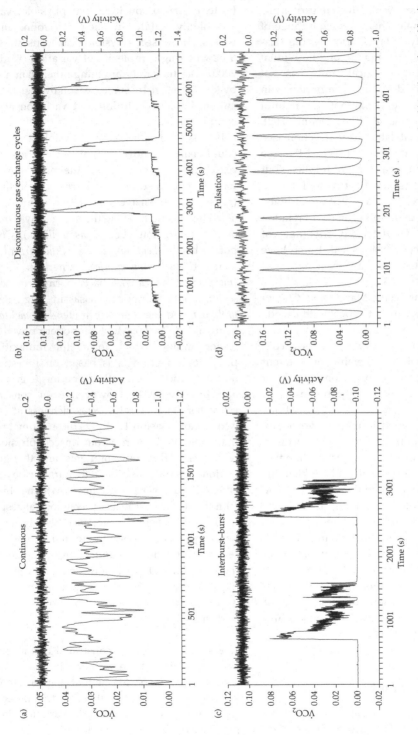

Figure 1.2 Gas exchange patterns shown by *Perisphaeria* sp. (Blattaria, Blaberidae) at rest: (a) continuous gas exchange, (b) discontinuous gas exchange cycles, (c) interburst–burst, (d) pulsation. In each case, $\dot{V}CO_2$ is shown as the lower curve (left axis) and activity as the upper curve (right axis). Activity is interpreted as the variance of activity about the mean value, rather than the absolute value of this activity, and it is negligible.

Source: Marais and Chown (2003).

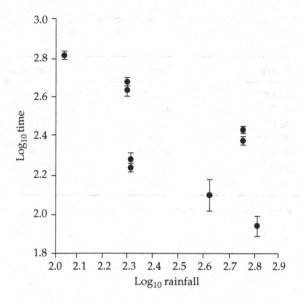

Figure 1.3 Variation in survival time (mean ± SE) under dry conditions in nine keratin beetle species across a rainfall gradient in southern Africa.

Source: Data from Le Lagadec *et al*. (1998).

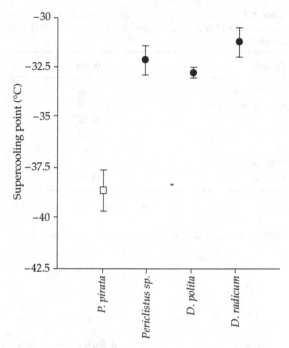

Figure 1.4 Supercooling points (or crystallization temperatures) (mean ± SE) for gall wasp (Hymenoptera, Cynipidae) prepupae overwintering above (open square) and below (filled circles) the snow at a single site (Sudbury, Ontario, Canada).

Source: Modified from Williams *et al*. (2002).

possible if physiological characteristics were species specific. Moreover, these findings are not unusual, at least as far as can be ascertained from studies of the levels at which life history traits are partitioned in vertebrates (Read and Harvey 1989), and from other investigations at lower levels for insects (Chown *et al*. 1999).

2. Even though there is a continuum of variation in many traits, clear physiological strategies can be identified. For example, despite considerable variation in the responses of insects to cold, insects can be clearly categorized as those that are freezing tolerant, those that are freeze intolerant, and one (but probably more) which survive subzero temperatures using cryoprotective dehydration (Sinclair *et al*. 2003*b*). Similar categorizations have been developed for responses to hypoxia, osmoregulation, and thermoregulation.

3. Insects do not occupy all of the available physiological 'space'. In other words, evolution has restricted occupation of all permutations of complexity to a limited diversity of workable solutions given the physical laws and biotic history

of the planet. For example, survival of low temperatures in insects is mediated by polyhydric alcohols and antifreeze proteins (Denlinger and Lee 1998), but not by molar quantities of salts or by the development of a subcutaneous fat layer (the latter being a solution that is open to and has been used by a variety of mammals).

4. Large-scale, geographic variation in traits can be detected, despite all the variation associated with feeding status, season, age, and taxon. For example, water loss rates vary with total rainfall (Addo-Bediako *et al*. 2001), extreme lower lethal limits decline with proximity to the poles (Addo-Bediako *et al*. 2000), and there is a weak, though significant, negative relationship between standard metabolic rate and environmental temperature (Fig. 1.5). This variation would not be detectable if insects showed a bewildering, non-understandable array of responses to the environment.

Table 1.1 Distribution of variance in supercooling point (SCP) of freezing tolerant and freeze intolerant insects, lower lethal temperature (LLT) of freezing tolerant insects, upper lethal temperature (ULT), critical thermal maximum (CT_{max}), metabolic rate, and water loss rate in insects

Variable	Order	Family	Genus	Species
Freeze intolerant SCP	18.68*	32.47**	33.14**	11.71
Freezing tolerant SCP	13.18	40.85**	44.12**	1.85
Freezing tolerant LLT	20.26	46.87**	0.06	32.81
Upper lethal temperature	0.72	46.61**	29.89**	22.78
CTmax	3.8	12.62	57.72**	25.86
Metabolic rate	21**	31**	28**	20
Water loss rate	13.95	24.51**	29.56**	31.98

Note: Tabulated values are percentage of the total variance accounted for at each successive level. The species level includes the error term in the data. *$p < 0.05$; **$p < 0.01$.

Source: Chown *et al.* (2002*a*).

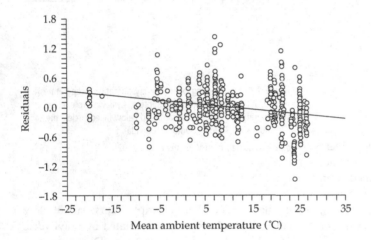

Figure 1.5 Scatterplot showing the negative relationship in insects between mean annual temperature and metabolic rate, expressed here as the residuals from a generalized linear model including body mass, trial temperature, respirometry method, and wing status. The fitted line serves to illustrate the trend.

Source: Addo-Bediako *et al.* (2002). *Functional Ecology* **16**, 332–338, Blackwell Publishing.

Nonetheless, the fact that there is variation about the major physiological themes has several important consequences. Most significant among these is that a comprehensive understanding of the ways in which insects respond physiologically to the world around them cannot be achieved solely by the study of model organisms (Feder and Mitchell-Olds 2003). The grounds for this view are not only rooted in the arguments provided above, but also go beyond these. In the first instance, it is clear that spatial variation among species at large geographic scales means that more than a handful of species must be investigated if this variation and its ecological implications are to be comprehended (Section 1.3). Indeed, most model taxa share a set of characteristics that are certainly not representative of all species. Model taxa tend to be widely distributed, fast-growing, relatively generalist in their food preferences, and not particularly fussy about the conditions under which they will survive, develop, and reproduce. If this were not the case, they would not be handy laboratory subjects, or model organisms. These very characteristics make them different from most other species.

Among the insects, most species are narrowly distributed, many feed on a limited range of host plants, and comparatively slow growth rates and long generation times are not unusual. Moreover, the large majority of species are less than happy to make anything but the most salubrious of laboratory environments their home. These characteristics are reflected in their physiologies too, and consequently they contribute substantially to physiological diversity. That there are likely to be important differences in the physiologies of common (and weedy) insects, and those that are rarer (and less ruderal) has long been appreciated. For example, both Lawton (1991) and Spicer and Gaston (1999) have argued that a comprehensive understanding of physiological diversity requires investigations of rare and common species. They have also suggested that such investigations could contribute substantially to our understanding of why some species are rare and others common, and why among the rare species some can survive rarity while others decline out of it and into extinction (Gaston 1994). These differences might also have potentially significant consequences for understanding the likely responses of species to environmental change (Hoffmann *et al.* 2003*a*).

Furthermore, while laboratory selection using model organisms is a useful approach for investigating the evolution of physiological attributes, and one that can circumvent many of the problems associated with comparative studies (Huey and Kingsolver 1993; Gibbs 1999), it is not without its problems, including those of repeatability and laboratory adaptation (Harshman and Hoffmann 2000; Hoffmann *et al.* 2001*a*). Among the insects, most laboratory selection experiments have also been based on a single insect order, the Diptera, and indeed often on a single family, the Drosophilidae. Because variation in many physiological characteristics is partitioned at the family and order levels (Chown *et al.* 2002*a*), results based on a single family or order might not be reliably generalized across the insects. Therefore, the generality of the findings based on model organisms must obviously be confirmed by investigations of physiological traits in selected higher taxa, and especially in non-model organisms that are often unwilling inhabitants of the laboratory (Klok and Chown 2003). If such investigations are undertaken within a modern comparative framework, they can complement and validate laboratory selection experiments (Huey and Kingsolver 1993), and supplement comparative work on other taxa. Thus,

by broadening the scope of comparative work (and laboratory selection), a comprehensive understanding of physiological variation can be achieved (Kingsolver and Huey 1998).

Investigations at a variety of scales are required to develop a full picture of the patterns of physiological variation and the mechanisms underlying them (Chown *et al.* 2003). These range from mechanistic molecular studies (Flannagan *et al.* 1998; Jackson *et al.* 2002; Storey 2002), through intraspecific comparisons, to broad-scale comparative analyses. To some extent, this approach presupposes that the question being posed determines the scale of the study (as does the magnification used in a microscopic examination—10 000× is just no good for understanding variation in the body form of dung beetles; Chown *et al.* 1998). Combining different-scale investigations into a single study can also be particularly powerful, as recent work on the likely responses of *Drosophila birchii* to environmental change has shown. This species shows geographic variation in desiccation resistance that tracks environmental variation in water availability. However, laboratory selection experiments have indicated that populations at the range margin are unlikely to be able to further alter their resistance owing to a lack of genetic variation in this trait (Hoffmann *et al.* 2003*a*).

1.3 Diversity at large scales: macrophysiology

While it is clear that studies at a variety of scales are required to understand physiological variation and its causes, studies at large geographic scales (in the ecologist's sense of large scale—that is, covering large geographic areas such as whole continents) are comparatively recent. Similar approaches were occasionally adopted in the past (Scholander *et al.* 1953), and qualitative comparisons of insect responses across large scales have also been undertaken since then (e.g. Zachariassen *et al.* 1987). However, extensive quantitative studies of broad-scale physiological variation are new.

The impetus for this work has come, first, from macroecology: the investigation of large-scale variation in species richness, and species abundances, distributions, and body sizes, through space and time (Brown and Maurer 1989; Gaston and Blackburn 2000). One of the many strengths of macroecology is its ability to reveal 'where the woods are, and why, before worrying about the individual trees' (Lawton 1999). In doing so, macroecological studies make several assumptions, some of which have to do with large-scale variation in physiological responses. It is these assumptions that have prompted large-scale investigations of variation in insect physiological traits, which likewise worry less about the trees than the woods.

For example, one of the explanations proposed for the latitudinal gradient in species richness relies on the Rapoport effect, or the increase in the latitudinal range sizes of species towards higher latitudes (Stevens 1989). The mechanism that is apparently largely responsible for producing this macroecological pattern is straightforward. To survive at higher latitudes individual organisms need to be able to withstand greater temporal variability in climatic conditions than at lower latitudes. This idea, known as the climatic variability or seasonal variability hypothesis, rests on two critical assumptions. First, that towards higher latitudes climates become more variable (Fig. 1.6). Second, that species at higher latitudes have wider climatic (physiological) tolerances than those at lower latitudes (Addo-Bediako et al. 2000). The second assumption plainly can only be examined

by adopting a large-scale comparison of physiological tolerances. Few studies have made such broad-scale comparisons, but in those cases where this has been done the physiological variability assumption appears to have been met (Gaston and Chown 1999a; Addo-Bediako et al. 2000) (Fig. 1.7).

A further example of macroecological questions that rely on information concerning geographic variation in physiological traits concerns spatial variation in body size (Gaston and Blackburn 2000). The mechanisms underlying body size variation, at either the intraspecific or interspecific levels, are firmly rooted in assumptions concerning geographic variation in physiological traits. These assumptions generally have (incorrectly) to do with thermoregulation, or correctly in the case of insects, with interactions between the temperature dependence of growth and development, and resource availability (Chown and Gaston 1999).

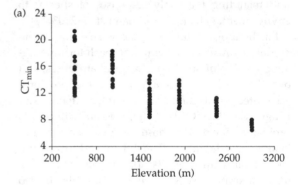

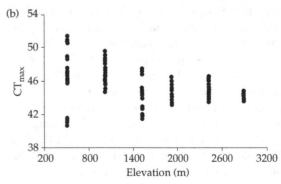

Figure 1.7 The relationship between elevation and (a) critical thermal minimum (CT_{min}), and (b) critical thermal maximum (CT_{max}) for dung beetles collected at six localities across a 2500 m elevational range in southern Africa.

Source: Gaston and Chown (1999). *Oikos* **86**, 584–590, Blackwell Publishing.

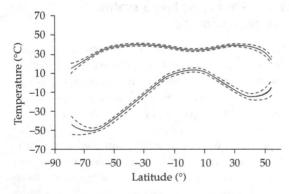

Figure 1.6 Best fit polynomial regression lines (±95%) showing the relationship between latitude and absolute maximum (top line) and absolute minimum (bottom line) temperatures across the New World (negative latitudes are north of the equator).

Source: Gaston and Chown (1999). *Oikos* **86**, 584–590, Blackwell Publishing.

Impetus for large-scale, geographic studies has also come from renewed interest in the utility of this approach for understanding physiological diversity. Indeed, it provides the only means to test several important hypotheses concerning physiological variation (such as interspecific metabolic cold adaptation; Addo-Bediako *et al.* 2002), and a useful way to address several others, such as the influence of climatic variability on physiological traits (Lovegrove 2000). In the latter case, geographically extensive studies have demonstrated that broad-scale environmental variation can influence insect physiological traits in ways that have previously gone unrecognized. Perhaps the most significant examples in this regard are the likely influence that climatic variability has had on hemispheric variation in thermal tolerances (Sinclair *et al.* 2003c) and rate-temperature relationships (Addo-Bediako *et al.* 2002; Chown *et al.* 2002a).

Large-scale studies have been given additional momentum by the recent development of nutrient supply network models as a means to account for the apparent prevalence of three-quarter scaling of metabolic rate (and other physiological rates) in all organisms (West *et al.* 1997). The extent to which the 'quarter-power' scaling assumed by these models applies to all organisms has been questioned (Dodds *et al.* 2001), and their utility for explaining global variation in species richness (Allen *et al.* 2002) has also come under considerable scrutiny. However, the models have rekindled interest in and excitement about the mechanisms underlying the scaling of physiological traits, and the broader ecological implications of physiological trait variation.

These examples make it clear that macrophysiology, or the investigation of variation in physiological traits over large geographic and temporal scales, and the ecological implications of this variation (Chown *et al.* 2004), are now an essential component of physiological investigations.

1.4 Growing integration

By the standards of only a few decades ago, the variety of approaches that can now be used to explore ecological success and its evolution, and consequently the mechanisms underlying patterns in biodiversity, is remarkable. These include functional genomics (Bettencourt and Feder 2002; Anderson *et al.* 2003; Feder and Mitchell-Olds 2003), genetic engineering (Feder 1999), comparative biochemistry (Mangum and Hochachka 1998; Storey 2002), laboratory selection (Gibbs 1999), evolutionary physiology (Feder *et al.* 2000a), and macrophysiology (Chown *et al.* 2004). They also include integrated investigations in which the need to simultaneously explore several different characteristics at a variety of levels is recognized, irrespective of whether these levels include behavioural, morphological, or life history traits (Chai and Srygley 1990; Kingsolver 1996; Jackson *et al.* 2002; Ricklefs and Wikelski 2002; Huey *et al.* 2003). Moreover, the diversity of approaches and opportunities for their integration seem set to continue their expansion as sequences of entire genomes become available for increasingly large numbers of species (Levine and Tjian 2003), and transcription profiling becomes more common. This situation is very different to the one which characterized much of twentieth-century physiological ecology, when physiology and ecology largely developed along separate paths (Spicer and Gaston 1999).

The benefits of integration of studies across the ecological and genetic hierarchies (Brooks and Wiley 1988) are likely to be profound. Such work will not only provide comprehensive understanding of the evolution of wild organisms *in situ*, but will also facilitate prediction of the outcomes of unintentional large-scale experiments such as climate change, biotic homogenization, and habitat destruction (Feder and Mitchell-Olds 2003). This is, perhaps, best illustrated by work on the adaptation of *Drosophila* to environmental extremes (Hoffmann *et al.* 2003b; Lerman *et al.* 2003), and the likely consequences of these responses for *Drosophila* diversity (Hoffmann *et al.* 2001b; Hoffmann *et al.* 2003a). Here, the distinction between classical model organisms and wild species is being blurred, as Feder and Mitchell-Olds (2003) argued it should be. Moreover, this integrated approach is genuinely able to address some of the major concerns precipitated by human manipulation of the environment, such as how different groups of species (e.g. generalists versus specialists) might respond to a changing environment (Hill *et al.* 2002), and how dispersal ability itself is likely to evolve (Zera and Denno 1997; Davis and Shaw 2001; Thomas *et al.* 2001).

1.5 This book

Given the growing integration of insect physiological ecology, the objectives of this book are threefold. First, to examine interactions between insects and their environments from a physiological perspective that integrates information across a range of approaches and scales. This will be done first by describing physiological responses from the molecular level through to that of the individual. Clearly, the degree to which this can be achieved varies with the trait in question. For example, it is relatively straightforward to develop this approach for thermal tolerance because of the immense amount of work at the sub-individual level in *Drosophila*, where the genes underlying thermal responses are known and their locations on the chromosomes have been identified (Bettencourt and Feder 2002; Anderson *et al.* 2003). However, in the case of other traits, such as interactions between gas exchange and metabolism (Chown and Gaston 1999), molecular level work lags far behind the level of understanding available at the tissue, organ, and whole-individual levels.

The second objective is to demonstrate that evolved physiological responses at the individual level are translated into coherent patterns of variation at larger, even global scales. Such variation cannot be detected using the small-scale, mechanistic approaches that are the cornerstone of much physiology. However, modern statistical and computer-based visualization methods (such as Geographic Information Systems—see Liebhold *et al.* 1993), make this large-scale approach relatively straightforward. Crucially, the value of applying this technique depends on the quality and spatial extent of the available data, a theme that is explored towards the end of the book. Because we are regularly concerned with taxonomic variation in traits, we provide two modern phylogenies as a context within which the animals and traits we

discuss can be viewed—one for the arthropods (Fig. 1.8), and one for the insect orders (Fig. 1.9).

The third objective is realized as something of an epiphenomenon. This is to draw attention to the ways in which methods and measurement techniques might have an influence on the conclusions drawn by a particular study. It is becoming increasingly apparent that what is done in a given experiment to a large extent determines the outcome. To some, this might seem obvious as the rationale for experimentation. However, these effects can be subtle, and can reflect investigations of different traits when this was not the initial intention. The best example of this physiological uncertainty principle is the measurement of heat shock, where ostensibly similar methods provide information on different traits.

We begin in Chapter 2 with nutritional physiology and ecology. All organisms require a source of energy, which can then be partitioned between the multitude of tasks that allow them to respond to environmental challenges and to strive to contribute their genes to the next generation. Here, we also examine development and growth because of its dependence on food quantity and quality, but take cognizance of the fact that even under ideal nutritional circumstances, insects can manipulate growth and development rates in response to other environmental cues (Nylin and Gotthard 1998; Margraf *et al.* 2003). Chapter 3 concerns metabolism and gas exchange. In general, insects make use of aerobic metabolism to catabolize substrates, and as a consequence, they have developed a sophisticated system for gas exchange that also has attendant problems of water loss and the likelihood of enhanced oxidative damage. Nonetheless, insects are also adept at living under hypoxic conditions, and metabolism under these circumstances is explored. Variation in the costs of living owing to variation within individuals depending on activity, between individuals as a consequence of size, and between populations and species is also explored.

Because water availability and temperature are two of the most significant environmental variables influencing the distribution and abundance of insects (Rogers and Randolph 1991; Chown and Gaston 1999), Chapters 4, 5, and 6 explore the ways in which insects respond to these abiotic variables.

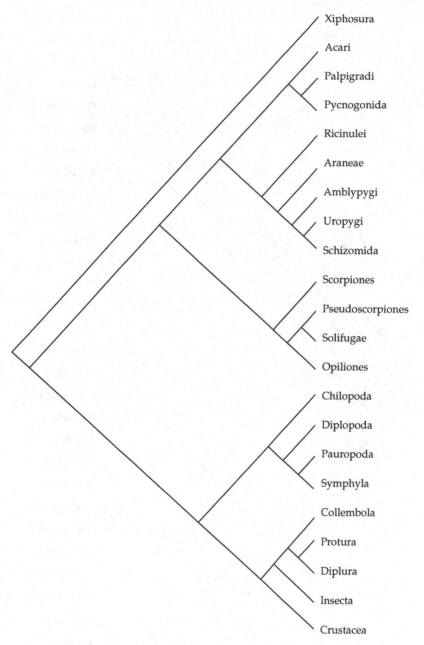

Figure 1.8 Composite phylogeny for the major arthropod taxa.
Note: Several of the clades are unstable, and relationships between some major taxa have yet to be resolved.
Source: Compiled from Giribet *et al.* (2001, 2002).

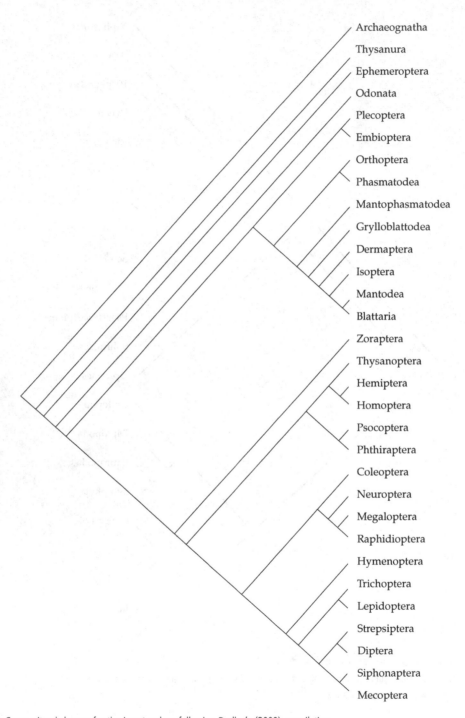

Figure 1.9 Composite phylogeny for the insect orders, following Dudley's (2000) compilation.

Note: The Mantophasmatodea is placed between the clade including the Phasmatodea and the one including the Grylloblattodea, based on the interpretations provided by Klaas *et al.* (2002), although they note that the phylogenetic placement of this new order has not been resolved.

Chapter 4 concerns not only the routes of water loss from insects, but also the ways in which water is obtained, particularly in species that have different feeding habits and which occupy different environments. Moreover, because osmoregulation is inseparable from water balance, the mechanistic basis of osmoregulation is also examined. This chapter is concluded with a discussion of the considerable insights that have been provided from laboratory selection and comparative studies of *Drosophila* (Gibbs *et al.* 2003*a*). Chapter 5 deals with responses of insects to temperature. Because the mechanisms underlying responses to high and low temperatures, but particularly heat and cold shock, are similar in several respects, responses to high and low temperatures are considered together. In Chapter 6, the focus is on behavioural and physiological thermoregulation and its consequences. Because this field has been repeatedly reviewed, rather than focusing on the well-known examples, we draw attention particularly to recent advances in understanding of thermoregulation. We also discuss the ecological implications of thermoregulation, especially in terms of competition and predation.

In Chapter 7 we synthesize information on the spatial extent of the physiological information that is available on insects. We highlight the fact that large-scale studies, in combination with mechanistic work, suggest that responses to temperature in insects might be different to those in other ectotherms. We also show that understanding of the evolution of body size, a trait that is of considerable ecological and physiological significance (Peters 1983), cannot be separated from investigations of several physiological traits. Moreover, these traits not only act in concert to determine body size, but body size in turn constrains these physiological variables. We conclude by drawing attention to the likely responses of insects to global climate change, and the role of an integrated insect physiological ecology in providing a basis for understanding and predicting these responses.

Before moving on, we offer a word of caution. Those readers who are enthusiastic about hormonal regulation, muscle physiology, neuronal functioning, sensory perception, sclerotization, the physics of flight, and the intricacies of insect clocks and diapause, will not find these topics discussed here. We make reference to them where appropriate, and acknowledge that insects would not be capable of interacting with the environment without systems that regulate internal function and coordinate external responses. However, our primary objectives are concerned with other issues. We urge disappointed readers to examine the excellent reviews of such topics provided by Nijhout (1994), Chapman (1998), Field and Matheson (1998), Sugumaran (1998), Dudley (2000), Denlinger *et al.* (2001), Nation (2002), Denlinger (2002), and Merzendorfer and Zimoch (2003), and then to come back to this book.

CHAPTER 2

Nutritional physiology and ecology

An understanding of insect ecology has been hampered by an inadequate knowledge of nutritional physiology.

Scriber and Slansky (1981)

Diverse insect diets are associated with entirely different constraints: liquid diets come with a weight or volume problem, solid diets require mechanical breakdown without damage to the gut, plant diets are poor in nutrients, and animal meals are unpredictable in time and space (Dow 1986). Most species of holometabolous insect could be represented in Fig. 2.1 by two linked circles, as a result of vastly different diets in the larval and adult stages. Folivory necessitates an increase in mass of the gut and its contents, with longer retention time, which is incompatible with flight (Dudley and Vermeij 1992). Although this applies to relatively sessile caterpillars, they metamorphose into nectar-feeding adults. About half of known insects are phytophagous, and among these some feeding guilds have been relatively well studied, in particular the leaf-chewers, which are mostly larvae. The nutritional ecology of immature insects was the subject of a classic review by Scriber and Slansky (1981). The present chapter is inevitably biased towards grasshoppers and caterpillars and, to a lesser extent, cockroaches and various fluid feeders. Recent technical advances and the advent of molecular biology have made it possible to study in some depth the nutrition of aphids, another group of agricultural pests, and their simpler diet means that synthetic diets are closer to the real thing. We consider the physiological constraints on feeding behaviour and the physiology of digestion and absorption, before turning to the difficulties of plant feeding and the longer-term consequences of feeding for

growth, development, and the life histories of insects.

A major theme of this chapter is compensatory feeding. In spite of the enormous variation in the quality of plant food, insects obtain their requirements by means of flexible feeding behaviour and nutrient utilization (Slansky 1993). There are three basic categories of compensatory responses shown by phytophagous insects (Simpson and Simpson 1990): *increased consumption* in order to obtain more of a limiting nutrient such as nitrogen, *dietary selection* of a different food to complement a limiting nutrient, or *increased digestive efficiency* to make the best use of a nutrient. The mechanisms of compensatory feeding have been studied in some detail for the major nutrients, proteins and carbohydrates. To avoid difficulties in interpreting experiments, the use of artificial diets is essential, in spite of their ecological limitations (Simpson and Simpson 1990). Another pervasive theme is nitrogen limitation. Insect herbivores tend to be limited by nitrogen because their C:N ratio is so much lower than that of the plants they eat (Mattson 1980).

It is feeding that makes insects into agricultural pests and disease vectors, although the choice of species and problems for research has often been very selective as a result (Stoffolano 1995). Knowledge of food consumption and utilization is of great importance in managing problem insects, and consequently there is an enormous literature on basic and applied insect nutrition and nutritional ecology. As examples of major reviews, the more

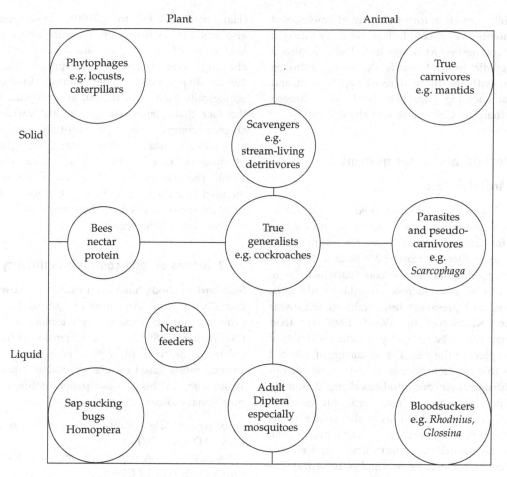

Figure 2.1 Classification of insect diets according to Dow (1986), based on plant/animal and liquid/solid dichotomies.

Note: Many insects can be placed in borderline positions, for example, adult female mosquitoes which feed on both nectar and blood.

Source: Reprinted from *Advances in Insect Physiology*, **19,** Dow, 187–328. © 1986, with permission from Elsevier.

mechanistic aspects are covered in nine chapters of volume 4 in the 1985 series *Comprehensive Insect Physiology, Biochemistry, and Pharmacology* (Kerkut and Gilbert 1985), regulation of insect feeding behaviour by Chapman and de Boer (1995), and the ecological context is represented in the 1000-page text of Slansky and Rodriguez (1987), with emphasis on feeding guilds, and by a substantial book on caterpillar foraging (Stamp and Casey 1993). Schoonhoven *et al.* (1998) list two pages of books and symposium proceedings devoted to insect–plant interactions. The treatment that follows is necessarily extremely selective.

Two areas of research in particular are providing new opportunities and motivation to investigate the effects of plant quality on insect herbivores (Awmack and Leather 2002). These are global climate change, a long-term experimental system involving gradual changes in host plant quality, and the development, since the mid-1990s, of transgenic plants expressing genes for insect resistance (first *Bacillus thuringiensis* toxin, then antinutrient proteins). Transgenic plants expressing antinutrient proteins permit direct measurement of the costs and benefits of plant defences. Very recently, the emerging field of elemental stoichiometry (Sterner and Elser 2002) is adding a new dimension to insect nutritional ecology. Fagan *et al.* (2002) have shown that insect predators contain on average 15 per cent more nitrogen than herbivores, even after correction for phylogeny, allometry, and gut contents (which could dilute the body nitrogen content of herbivorous species). There is also a phylogenetic trend towards decreasing nitrogen content, with the more derived Lepidoptera and Diptera showing significantly lower nitrogen values than the Coleoptera and Hemiptera.

2.1 Method and measurement

2.1.1 Artificial diets

Artificial diets are widely used in nutritional studies. Often they are semi-synthetic and contain crude fractions of natural diets; for example, the widely used diet for rearing *Manduca sexta* (Lepidoptera, Sphingidae) larvae contains wheat germ, yeast, casein, and sucrose, together with salts, vitamins, and preservatives, combined with agar in water (Kingsolver and Woods 1998). For food specialists it may be necessary to include extracts of the host plant in the diet. The advantage of artificial diets is that single nutrients or allelochemicals can be omitted or their concentrations changed, and the effect on performance measured. An essential nutrient can be detected from the effects of its deletion on growth, development, or reproduction, but the determination of nutritional requirements tends to be laborious. Protein and carbohydrate are the major macronutrients, so lipids are generally minimal components of artificial diets, even those for wax moths (Dadd 1985). In compensatory feeding studies, animals may respond differently to diets diluted with water or indigestible agar (Timmins *et al.* 1988).

Artificial diets have certain limitations. They are based on purified proteins, such as the milk protein casein, which are probably easier to digest than plant proteins because they contain little secondary structure and are not protected by cell walls (Woods and Kingsolver 1999). Artificial diets are rich, and caterpillars raised on them have much higher fat contents than those fed on leaves (Ojeda-Avila *et al.* 2003). Laboratory selection experiments using *Drosophila* are influenced by the abundance of food, so that the responses of flies to various forms of selection have tended to involve energy storage rather than energy conservation

(Harshman and Hoffman 2000). These diets are also much softer than plant material, and this can lead to a reduction in the size of the head and chewing musculature in caterpillars (Bernays 1986*a*). Experiments using natural forage are ecologically more realistic, but are complicated by the fact that plant tissue is highly variable in chemical composition and levels of nitrogen, water, and allelochemicals tend to covary. Nitrogen and phosphorus also covary in plant tissue (Garten 1976). The use of excised leaves is not recommended in assays of herbivory, because induced plant defences may reduce their nutritional quality (Olckers and Hulley 1994).

2.1.2 Indices of food conversion efficiency

Standard methods have been extensively used for quantifying food consumption, utilization, and growth in insects, especially phytophagous larvae (Waldbauer 1968; Scriber and Slansky 1981). The efficiency of food utilization is assessed using various ratios based on energy budget equations. Waldbauer, in his classic paper, defined three nutritional indices:

Approximate digestibility (or assimilation efficiency) $AD = (I - F)/I$;
Efficiency of conversion of ingested food (or growth efficiency) $ECI = B/I$;
Efficiency of conversion of digested food (or metabolic efficiency) $ECD = B/(I - F)$,

where $I =$ dry mass of food consumed, $F =$ dry mass of faeces produced, and $B =$ dry mass gain of the insect. Performance is expressed in terms of relative (i.e. g per g) rates of consumption (RCR) and growth (RGR). Various interconversions between nutritional indices and performance rates are possible, for example, $RGR = RCR \times AD \times ECD = RCR \times ECI$. An insect may maintain its growth rate over various combinations of these parameters because there are trade-offs between rates and efficiencies, for example, a higher RCR lowers retention time and thus AD. Slansky and Scriber (1985) discussed the methodology and summarized an enormous amount of data on the nutritional performance of insects in different feeding guilds. Slansky (1993) recommended measuring food consumption based on fresh weight as well as dry weight, otherwise compensatory feeding (see below) may not be evident when the foods differ in water content. Errors resulting from inaccurate measurement of food water content (especially leaves) are common and potentially serious.

Dry mass measurements (most of the data) can be converted for calculation of energy or nitrogen budgets (Wightman and Rogers 1978). ECI and ECD of a phytophage will be higher when expressed in terms of energy content than when expressed in dry mass because insect tissue has a higher energy content than plant tissue (Waldbauer 1968).

2.1.3 Use of a geometric framework

Ratio analyses in ecophysiology are problematic (Packard and Boardman 1988; Raubenheimer 1995; Beaupre 1995), and statistical problems can be avoided by direct analysis of measured variables. This approach has been convincingly advocated over the past decade by Simpson and Raubenheimer (Simpson and Raubenheimer 1993*a*; Simpson *et al.* 1995; Raubenheimer and Simpson 1999) for nutritional analysis in insects at both ingestive and post-ingestive levels. Their geometric approach has been valuable for demonstrating how animals eating unbalanced or suboptimal foods compromise between intakes of different nutrients, and is briefly explained here.

The concept of an 'intake target' provides a new way of looking at the regulation of nutrient intake (Simpson and Raubenheimer 1993*a*). Intake targets, which vary with growth or reproduction, are defined as the optimal amount and balance of nutrients that must be ingested for post-ingestive processes to operate at minimal cost to fitness. In the simplest case of two nutrients (such as protein and carbohydrate), with each represented as an axis on a two-dimensional graph, an insect given a single food type consumes a fixed proportion of nutrients so that its intake lies on a line passing through the origin, termed a 'rail' to suggest movement in a fixed direction (Fig. 2.2). The insect will not be able to achieve its intake target for the two nutrients if the rail does not pass through the target. It may have to compromise by eating an

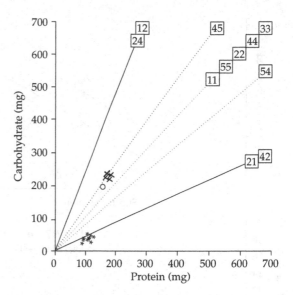

Figure 2.2 Geometric approach to nutrient intake in fifth-instar nymphs of *Locusta migratoria* fed a wide range of artificial diets.

Note: Nutrient consumption is shown as a bivariate plot of protein and carbohydrate consumption. Crosses indicate the intake target reached in experiments with various choices of foods (the circle is a separate estimate of the intake target). Asterisks indicate the growth target reached in no-choice experiments on different diets. Boxes at the end of the rails give the proportions of carbohydrate to protein. The intake target is close to a 1 : 1 ratio.

Source: Simpson and Raubenheimer (1993). *Philosophical Transactions of the Royal Society of London B*. **342**, 381–402, The Royal Society Publications.

excess of one nutrient or insufficient amounts of the other. But if the insect is allowed to choose between two foods containing different nutrient ratios (i.e. defining different rails), it will be able to attain any target lying within the nutrient space between the rails. This can be demonstrated by offering various combinations of paired diets.

Alternatively, nutrient balance can be achieved post-ingestively by removing nutrients which are in excess of metabolic requirements: this enables an insect to move across nutrient space from the intake target to the growth target (Fig. 2.2). This is also important in cases when the unbalanced foods are not complementary and it is impossible to reach the intake target. Post-ingestive aspects of nutrition can be examined by constructing utilization plots of nutrient output versus intake; a change in slope indicates the point above which ingested nutrient is not utilized (Fig. 2.3). Bicoordinate utilization plots

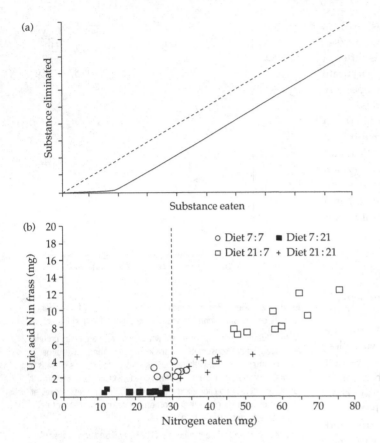

Figure 2.3 Utilization plots of nutrient output versus intake. (a) Hypothetical plot in which the dotted line represents total elimination of a nutrient, while the solid line shows its utilization up to a certain level, with elimination occurring beyond this level. (b) Utilization plot for nitrogen in fifth instar *Locusta migratoria* fed four different diets (7 : 7 indicates a diet containing 7% protein and 7% digestible carbohydrate). When nitrogen consumption exceeded 30 mg, uric acid excretion was used to remove the excess.

Source (a): *Regulatory Mechanisms in Insect Feeding*, 1995, Simpson *et al.*, pp. 251–278. With kind permission of Kluwer Academic Publishers. *Source* (b): Zanotto *et al.* (1993). *Physiological Entomology* **18**, 425–434, Blackwell Publishing.

are the geometrical representation of analysis of covariance (ANCOVA) designs and are statistically preferable to the widely used ratio-based indices (Raubenheimer 1995). Specific examples of both behavioural regulation of food intake and physiological regulation of its utilization are described in Section 2.2.3 below.

Some studies have used both old and new methods, represented by nutritional indices and bicoordinate plots, to assess nutritional performance (Chown and Block 1997; Ojeda-Avila *et al.* 2003). A plot of growth rate against consumption rate gives a visual assessment of ECI, while growth against absorption gives ECD.

2.2 Physiological aspects of feeding behaviour

Regulatory mechanisms involved in insect feeding were comprehensively reviewed by Chapman and de Boer (1995). We will briefly examine the

inter-relationships between feeding and digestion in caterpillars, the regulation of meal size in fluid feeders, and finally the regulation, especially in locusts, of intake of protein and digestible carbohydrate. These are all relatively short-term aspects, whereas diet switching, as defined by Waldbauer and Friedman (1991), refers to long-term changes in diet that occur between larva and adult or between instars in some herbivores.

2.2.1 Optimal feeding in caterpillars

Inter-relationships between consumption and digestion have been explored in detail in caterpillars (for excellent reviews see Reynolds 1990; Woods and Kingsolver 1999). The experimental caterpillar is commonly the fifth instar of the tobacco hornworm *M. sexta* (Lepidoptera, Sphingidae); appropriately, *Manduca* means 'the chewer'. During its last instar this species increases in mass at the rate of 2.7 g per day, which is faster than the growth rates of similar-sized altricial birds and is achieved without the benefit of endothermy (Reynolds *et al.* 1985). The gut and its contents form 39 per cent of the mass of the caterpillar throughout the feeding period of the fifth instar. The transformation of leaf protein into insect tissue to achieve such rapid growth can be divided into four steps (Woods and Kingsolver 1999): consumption of leaves, digestion of protein into small peptides and amino acids, absorption of amino acids across the midgut epithelium, and construction of tissue. The last three steps are post-ingestive events which influence, and are influenced by, the rate of consumption.

In the continuous flow digestive system of a caterpillar, gut passage rates are equal to rates of consumption. Gut passage rates involve a trade-off between fast processing and thorough processing. Woods and Kingsolver (1999) used a chemical reactor model to predict the concentration profiles of proteins and their breakdown products along the midgut, and found that an intermediate consumption rate gave the highest rate of absorption. Reynolds (1990) reached the same conclusion using a model of optimal digestion (Sibly and Calow 1986): AD is optimized at the optimal retention time, which then determines the rate of consumption. Caterpillars, therefore, restrain food intake to an optimal level which maximizes the rate of nutrient uptake and the rate of growth (Reynolds *et al.* 1985). From measurements of food passage rates, midgut dimensions, proteolytic activity in the lumen (V_{max}), and the protein concentration giving half-maximal rates (K_m), Woods and Kingsolver (1999) predicted that protein is digested rapidly in the anterior midgut but absorption of breakdown products may be a limiting step. However, Reynolds (1990) measured rapid uptake of labelled amino acids in the anterior midgut, which would suggest post-absorptive rather than absorptive constraints on growth. Further studies should emphasize caterpillars eating leaves, because plant proteins differ from those in artificial diets, and undigested protein in leaf fragments will extend further along the midgut (Woods and Kingsolver 1999). Most caterpillars feed by leaf-snipping (their mandibles have cutting but not grinding surfaces) and few of the cells in the ingested tissue are crushed. However, AD values for carbohydrate and protein are surprisingly high, suggesting that nutrients are extracted when the cell walls become porous, and this digestive strategy is apparently as efficient as that of grasshoppers, which crush leaf tissues (Barbehenn 1992).

Evidence for restrained food intake also comes from behavioural observations on temporal patterns of feeding in caterpillars. The combination of meal durations and meal frequencies determines the proportion of time an insect spends feeding. This, combined with the instantaneous feeding rate, gives the overall consumption rate (Slansky 1993). The proportion of time spent feeding by *M. sexta* larvae is up to 80 per cent on tobacco leaves, compared to 25 per cent on artificial diet, the difference being due to relative water contents, although growth rates are identical on both diets (Reynolds *et al.* 1986). Bowdan (1988) examined the microstructure of feeding on tomato leaves using an electrical technique, and showed that larger caterpillars ate more by increasing bite frequency and the length of meals, but meal frequency was unchanged. The periods of inactivity even at high rates of consumption, and the compensatory feeding which occurs when artificial diet is diluted with water or cellulose (Timmins *et al.* 1988), confirm that caterpillars could consume more food than they do.

The feeding rhythms of caterpillars vary greatly because they depend on ecological factors as well as digestive processes. The caterpillar's task is to maximize growth rate while avoiding risk, and the risk from predators and parasitoids can be great. Bernays (1997) quantified the risk during continuous observation of two caterpillar species, and mortality was so high during feeding that there must be strong selection for rapid food intake. Predation will also increase considerably on nutritionally poor plants. Caterpillars undergo spectacular changes in body size with growth, with major effects on feeding ecology, behaviour, and predator assemblages (Reavey 1993; Gaston *et al.* 1997). Minimizing risk can involve changes in feeding habit as individuals grow: as body size increases there is a general trend from concealed feeding (leaf miners or gall formers, which suffer from space constraints) to spinning or rolling leaves, to external feeding in late instars. In bigger caterpillars the surface area of gut for absorption is relatively less in relation to the volume of gut contents. Rates tend to decrease with increasing body size but efficiencies do not (Slansky and Scriber 1985). AD does not vary with size in *M. sexta*, remaining about 60 per cent throughout the fifth instar, but retention time increases. Correction of the body mass component in nutritional indices for the presence of food in the gut reduces their value only slightly (Reynolds *et al.* 1985).

2.2.2 Regulation of meal size: volumetric or nutritional feedback

The physiological regulation of meal size can potentially include sensory stimuli (either positive or negative), volumetric feedback via stretch receptors in the gut or body wall, haemolymph composition (osmolality or the concentration of individual nutrients), available reserves, and neuropeptides, many of which are known to affect contractile activity of the gut (Gäde *et al.* 1997). Unravelling causal relationships is far from simple. For chewing insects, the most detailed information comes from acridids, in which volumetric feedback from stretch receptors in the gut is important in terminating a meal (reviewed by Simpson *et al.* 1995). These receptors are located in both the crop and ileum, those in the latter being stimulated by the remains of the previous meal. It is also likely that rapid changes in haemolymph osmolality and nutrient concentration inhibit further feeding. Locusts fed high-protein diets exhibited much greater increases in haemolymph osmolality and amino acid concentrations during a meal than those on low-protein diets, and the result was a longer interval until the next meal (Abisgold and Simpson 1987). Feeding stops when inhibitory feedbacks force excitation below the feeding threshold, and increasing inhibition during a meal is reflected in declining ingestion rates (Simpson *et al.* 1995). Volumetric feedbacks are less obvious in caterpillars, which lack the capacious crop of acridids, and meal size in *Manduca* may depend on feedback from nutrients in the gut lumen. Injection of soluble diet extract into the midgut lumen inhibited feeding, while an injection of xylose

solution of the same osmolality did not (Timmins and Reynolds 1992).

Carbohydrate feeding is best understood in Diptera, although it is also fundamental to the aerial success of adult Lepidoptera and Hymenoptera, all three orders depending on a variety of liquid carbohydrate resources as immediate energy for flight (Stoffolano 1995). These insects have evolved an expandable and impermeable crop (diverticular in Diptera and Lepidoptera, linear in the Hymenoptera) located in the abdomen. The blowfly *Phormia regina* (Calliphoridae) has been used as an experimental model, and it is clear that information from abdominal stretch receptors ends the meal. Not surprisingly, the regulation of feeding behaviour has also been thoroughly investigated in blood feeders, especially mosquitoes, where nectar meals are directed to the crop and blood meals to the midgut but both kinds of meal are terminated by abdominal distension (reviewed by Davis and Friend 1995).

Meal quality and feeding regime also influence crop filling. The Australian sheep blowfly *Lucilia cuprina* ingests greater volumes of dilute glucose solutions by taking larger and more frequent meals (Simpson *et al.* 1989). Crop volume at the end of a meal was similar in *L. cuprina* ingesting 0.1 and 1.0 M glucose, due to volumetric inhibition. However, the flies maintained on 1.0 M glucose had fuller crops at the beginning of a meal because dilute solutions empty from the crop more rapidly (see Section 2.3.3) and can be ingested in greater quantity. Evaporative losses caused by bubbling behaviour in *Rhagoletis pomonella* (Diptera, Tephritidae) reverse the volumetric inhibition, permitting feeding to continue on dilute solutions (Hendrichs *et al.* 1992). Some confusion in the literature has arisen because researchers have used insects in very different nutritional states. Feeding behaviour varies greatly between insects fed *ad libitum* and those which are deprived of food and then offered single meals, as elegantly demonstrated by Edgecomb *et al.* (1994) for *Drosophila melanogaster* feeding on sucrose–agar diets. Flies fed *ad libitum* maintained much smaller crop volumes than food-deprived flies fed a single meal, and responded differently to sucrose concentrations up to 0.5 M (Fig. 2.4). In general, the volumes of sugar solution

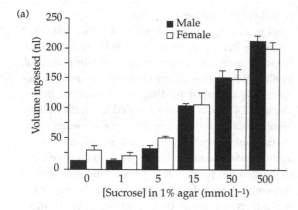

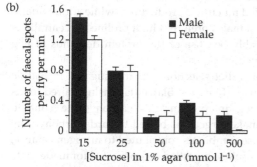

Figure 2.4 Responses of *Drosophila melanogaster* to diet concentration depend on feeding regime. (a) Volume ingested increases with sucrose concentration in flies deprived of food for 24 h and then fed for 5 min. Each column is the mean ± SE from four trials involving 20 flies each. (b) Excretion rate, measured as the number of faecal spots per min, decreases with sucrose concentration in flies fed *ad libitum*. Each column represents the mean ± SE from 5 trials of 50 flies each.

Source: Edgecomb *et al.* (1994).

ingested by insects are positively correlated with concentration in previously starved individuals offered single meals, but insects feeding *ad libitum* show compensatory feeding and the volume imbibed is then negatively correlated with concentration.

Regulation of load size may be more complex in social insects, which begin foraging with empty crops. Recently, Josens *et al.* (1998) investigated nectar feeding in the ant *Camponotus mus* by weighing foragers as they crossed a small bridge between the colony and the foraging arena, then weighing them again on the return trip. Crop load increased with increasing sucrose concentration

to a maximum at about 1.5 M, then diminished because of viscosity effects. Workers carried up to 60 per cent of their own weight in the crop, but the loads were partial for either dilute or very concentrated solutions, when motivational state of the ants or physical properties of the solution played a role, respectively.

The central role of haemolymph in nutritional homeostasis was highlighted by Simpson and Raubenheimer (1993b). The haemolymph provides a continuous reading of the insects' nutritional and metabolic state: it integrates information on the time since the last meal, its size and quality, as well as current and recent demands by tissues. Feeding may be inhibited by high haemolymph osmolality or, more accurately, by high concentrations of specific nutrients such as amino acids or sugars. This feedback results from both previous and current meals. Haemolymph nutrient concentrations change during the course of a meal, as a result of nutrient absorption and secretion of dilute saliva. Movement of water from haemolymph to gut lumen can also be expected initially, down an osmotic gradient created by hydrolysis of macromolecules. Reynolds and Bellward (1989) showed that midgut water content of *M. sexta* was regulated between 87 and 91 per cent, even when dietary water content was much lower. Injections which increase haemolymph osmolality or amino acid concentration mimicked the effect of a high-protein meal and delayed the next meal in locusts (Abisgold and Simpson 1987). These authors did not find the same effect with high and low carbohydrate diets, probably because the products of carbohydrate digestion are rapidly removed from the haemolymph after absorption.

Hormonal involvement in feeding regulation is also likely. The diffuse endocrine cells of the locust midgut are more dense in the ampullae at the Malpighian tubule–gut junction, where they are perfectly positioned to monitor three key fluids; the midgut luminal contents, tubule fluid, and haemolymph (Zudaire *et al.* 1998). FMRFamide-like peptides are thought to be involved in digestive processes, and FMRFamide-like immunoreactivity of the ampullar endocrine cells was correlated with food quality, increasing as protein/carbohydrate composition of the diet shifted away from optimal.

Stoffolano (1995) has referred to the midgut as the least studied, but largest, endocrine tissue in insects.

2.2.3 Regulation of protein and carbohydrate intake

Waldbauer and Friedman (1991) defined self-selection of optimal diets as a continuous regulation of intake involving frequent shifts between foods. The fact that insects perceive nutritional deficiencies, and alter behaviour to correct them, has been clearly illustrated by application of the geometrical approach to protein and carbohydrate intake in *Locusta migratoria*. Many aspects of nutritional regulation in this species stem from interactions between these two macronutrients (Raubenheimer and Simpson 1999). Animals given a balanced diet, or two or more unbalanced but complementary diets, can satisfy their nutrient requirements (arrive at the same point in nutrient space) and achieve similar growth performances (Fig. 2.2). It must be emphasized that the nutritional needs reflected in intake targets are not static. A flight of 2 h duration moves the intake target of adult *L. migratoria* towards increased carbohydrate levels, and targets also vary as requirements change throughout development (Raubenheimer and Simpson 1999).

When fed unbalanced diets and prevented from reaching their intake targets, the grass-feeding species *L. migratoria* is less willing to eat an unwanted nutrient than the polyphagous *Schistocerca gregaria*, perhaps because the latter has a better chance of encountering new host plants of different composition (Raubenheimer and Simpson 1999). Put another way, the amount eaten of the unbalanced food should reflect the probability of encountering an equally and oppositely unbalanced food. This is supported by comparisons of nutritional regulation in solitary and gregarious phases of *S. gregaria* (Simpson *et al.* 2002). Solitary locusts are less mobile and encounter fewer host plants, and so experience less nutritional heterogeneity. Gregarious locusts, not subject to these constraints, ingest more of the excess nutrient in unbalanced foods. Locusts respond rapidly to nutritional deficiencies: compensatory selection for either protein or carbohydrate was evident after a single deficient meal in *L. migratoria*, but not in another relatively sessile herbivore, *Spodoptera littoralis*, in which the response may take longer to develop (Simpson *et al.* 1990). Interactions between nutrients and allelochemicals in locust feeding are considered below (Section 2.4.3). On a more detailed level, *Phoetaliotes nebrascensis* grasshoppers are able to select individual amino acids from a background mixture of amino acids and sucrose applied to glass fibre discs (Behmer and Joern 1993, 1994). This selection is determined by nutrient requirements: Nymphs but not adults preferred diets high in phenylalanine (needed for cuticle production), while adult females but not males preferred high proline concentrations (probably because of the protein demands of egg production).

When diet selection was investigated in *Blatella germanica* (Blattaria, Blatellidae) using paired foods differing in protein and carbohydrate content, the intake target was biased towards carbohydrate because symbionts contributed to nitrogen balance, and cellulose digestion compensated for inadequate levels of soluble carbohydrate in diluted diets (Jones and Raubenheimer 2001). Kells *et al.* (1999) investigated the nutritional status of the same cockroach species in the 'field' (low income apartments). In spite of the reputation of cockroaches as successful generalists, the apartment diet was considered suboptimal (low in protein) compared to rodent chow because the uric acid content of field cockroaches was much lower. Stored uric acid is utilized by symbiont bacteria (see Section 2.4.2).

Nutritional homeostasis involves not only long-term regulation of feeding, but also of post-ingestive utilization. The geometric approach draws attention to the fact that regulating intake of one nutrient often involves ingesting, and then removing, excesses of another. There is so far little evidence in insects for pre-absorptive removal of excess nutrients by the most likely mechanisms of increasing gut emptying rate, decreasing enzyme secretion, or reducing absorption rates (for references see Simpson *et al.* 1995). Instead, the major site of differential regulation appears to be post-absorptive. Nymphs of *L. migratoria* feeding on unbalanced foods remove excess nitrogen by increased uric acid excretion (Fig. 2.3b) and excess carbon by increased CO_2 output, that is, 'wastage' respiration (Zanotto *et al*. 1993, 1997). Use of chemically defined diets with sucrose radiolabelled in either the glucose or fructose moiety has shown that the pea aphid, *Acyrthosiphon pisum*, preferentially assimilates and respires the fructose from ingested sucrose, while converting the glucose into oligosaccharides which are excreted in the honeydew (Ashford *et al*. 2000).

2.3 Digestion and absorption of nutrients

Some digestion may occur in the crop, as a result of salivary enzymes or midgut enzymes moving anteriorly (as in beetles, Terra 1990), but most biochemical transformation occurs in the midgut. Occasionally the midgut has a storage function, like the anterior midgut of *Rhodnius prolixus* (Hemiptera, Reduviidae) (exploited by researchers wishing to obtain blood non-invasively from bats, Helverson *et al*. 1986). Dow's (1986) review brought together information on ultrastucture, ion transport, enzymes, and detoxification for midguts from all main insect feeding types shown in Fig. 2.1. In addition to the hindgut (Section 4.1.3), the midgut is a major site of ion regulation (which is fundamental to nutrient absorption) and the best understood transport epithelium in insects.

The midgut, as a primary interface between insect and environment, is a target for insect control. Insecticides based on Bt toxin from the soil bacterium *Bacillus thuringiensis* are highly effective against certain pests, particularly larvae of Lepidoptera, Coleoptera, and Diptera, and transgenic crops expressing Bt genes are now in widespread use. The toxin, which is activated by midgut proteases, is thought to bind to receptor proteins on the columnar cell microvilli; it then undergoes a conformational change and inserts into the membrane in aggregates, forming pores that result in osmotic lysis and disintegration of the epithelium (Pietrantonio and Gill 1996). Assays of osmotic swelling in membrane vesicles from the midguts of target insects can be used to measure susceptibility to the toxins produced by different strains of Bt (Escriche *et al*. 1998). Young, actively feeding larvae are most affected, and the success of Bt products as

microbial insecticides is due to their specificity to the target insects. Possible harmful effects on nontarget species, including natural enemies, are the focus of active research (Zangerl *et al*. 2001; Dutton *et al*. 2002).

2.3.1 Digestive enzymes and the organization of digestion

Digestive enzymes

Insect digestive enzymes are all hydrolases, show general similarities to mammalian enzymes and are classified using standard nomenclature based on the reactions they catalyse (Applebaum 1985; Terra and Ferreira 1994; Terra *et al*. 1996*a*). The biochemistry and molecular biology of purified enzymes is currently an active field. Molecular biology is now used as an alternative to exhaustive protein purification, especially for the large arrays of serine proteinase genes present in disease vectors and other insects (Muharsini *et al*. 2001). For example, there are about 200 genes encoding serine proteinases in the genome sequence of *D. melanogaster* (Rubin *et al*. 2000).

Serine proteinases (of which the trypsins and chymotrypsins are well characterized) hydrolyse internal peptide bonds, while carboxypeptidases and aminopeptidases remove terminal amino acids. The term protease, thus, includes both proteinases (endopeptidases) and exopeptidases. Serine proteinases of blood-sucking insects are important in vector-parasite relationships: infection requires that the parasite survive protease activity in the midgut. For phytophages, naturally occurring proteinase inhibitors are important secondary plant compounds (Section 2.4.3), and when chewing breaks up plant cell walls some of the plant enzymes released are also active in the gut lumen (Appel 1994). In bruchid beetle larvae, which specialize on a diet of legume seeds, reduced levels of proteases are complemented by additional enzymes obtained from the seeds (Applebaum 1964). The main digestive proteinases of Bruchidae are not serine but cysteine proteinases, which require a lower pH and are common in the midguts of Hemiptera and some Coleoptera. The distribution of cysteine proteinases in beetles has a phylogenetic basis. These enzymes first appeared in an

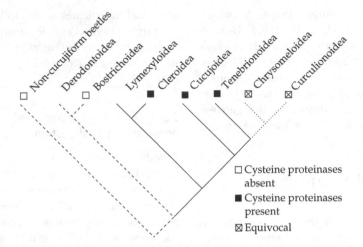

Figure 2.5 Hypothesized cladogram showing the distribution of cysteine proteinases in the major superfamilies of Coleoptera.

Note: The cladogram is based on data for 52 species representing 17 families. Some secondary loss of cysteine proteinases has occurred in the more derived and phytophagous superfamilies.

Source: Reprinted from *Comparative Biochemistry and Physiology B*, **126**, Johnson and Rabosky, 609–619. © 2000, with permission from Elsevier.

ancestor of the series Cucujiformia (Fig. 2.5), perhaps in response to a seed diet rich in trypsin inhibitors, and in total occur in five related superfamilies (Johnson and Rabosky 2000). Cysteine proteinases have been secondarily lost in some Cerambycidae (superfamily Chrysomeloidea), suggesting reversion to a digestive strategy utilizing serine proteinases. Although polymer digestion is considered unnecessary in aphids, Cristofoletti *et al.* (2003) have recently demonstrated substantial cysteine proteinase activity in the midgut of the pea aphid, *A. pisum*.

Carbohydrases fall into two broad categories according to whether they are active on polysaccharides or on smaller fragments. Amylases (active on starch or glycogen) and cellulases (see Section 2.4.1) cleave internal bonds in polysaccharides. Lysozyme is involved in the digestion of bacterial cell walls in the midgut of cyclorrhaphan Diptera. The second category includes glucosidases and galactosidases, which hydrolyse oligosaccharides and disaccharides. Use of the term 'sucrase' does not differentiate between sucrose hydrolysis by α-glucosidases or less common β-fructosidases. Trehalase, which hydrolyses trehalose into glucose, is widespread in insect tissues and, in the midgut, may counteract back-diffusion

of trehalose into the lumen (first suggested by Wyatt 1967).

Insect lipases are less easily studied than proteases and carbohydrases because of their lower activities. Moreover, reaction between the enzymes, which are water soluble, and their substrates, which are not, requires a suitable emulsion. These are difficult to prepare experimentally (Applebaum 1985). Phospholipases act on membrane lipids and cause cell lysis. Triacylglycerols are major lipid components of the diet and are hydrolysed to fatty acids and glycerol. In general, lipid digestion is poorly understood in insects (Arrese *et al.* 2001).

Regulation of enzyme levels

Do the levels of digestive enzymes vary according to the quantity or quality of food? Continuous and discontinuous feeders will obviously have different requirements regarding control of digestive enzyme secretion. Ultrastructural evidence indicates that secretion usually follows soon after synthesis, even in discontinuous feeders, although there are examples of storage of enzymes in the latter group (Lehane and Billingsley 1996). Synthesis and secretion are controlled in two ways: 'secretagogue' stimulation according to the amount of relevant substrate in the gut, or hormonal regulation. These are not necessarily mutually exclusive, and may operate on different time scales (Applebaum 1985). Unfortunately, experimentally distinguishing these types of control is difficult. The best evidence for hormonal influences comes from mosquitoes (Lehane *et al.* 1995). Because the term 'secretagogue' can be confusing, the latter authors have proposed that direct interaction of a component of the meal with enzyme-producing cells should be termed a prandial mechanism. They also distinguish between paracrine and endocrine mechanisms: a paracrine effect is a local hormonal effect on neighbouring cells. The diffuse endocrine system of the insect midgut remains a major obstacle to differentiating between prandial and endocrine control.

Many studies have investigated the effect of proteins on protease levels in haematophagous Diptera, because of their large and infrequent blood meals, and their role as disease vectors. Diverse soluble proteins stimulate trypsin secretion into the incubation medium of midgut homogenates of the stable fly *Stomoxys calcitrans* (Diptera, Muscidae), and the effect is concentration-dependent (Blakemore *et al.* 1995). This method distinguishes between effects on synthesis and on secretion, because new synthesis over the time scale of the incubations is considered negligible. Insoluble proteins, small peptides and amino acids do not stimulate trypsin secretion. Regulation of levels can vary within an enzyme family. In female *Anopheles gambiae* (Diptera, Culicidae), for example, some trypsins are constitutively expressed, while others, produced in larger amounts, are induced by blood feeding (Müller *et al.* 1995). Rapid advances at the molecular level are being driven by interest in the regulation of serine proteinases of blood-sucking insects (Lehane *et al.* 1995). This interest extends to midgut immunity through the recently discovered defensin family of peptides. Hamilton *et al.* (2002) have shown that midgut defensins of *S. calcitrans* are colocalized with a serine proteinase during storage, and that the complex dissociates on secretion into the lumen, the defensins protecting the stored blood meal from bacterial attack. Insects adapt to proteinase inhibitors in their diet by hyperproduction of proteinases or by switching to novel proteinases that are insensitive to these plant defences (see Section 2.4.3).

Peritrophic matrix and the organization of digestion

The midgut cells of most insects (Hemiptera excluded) secrete a multilayered peritrophic matrix (although frequently called a peritrophic membrane, it lacks cellular structure) consisting of chitin, proteins, and proteoglycans. This functions as a physical barrier to protect the epithelium from mechanical abrasion, toxic plant allelochemicals, and pathogens, and also allows compartmentalization of the gut lumen and the spatial organization of digestive processes. Large macromolecules are hydrolysed by soluble enzymes inside the peritrophic matrix, until they are small enough to diffuse into the space between the peritrophic matrix and midgut epithelium, where digestion is completed by membrane-bound enzymes which may be integral proteins of the microvillar membranes (Terra *et al.* 1996*b*). Structural strength is provided by the meshwork of chitin fibrils, while permeability properties are determined by pore diameters in the gel-like matrix. Labelled dextrans with diameters ranging from 21 to 36 nm penetrate the peritrophic matrix of several species of Lepidoptera and Orthoptera (Barbehenn and Martin 1995), so size exclusion does not explain impermeability to digestive enzymes or to tannins, and other properties of the matrix may be involved.

The importance of midgut compartments in the efficient, sequential breakdown of food was first demonstrated in *Rhynchosciara americana* (Diptera, Sciaridae) by assaying enzyme activities in different luminal compartments and midgut tissue (Terra *et al.* 1979). It is also thought that countercurrent flux of fluid assists in both the absorption of nutrients and the recycling of digestive enzymes. Countercurrent fluxes may result from fluid secretion in the posterior midgut or the anterior movement of primary urine from the Malpighian tubules, the result being that fluid moves in an anterior direction outside the pertriophic matrix, and is absorbed in the anterior midgut or caecae. The anatomical differences vary with phylogeny. The evidence for compartmentalization of digestive processes in the major insect orders has been thoroughly reviewed by Terra and colleagues (Terra 1990; Terra and Ferreira 1994; Terra *et al.* 1996*b*). Countercurrent movement of gut fluids occurs in Orthoptera, but only in animals deprived of food (Dow 1981), and it is considered unlikely in continuously feeding caterpillars (Dow 1986; Woods and Kingsolver 1999). Terra and colleagues argue that phylogeny is more important than feeding habits in determining the enzymes present and the organization of digestion, and it is possible that most insect species possess a full complement of digestive enzymes, although the relative amounts vary with diet (Terra *et al.* 1996*a*). The fast gut passage rates of many insects suggest potential costs in terms of digestive enzyme loss, but countercurrent fluxes may displace enzymes anteriorly and aid in their recycling (Terra *et al.* 1996*b*).

2.3.2 Gut physicochemistry of caterpillars

Conditions in the gut lumen can vary dramatically among insect herbivores (Appel 1994). The most extreme conditions (and most expensive to maintain) are found in caterpillars. Dow (1984) recorded pH values over 12, the highest known in any biological system, in the anterior and middle regions of caterpillar midgut (Fig. 2.6). The large volume of the midgut compartment suggests substantial acid–base transport to regulate this extreme pH. Ion transport in the midgut of *M. sexta* has been studied intensively using electrophysiological techniques, initially because of its potent K^+-transporting ability (Klein *et al.* 1996). *Manduca* midgut is a model tissue (the frog skin of invertebrates), possessing the advantages of large size, commercial availability of insects and synthetic diet, and ease of making *in vitro* preparations. Caterpillar midgut was also the first animal tissue in which proton pumps were identified and found to energize secondary active transport (Wieczorek *et al.* 1991). Vacuolar-type proton ATPases (V-ATPases) are highly conserved enzymes located in bacterial, yeast, and plant plasma membranes but are now also known to occur in many animal plasma membranes (Harvey *et al.* 1998; Wieczorek *et al.* 1999). They are coupled to antiporters: in caterpillar midgut the V-ATPase is coupled to a $K^+/2H^+$ antiporter to produce what was long assumed to be a primary K^+ pump (Harvey *et al.* 1998). The V-ATPase is confined to the apical membrane of the goblet cells (shown by

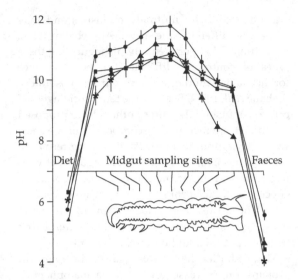

Figure 2.6 Midgut pH profile for four larval lepidopterans, with values for food and faeces shown for comparison.

Note: In all cases, haemolymph pH was 6.7. Species: circles, *Acherontia atropos* (Sphingidae); triangles, *Manduca sexta* (Sphingidae); squares, *Lichnoptera felina* (Noctuidae); and asterisks, *Lasiocampa quercus callunae* (Laslocampidae). Mean ± SE, $n = 4$.

Source: Dow (1984).

immunohistochemistry using plasma membrane fractions) and is responsible for a large lumen-positive apical voltage in excess of 150 mV. It serves to alkalinize the midgut lumen, maintaining favourable pH conditions for enzyme activity, and energizes amino acid uptake via K^+-amino acid symport (see below). Alkalinization results from the stoichiometry of the $K^+/2H^+$ antiporter and the high voltage generated by the V-ATPase, the result being net K^+ secretion and net H^+ absorption (Wieczorek *et al.* 1999). A model of midgut transport processes is presented in Fig. 2.7.

Leaf-eating caterpillars have a diet very low in Na^+, and their low haemolymph Na^+ concentration precludes use of a sodium pump as primary energizer (goblet cells lack any detectable Na^+/K^+-ATPase). However, carnivorous and some herbivorous insects may use the Na^+/K^+-ATPase to drive absorption of fluid and organic solutes, as vertebrates do (Klein *et al.* 1996). In both cases secondary processes are coupled to a primary ion transport ATPase, but the V-ATPase and

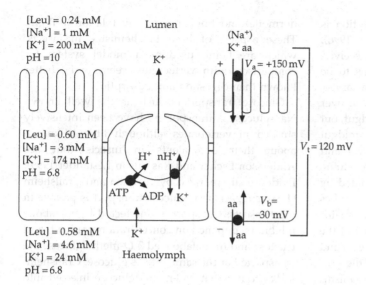

[Leu] = 0.24 mM
[Na$^+$] = 1 mM
[K$^+$] = 200 mM
pH =10

Lumen

(Na$^+$)
K$^+$ aa

K$^+$

V_a = +150 mV

[Leu] = 0.60 mM
[Na$^+$] = 3 mM
[K$^+$] = 174 mM
pH = 6.8

H$^+$ nH$^+$

V_t = 120 mV

ATP   ADP   K$^+$

aa V_b = −30 mV

[Leu] = 0.58 mM
[Na$^+$] = 4.6 mM
[K$^+$] = 24 mM
pH = 6.8

K$^+$

Haemolymph

aa

Figure 2.7 Model of amino acid absorption in the midgut of a caterpillar (*Philosamia cynthia*).

Note: A goblet cell is shown between two columnar absorptive cells. The left columnar cell shows leucine, Na$^+$ and K$^+$ concentrations and pH values in the lumen, cell and haemolymph. The goblet cell shows mechanisms involved in K$^+$ transport. The right columnar cell shows mechanisms involved in amino acid (aa) absorption and apical (V_a), basal (V_b) and transepithelial (V_t) electrical potential differences. Basal exit mechanisms for amino acids are less well known.

Source: Biology of the Insect Midgut, 1996, pp. 265–292, Sacchi and Wolfersberger, with kind permission of Kluwer Academic Publishers.

Na$^+$/K$^+$-ATPase are located on apical and basal membranes, respectively.

The plasma membrane V-ATPase of *M. sexta* is well characterized (Harvey *et al.* 1998). When feeding ceases in preparation for a larval–larval moult, downregulation of the V-ATPase is thought to be achieved by reversible dissociation of the peripheral ATP-hydrolysing complex from the membrane-bound H$^+$-translocating complex (Sumner *et al.* 1995). Expression of V-ATPase genes is also downregulated at this time, under the control of ecdysteroids (Reineke *et al.* 2002). This economy is necessary because 10 per cent of larval ATP production is consumed by midgut K$^+$ transport, that is, by the V-ATPase.

The pH of the midgut lumen varies with phylogeny and feeding ecology, and extreme alkalinity occurs in several orders besides Lepidoptera (Clark 1999; Harrison 2001). Extreme pH has complex effects on the activity of ingested allelochemicals (Section 2.4.3 and see Appel 1994). For caterpillars, a disadvantage of high gut pH is that it facilitates activation of Bt toxin (Dow 1984). The midgut of mosquito larvae is highly alkaline, probably through similar molecular mechanisms, and this characteristic might provide a basis for disease vector control just as it has for control of agricultural pests (Harvey *et al.* 1998). Insect acid–base physiology was reviewed by Harrison (2001).

2.3.3 Absorption of nutrients

Gut absorption was reviewed by Turunen (1985). Absorption includes transport across both apical and basal membranes, but there is more information on apical mechanisms. These are usefully studied by using purified plasma membranes which form sealed vesicles containing fluid of known composition, and can be pre-loaded with ions or amino acids (Sacchi and Wolfersberger 1996).

Leucine absorption in the midgut of *Philosamia cynthia* larvae (Lepidoptera, Saturniidae) has been well studied, and a model for leucine uptake by columnar cells is shown in Fig. 2.7. Goblet and columnar cells cooperate in ionic homeostasis and absorption of nutrients: the V-ATPase and K$^+$/nH$^+$ antiporter on the apical membrane of the goblet cells energize K$^+$-amino acid symporters on the microvilli of columnar cells. Amino acid transport is coupled to the movement of K$^+$ down its electrochemical gradient from lumen to cell. This contrasts with the Na$^+$-cotransport system of vertebrates and some insects (e.g. cockroaches), involving the basolateral Na$^+$/K$^+$-ATPase and apical transport proteins in the same cell (Sacchi and Wolfersberger 1996). The affinity of the symporter for Na$^+$ is about 18 times that for K$^+$, but since the luminal K$^+$ concentration is 200 times higher (Fig. 2.7), most amino acid absorption is K^+-dependent (Sacchi and Wolfersberger 1996). Transport of neutral amino acids has received the most attention, and this model seems to be generally applicable to other lepidopteran larvae. Neutral amino acid-K^+ symport is effective over most of the pH range found in *M. sexta* midgut, but occurs mostly in the posterior third of the midgut where luminal pH is less extreme (Sacchi and Wolfersberger 1996). The latter review of amino acid absorption in insect midgut was updated by Wolfersberger (2000), including molecular studies of the symporters involved. The literature remains unbalanced in favour of large caterpillars, but the focus has shifted from *P. cynthia* to *Bombyx mori* and *M. sexta*, which can be reared throughout the year on artificial diets. Several absorption mechanisms are evident in midguts of larval *B. mori*: the neutral amino acid-K^+ symport described above, and a less selective uniport system which facilitates diffusion of amino acids (Giordana *et al.* 1998; Leonardi *et al.* 1998). Another uniport system transports the dibasic amino acids arginine and lysine (Casartelli *et al.* 2001). The cDNA encoding a K^+-amino acid symporter from *M. sexta* midgut has been isolated and cloned, and the deduced amino acid sequence shows homology to mammalian amino acid transporters (Castagna *et al.* 1998).

Absorption of lipids requires solubilization in the layer of water adjacent to the absorptive cells, but the details are poorly understood in insects (Turunen and Crailsheim 1996). Nothing is known about the possible role of fatty acid transporters in the apical membrane of midgut cells (Arrese *et al.* 2001). After absorption, fatty acids are converted to diacylglycerols in midgut cells and released to a haemolymph lipoprotein called lipophorin: this is a transport protein which acts as a reusable shuttle and delivers diacylglycerols and other lipids to various tissues (Arrese *et al.* 2001). In the fat body the diacylglycerols are converted to triacylglycerols for storage, and in the larva of *M. sexta* they can be 30 per cent of the wet mass of this tissue. Lipophorin also transports diacylglycerols from fat body to flight muscles during sustained flight. Ryan and van der Horst (2000) recently reviewed lipid mobilization (in response to adipokinetic hormone) and lipid transport in relation to flight. These aspects of lipid biochemistry in insects, which are being used as a model system for comparison with vertebrates, seem to be better known than digestion and absorption.

Glucose transporters such as the well known Na^+-glucose cotransporters have been intensively studied in vertebrates, although little is known about their equivalents in insects (but see Andersson Escher and Rasmuson-Lestander 1999). Evidence summarized by Turunen and Crailsheim (1996) suggests that glucose transport is passive in most insects: transport is unaffected by metabolic inhibitors, depends on concentration gradient, and fructose and unmetabolized 3-O-methylglucose are transported at the same rate as glucose. Crailsheim (1988) found that 3-O-methylglucose injected into the haemolymph of honeybees became equally distributed between midgut lumen and haemolymph in 30 min. It is assumed that fructose transport across the gut wall is also passive, and fructose is then converted to glucose by hexokinase and phosphoglucoisomerase (Bailey 1975). In the fat body, trehalose is synthesized from glucose via hexose phosphates (also intermediates in glycogen synthesis). Like the transport disaccharide of plants (sucrose), it is a non-reducing sugar and less reactive than glucose (Candy *et al.* 1997). Treherne (1958) first showed the conversion of labelled glucose and fructose to trehalose, and pointed out its significance in maintaining a steep concentration gradient for absorption of monosaccharides. Water absorption from the midgut would also increase this concentration gradient (Turunen and Crailsheim 1996). Absorption of sugars is fast and complete. For example, female mosquitoes that have been flown to exhaustion will resume continuous flight within a minute of starting to feed on glucose solution (Nayar and Van Handel 1971). Unfed honeybees given labelled glucose incorporate it into trehalose within 2 min of feeding and there is no loss of label in the excreta (Gmeinbauer and Crailsheim 1993).

These last examples concern insects with initially empty crops. Crop-emptying is in fact, the limiting process for absorption of monosaccharides in insects, and its control has been attributed variously to the osmolality or sugar concentrations of food or haemolymph. Recently the control of crop-emptying has been carefully investigated in honeybees (*Apis mellifera carnica*), using unrestrained bees trained to collect defined amounts of sucrose solution (Roces and Blatt 1999; Blatt and Roces 2001, 2002). As in other insect species, crop-emptying rates measured by volume were inversely related to food concentration. Conversion to sugar transport rates showed that sugar left the crop at a constant rate, independent of food concentration but corresponding closely with the metabolic rate of the bees (Fig. 2.8). It is well known that the metabolic rate of honeybees depends on the reward rate at the food source (Moffat and Núñez 1997; see also Chapters 3 and 6). Haemolymph sugar homeostasis was maintained under all conditions except those involving dilute food and a high metabolic rate (induced by cold); haemolymph trehalose concentration then decreased. Crop-emptying is, therefore, adjusted to the energy demands of the bee, mediated by the trehalose concentration of its haemolymph.

We conclude this section on midgut physiology by considering phenotypic flexibility and whether there are reversible changes in gut surface area and absorption capacity depending on demand, as have been demonstrated in vertebrates (Diamond 1991; Weiss *et al.* 1998). Increased gut size helps to compensate for reduced food quality in grasshoppers (Yang and Joern 1994). Larval *M. sexta* reared on low protein diet allocate more tissue to midgut, although they still grow more slowly (Woods 1999). Woods and Chamberlin (1999) measured proline transport in the posterior midgut of *M. sexta* and found no response to dietary history (Fig. 2.9). They used flat sheet preparations bathed by asymmetrical salines designed to resemble *in vivo* conditions, and measured fluxes of ^{14}C-labelled L-proline, which was transported from lumen to haemolymph 15 times faster than in the reverse direction. However, leucine transport in brush border membrane vesicles from the midgut of *B. mori* (Lepidoptera, Bombycidae) is increased in starved larvae (Leonardi *et al.* 2001). In contrast to data from vertebrates (e.g. Weiss *et al.* 1998) showing upregulation of gut function in response to increased substrate levels, the compensatory responses described above in insects suggest

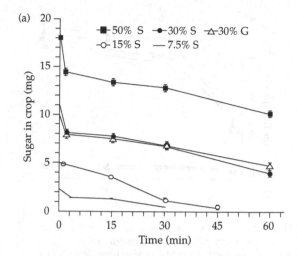

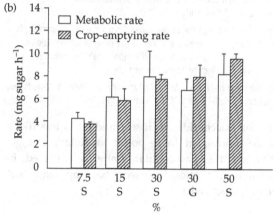

Figure 2.8 Crop-emptying rates and metabolic rates in honeybees collecting sugar solutions of different concentrations. (a) Sugar content of the crop as a function of time in bees which collected 30 µl of each sucrose (S) or glucose (G) concentration (means ± SD). (b) Comparison of crop-emptying and metabolic rates, both expressed as mg sugar h^{-1} (means ± SD). Only the rates for 7.5% sucrose were significantly different from the others.

Source: Reprinted from *Journal of Insect Physiology* **45**, Roces and Blatt, 221–229. © 1999, with permission from Elsevier.

possible upregulation of gut function in response to *declining* nutrients (Woods and Chamberlin 1999). Decreased secretion of digestive enzymes after exposure to low substrate concentrations, or decreased absorption of the products of hydrolysis, does not seem an efficient way to maximize the value of low-nutrient diets (Simpson *et al.* 1995). Passive absorption of nutrients provides a mechanism for matching the rate of absorption to the rate of hydrolysis (Pappenheimer 1993), so absorption

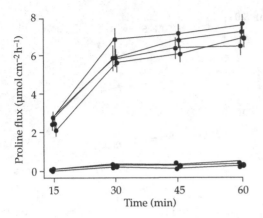

Figure 2.9 Rates of proline transport by posterior midguts of *Manduca sexta* caterpillars (fifth instar) reared on high or low protein diets and then transferred to high or low protein diets for 24 h. Mean ± SE of 6–7 midguts.

Note: The four high traces are fluxes from lumen to haemolymph, and the four low traces are fluxes in the other direction. Lower values at 15 min are due to equilibration of labelled proline. There was no effect of rearing or test diets.

Source: Reprinted from *Journal of Insect Physiology* **45**, Woods and Chamberlin, 735–741. © 1999, with permission from Elsevier.

of monosaccharides in insect midgut is less likely to be modulated in response to dietary change. Whether the paracellular route is involved in nutrient absorption in insects has apparently not been considered.

2.4 Overcoming problems with plant feeding

Half of all insect species feed on living plants. Although few orders have overcome the evolutionary hurdles of plant feeding, those few (especially Orthoptera, Hemiptera, and Lepidoptera) have been extremely successful (Southwood 1972; Farrell 1998). Apart from mechanical obstacles to feeding, herbivores must contend with indigestible cellulose, nutrient deficiencies (especially low nitrogen), and allelochemicals. Often, maximizing nutrient intake while minimizing secondary compound intake requires complex foraging decisions and can prevent the insect from reaching its intake target (Behmer *et al.* 2002). Many of the solutions involve interactions with microorganisms. There is a vast literature in this area and our treatment is highly selective.

2.4.1 Cellulose digestion: endogenous or microbial?

Caterpillars and most other insect folivores are unable to use the huge proportion of plant energy locked up in cellulose. Martin (1991), in reviewing the evolutionary ecology of cellulose digestion, suggests that it is rare because insect herbivores are usually limited by nitrogen or water and not by carbon, so they would derive no particular benefit from exploiting the energy in cellulose. Cellulose digestion is much more likely in wood-feeding (xylophagous) insects or omnivorous scavengers with nutritionally poor diets, especially those whose guts are colonized by microorganisms. The best known cellulose-digesting insects are termites and the closely related cockroaches.

The traditional assumption is that cellulose-digesting enzymes are derived from protozoa or bacteria residing in the hindgut, or fungi ingested with the diet; this assumption is consistent with the independent evolution of the capacity in different taxa (Martin 1991). However, the contribution of endogenous enzymes has been strongly debated (Slaytor 1992). Because insects are almost universally associated with microorganisms, it has been difficult to refute the long-standing hypothesis of derived enzymes, even though most symbionts are located in the hindgut, endogenous cellulase is present in salivary glands and midgut, and cockroaches have far smaller symbiont populations than do termites. Lower termites appear to utilize both endogenous and protozoal cellulases; their specialized hindgut has chambers containing large populations of protozoa, which break down cellulose to glucose and ferment it to short-chain fatty acids, mainly acetate, propionate and butyrate. Higher termites (the majority of termite species) lack hindgut protozoa and appear to utilize endogenous enzymes, except for fungus-growing species in the subfamily Macrotermitinae which acquire additional cellulases from the fungus (*Termitomyces* sp.) that they cultivate and consume.

Cellulase is an enzyme complex capable of converting crystalline cellulose to glucose (Slaytor 1992). In termites and cockroaches its main components are endo-β-1,4-glucanases, which cleave glucosidic bonds along a cellulose chain, and β-1,4-glucosidases, which hydrolyse cellobiose to glucose. Insects apparently lack an exoglucanase active against crystalline cellulose (Martin 1991), but their endoglucanases possess some exoglucanase activity and can be present in large quantities, as in *Panesthia cribrata* (Blattaria, Blaberidae), which feeds on rotting wood (Scrivenor and Slaytor 1994). An endogenous insect cellulase, endo-β-1, 4-glucanase from the lower termite *Reticulitermes speratus* (Rhinotermitidae), was identified by Watanabe *et al.* (1998). It is secreted in the salivary glands, along with a β-glucosidase, and produces glucose from crystalline cellulose. Unhydrolysed cellulose is then fermented to acetate by hindgut protozoa, and this double action could account for the high efficiencies of cellulose digestion mentioned by Martin (1991).

The Macrotermitinae cultivate symbiotic fungi on combs constructed from undigested faeces, and consume fungus nodules and older comb. Besides enzymes, they acquire concentrated nitrogen, because fungi contain reduced quantities of structural carbohydrates (Mattson 1980). The fungus comb in newly founded termite colonies is inoculated with spores carried by alates or collected by foragers (Johnson *et al.* 1981). Genetic techniques have recently been applied to the evolutionary histories of fungus-farming in ants, termites, and beetles (see Mueller and Gerado 2002). Fungiculture has evolved several times independently: multiple origins are evident in certain beetles such as cerambycid larvae, but only a single origin each in ants and termites. This sophisticated form of cellulose digestion has enabled leaf-cutting ants and the Macrotermitinae to become dominant herbivores and detritivores in tropical ecosystems, and fungus-growing beetles are major forestry pests. The emphasis in the literature has been on cellulose digestion, but this is still in dispute where leaf-cutting ants are concerned (Abril and Bucher 2002). Other polysaccharides may be important: leaf-cutting ants and their symbiotic fungi together possess the enzymes necessary to degrade the xylan and laminarin of hemicellulose (D'Ettorre *et al.* 2002). Workers of *Atta sexdens* obtain a large proportion of their nutritional needs from the extracellular degradation of starch and xylan by enzymes of fungal origin (Silva *et al.* 2003).

Scarabaeid dung beetles are extraordinarily successful on a food resource that is patchy and ephemeral, but also rich: dung, especially from ruminants, contains substantial nitrogen (Hanski 1987). Although both larvae and adults feed on dung of mammalian herbivores, only larvae digest cellulose, with the help of bacteria in a hindgut fermentation chamber, and reingestion of faeces. This ensures maximum utilization of the strictly limited amount of food in the brood ball (Cambefort 1991). Larger particles in fresh dung are indigestible plant fragments, which may be macerated by larvae. By contrast, adults have filtering mouthparts that reject such coarse particles. Recently, latex balls of various diameters, manufactured for calibration of Coulter Counter® instruments, were mixed with the preferred dung of 15 species of adult Scarabaeinae (size range 0.05–7.4 g) to determine the maximum size of ingested particles (Fig. 2.10): the range was only 8–50 µm (Holter *et al.* 2002). These very small particles have higher nutritional value because their large surface area to volume ratios promote microbial activity. Fine particles (<20 µm) have a much lower $C:N$ ratio than coarse particles (>100 µm) or bulk dung, and the value for fine particles resembles the $C:N$ ratio for bacteria (P. Holter and C.H. Scholtz, unpublished). Thus,

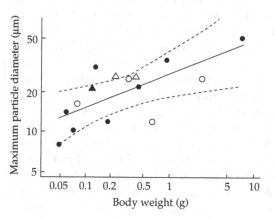

Figure 2.10 Maximum diameter of ingested particles as a function of body mass in 15 species of dung-feeding Scarabaeinae.

Note: Both scales are logarithmic. Empty symbols indicate species preferring rhino or elephant dung.

Source: Holter *et al.* (2002). *Ecological Entomology* **27**, 169–176, Blackwell Publishing.

even beetles preferring the very coarse dung of elephant or rhino are essentially liquid feeders which benefit from microbial protein.

2.4.2 Nitrogen as a limiting nutrient

Because animals consist mainly of protein but plants are mainly carbohydrate, it follows that nitrogen is a limiting nutrient for many herbivores (for a major review see Mattson 1980). Nitrogen in plant tissue ranges from 0.03 to 7.0 per cent dry mass, the highest concentrations occurring in actively growing or storage tissues. Assessment of the importance of nitrogen is complicated by the fact that nitrogen and water contents of foliage vary enormously and tend to vary together, especially when the proportion of structural carbohydrates increases in maturing leaves (Slansky and Scriber 1985). Performance indices of many insect-feeding guilds are strongly correlated with both nitrogen and water contents of the food, which is why larval feeding often occurs early in the growing season (Slansky and Scriber 1985). Insects respond to low nitrogen content by increasing food consumption or the efficiency of nitrogen use (=N gained/N ingested) (Slansky and Feeny 1977; Tabashnik 1982). Both responses are shown in Fig. 2.11 for larvae of *Pieris rapae* (Lepidoptera, Pieridae) feeding on a variety of wild and cultivated plants, mostly Cruciferae (Slansky and Feeny 1977). Much of the research on nitrogen limitation has concerned caterpillars, which accumulate nitrogen for adult reproduction (see Section 2.5.2), but positive responses to nitrogen in larvae may not always benefit pupal and adult stages of a species. For example, increased dietary nitrogen (via fertilizer treatment of food plants) decreases development time of *Lycaena tityrus* (Lepidoptera, Lycaenidae), but also increases pupal mortality and reduces adult size (Fischer and Fiedler 2000). Results from larval stages only should be treated with caution. Climate change has stimulated research on insect herbivory, although many contradictions remain. Plants grown in high CO_2 levels generally have lower foliar nitrogen concentrations (Watt *et al.* 1995), and negative effects on insect herbivores have been ascribed to dilution of plant nitrogen levels by cellulose and other carbon-based

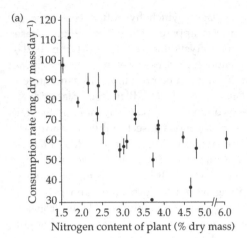

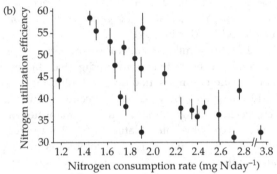

Figure 2.11 Responses of fifth instar *Pieris rapae* larvae to variation in nitrogen content of their food plants (mainly Cruciferae). (a) Rate of consumption (mg dry mass per day) as a function of plant nitrogen content (% dry mass). (b) Nitrogen utilization efficiency as a function of the rate of consumption of nitrogen (mg N per day).

Note: Data are means ± SEs, based on two experiments with various food plants and a nitrogen fertilization experiment.

Source: Slansky and Feeny (1977).

compounds, so that insects must compensate by eating more and their development is prolonged. The situation is more complex because allelochemicals may be affected, more carbon being available for allocation to defensive compounds (Coviella *et al.* 2002).

Although the focus has been on nitrogen as a limiting nutrient in terrestrial systems, Elser *et al.* (2000) have recently demonstrated that terrestrial plants are also poor in phosphorus. Stoichiometric analyses provide a more quantitative way of thinking about the differences between trophic levels, and the C : N and C : P ratios of terrestrial herbivores are 5–10-fold lower than those of foliage. Phosphorus content is inversely related to body mass in insects, and more recently derived orders tend to have lower nitrogen and phosphorus contents (Fagan *et al.* 2002; Woods *et al.* 2004).

Nitrogen is a general indicator of host plant quality, but because other phytochemicals and water vary simultaneously with nitrogen, causality is difficult to prove (Kytö *et al.* 1996; Speight *et al.* 1999; Karley *et al.* 2002). An exception is found in phloem feeders, whose relatively simple but nutritionally unbalanced food provides an excellent opportunity for testing mechanistic relationships between plant quality and insect performance. Nitrogen quality is also important, and this is measured as the concentrations of individual amino acids in phloem sap. Silverleaf whiteflies *Bemisia tabaci* (Aleyrodidae) feeding on cotton plants with and without fertilizer treatment differ greatly in free amino acid pools, especially the proportion of the non-essential amino acid glutamine (Crafts-Brandner 2002). Another rapid adjustment was in amino nitrogen excretion (but not honeydew production), which essentially stopped for whiteflies fed on low-nitrogen plants. Aphids are serious pests of potato crops, and Karley *et al.* (2002) compared several performance parameters of *Myzus persicae* and *Macrosiphum euphorbiae* on young and old potato plants, and on artificial diets mimicking their phloem sap. Decreased performance on older plants is due to changes in the amino acid profile of the phloem sap, especially a dramatic decline in glutamine levels. It is interesting that there was no significant correlation between the C:N ratio of plant tissue and the phloem sap sucrose:amino acid ratio (Karley *et al.* 2002). Incidentally, carbon is even more scarce than nitrogen in a xylem diet, and carbon retention by three species of leafhoppers (Homoptera, Cicadellidae) far exceeds nitrogen retention, excess nitrogen being excreted as ammonia (Brodbeck *et al.* 1993).

Bernays (1986*b*) compared the utilization of a wheat diet in a grasshopper and a caterpillar of similar size, reared under identical conditions. Values of AD were similar, but ECD was lower in the grasshoppers, and this was attributed to their large investment in cuticle mass. Cuticle consists mostly of protein and chitin, which are 16 and 7 per cent nitrogen by mass, respectively. Using rain forest beetles in Borneo, Rees (1986) found that adult Chrysomelidae carried significantly less exoskeleton in proportion to body mass than did representatives of several other beetle families, and attributed this to a shortage of nitrogen in their plant diet. However, this conclusion may be biased by phylogenetic relatedness and allometric considerations: the fraction of body mass in the skeleton increases with increasing body size (Schmidt-Nielsen 1984). The importance of recycling cuticular nitrogen was demonstrated recently in cockroaches eating and digesting their own exuviae (Mira 2000). This behaviour was more common in females, in insects reared on a low-protein diet, and in those deprived of their endosymbiotic bacteria. Mira also speculated that acquiring particular amino acids might be important, rather than nitrogen in general: phenylalanine is abundant in cuticle but scarce in plant tissues, and is selected by grasshoppers (Behmer and Joern 1993). Pea aphids (*A. pisum*) lose in their exuviae about 10 per cent of the total amino acids in the tissues of the adult aphid (Febvay *et al.* 1999).

Contribution of symbionts to nitrogen balance

Many insects possess microbial symbionts which assist with apparently unpromising or deficient diets. Their contributions are diverse and hold much potential in the area of pest management (Douglas 1998). The symbionts may be extracellular, like those in the gut lumen of termites, or intracellular, confined to large cells known as mycetocytes. Mycetocyte symbiosis is best known in Blattaria, Homoptera, Phthiraptera, and Coleoptera living on nutritionally poor diets such as wood, plant sap, or vertebrate blood. Nutritional benefits to the host are often assumed to involve nitrogen, but blood-feeding tsetse flies are provided with missing B vitamins and other insects with sterols (for review see Douglas 1989). Simpson and Raubenheimer (1993*a*) presented a phylogenetic analysis of the effect of mycetocyte symbionts on the ratio of protein to digestible carbohydrate required in insect diets, using data from an extensive literature on artificial diets. Insects with the lowest protein requirements in relation to carbohydrate were those with endosymbiotic bacteria which presumably contribute to nitrogen metabolism. Use of a geometric framework to investigate performance of newborn pea aphids, *A. pisum* (Homoptera, Aphididae), demonstrated an intake target based on an 8:1 ratio of sucrose to amino acids (Abisgold *et al.* 1994).

The best direct evidence that mycetocyte symbionts are important to nitrogen balance comes from aphids. The amino acid composition of phloem sap is unbalanced, and essential amino acids are synthesized by symbiotic bacteria of the genus *Buchnera* (Febvay *et al.* 1999; Douglas *et al.* 2001). These bacteria live inside the host mycetocytes (occupying most of the cell volume) and are transmitted vertically from a female to her progeny. The symbionts convert non-essential to essential amino acids, and also use dietary sucrose extensively in the synthesis of essential amino acids (Fig. 2.12), even when the diet resembles aphid tissues (and not phloem sap) in composition (Febvay *et al.* 1999). Aposymbiotic insects, in which heat or antibiotic treatment is used to eliminate the intracellular microorganisms, have been widely used to investigate interactions between partners. Aphid performance is dramatically reduced after treatment with antibiotics, especially when the insects are reared on diets from which individual amino acids have been omitted (Douglas *et al.* 2001). It has been suggested that symbionts of the silverleaf whitefly *B. tabaci* (Aleyrodidae) are responsible for production of trehalulose (Davidson

et al. 1994), although oligosaccharide synthesis is unchanged in aposymbiotic pea aphids (Wilkinson *et al.* 1997). Yeast-like endosymbionts in the brown planthopper *Nilaparvata lugens* (Delphacidae), a major pest of rice, have high uricase activity and may be recycling nitrogen (Sasaki *et al.* 1996).

Nitrogen recycling also occurs in cockroaches, but here the microbiology is more complicated because they have both a complex hindgut microflora and bacterial endosymbionts, which mobilize urate deposits in the fat body on low-nitrogen diets and convert them to essential amino acids. Nitrogen is often limiting in the diets of these opportunistic scavengers (Kells *et al.* 1999). Termites survive on diets with very high C:N ratios, but possess hindgut bacteria which contribute significantly to the nitrogen economy of their hosts by recycling uric acid nitrogen and by fixing atmospheric nitrogen (Breznak 2000). Nardi *et al.* (2002) have recently drawn attention to the substantial contribution that microbes in the guts of arthropod detritivores may be making to nitrogen fixation in terrestrial ecosystems. Molecular techniques have made it possible to study gut microbes and identify their nitrogenase enzymes without the necessity of culturing difficult organisms, which has hindered such studies in the past (Breznak 2000).

2.4.3 Secondary plant compounds

The molecular structure of secondary plant compounds is far better known than their modes of

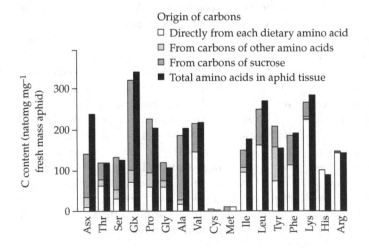

Origin of carbons
☐ Directly from each dietary amino acid
▨ From carbons of other amino acids
▩ From carbons of sucrose
■ Total amino acids in aphid tissue

Figure 2.12 Origin of the carbons of each amino acid in the pea aphid, *Acyrthosiphon pisum*, reared on an artificial diet with balanced amino acid composition as in aphid tissue.

Note: Carbons are derived from amino acids in the food, from other dietary amino acids after conversion, or from dietary sucrose. For each amino acid, the sum of the carbons from different sources is roughly the same as the amount recovered in aphid tissues (black bars).

Source: Febvay *et al.* (1999).

action, and is characterized by great diversity—especially in those plant taxa on which phytochemists have focused their attention (Jones and Firn 1991). In fact, Harborne (1993) considers the three primary areas of biological diversity to be angiosperms, insects, and secondary compound chemistry. Chemical defences have been classified according to various plant–herbivore theories (discussed by Speight *et al.* 1999). In broad terms, a distinction has been made between toxins (produced in small quantities by rare or ephemeral plants) and digestibility-reducing allelochemicals ('quantitative' or deterrent defences, produced in large quantities by long-lived or apparent plants). Mattson (1980) pointed out that toxic compounds are associated with nitrogen-rich plants such as legumes, and many of the toxins are nitrogen-based (e.g. alkaloids, cyanogenic glycosides, non-protein amino acids), although others (cardiac glycosides) are not (Harborne 1993). Moreover, proteinase inhibitors are nitrogen-based but not toxic. Digestibility-reducing allelochemicals, on the other hand, tend to occur in plants adapted to low nitrogen environments (such as *Eucalyptus*) and are carbon-based (tannins, terpenoids). According to the carbon–nutrient balance hypothesis of Bryant *et al.* (1983), the production of defences is costly to plants and involves a trade-off between growth and defence. Nitrogen-based defences increase in concentration when nitrogen availability to plants increases, while concentrations of carbon-based defences decrease (Kytö *et al.* 1996). More specifically, Haukioja *et al.* (1998) have proposed that protein synthesis, because of the requirement for phenylalanine, competes with synthesis of condensed tannins. Conflicting results are common in tests of the carbon–nutrient balance hypothesis (Hamilton *et al.* 2001) and other hypotheses generated to explain patterns in plant defences: see Cipollini *et al.* (2002) for an example involving measurement of many different variables in a thorough study of leaf chemistry and herbivory. The distinction between toxic and deterrent plant allelochemicals can be misleading: a recent review uses the term 'antinutrients' for natural products which reduce nutrient availability to insects (Felton and Gatehouse 1996). Reduction of nutrient availability may involve chemical modification of

the nutrient, formation of complexes with it, or interference with its digestion or absorption, and the effect of antinutrients can be overcome by providing the insects with supplemental nutrients.

Tannins and other phenolics
Feeding experiments involving these aromatic compounds require careful design and have shown some complex effects. For example, Raubenheimer and Simpson (1990) found no interactive effects of tannic acid and protein–carbohydrate ratios in *Locusta*, which is a grass-feeding oligophage, and *Schistocerca*, which is polyphagous and receives greater exposure to tannins in its diet. In the short term, tannic acid had a phagostimulatory effect on *Schistocerca*, and this must be distinguished from compensatory consumption of an inferior diet. More recent use of the geometric approach has shown that tannic acid is more effective as a feeding deterrent in *Locusta* than as a post-ingestive toxin, but its effect is markedly influenced by the proportions of protein and carbohydrate in the food (Behmer *et al.* 2002). Tannic acid has a stronger deterrent effect when foods contain a large excess of carbohydrate relative to protein, whereas locusts are willing to consume relatively large amounts of tannic acid included in protein-rich diets. These authors suggest that this is another factor favouring carbon-based defences in resource-poor habitats.

Eucalyptus species contain very high concentrations of phenols and essential oils (mixtures of terpenoids), yet these do not seem to affect feeding by chrysomelid beetles, which are far more constrained by the low nitrogen content of *Eucalyptus* foliage (Fox and Macauley 1977; Morrow and Fox 1980). This explains the apparent paradox of heavy grazing in spite of these quantitative chemical defences. Phenolics have varied effects, both inhibitory and stimulatory, on the performance of insects (Bernays 1981) and negative effects on feeding and growth may involve a variety of mechanisms, including oxidative stress. Felton and Gatehouse (1996) exclude tannins from their review of plant antinutrients.

Caterpillars maintain strongly alkaline midguts, and Berenbaum's (1980) survey of published gut pH values for 60 species showed that those feeding on woody plant foliage have higher gut pH than those feeding on herbaceous plants. Condensed tannins are characteristic of trees, and high pH may be advantageous in reducing the stability of tannin–protein complexes. The association between phylogeny, diet, and midgut pH was considered further by Clark (1999). Exopterygote insects have near-neutral midguts, and physicochemical conditions in grasshopper guts are apparently not influenced by patterns of host plant use (Appel and Joern 1998), although Frazier *et al.* (2000) suggest that some grasshopper diets pose significant acid–base challenges. In an excellent review, Appel (1994) has stressed the importance of the gut lumen of insect herbivores as the site of interaction of nutrients (often refractory), insect and plant enzymes, allelochemicals, and pathogens. These interactions are affected by widely differing pH, redox conditions, and antioxidant activities. Although alkaline pH weakens protein–tannin binding, this mode of action is now considered less likely than oxidation of phenols to reactive quinones (Appel 1994). Oxidation of allelochemicals to toxic metabolites is, in fact, favoured by high pH. However, low oxygen levels in the gut lumens of herbivores reduce the rates of oxidation of allelochemicals (Johnson and Rabosky 2000). Ascorbate, an essential nutrient for many insects, is an antioxidant that maintains phenols in a reduced state in the gut lumen, minimizing their negative effects, and the recycling of ingested ascorbate may be the biochemical basis of differing tolerances to tannins among caterpillar species (Barbehenn *et al.* 2001).

Antinutrient proteins

Research on plant allelochemicals has shifted primarily to antinutrient proteins, because of their enormous potential in plant biotechnology (Lawrence and Koundal 2002). Development of transgenic crops expressing genes for insect resistance was first based on expression of Bt toxins, but transgenic plants equipped with genes for proteinase inhibitors and lectins are providing interesting opportunities for collaboration between chemists, physiologists, and applied ecologists. Proteinase inhibitors are inducible plant defences that are synthesized in leaves as a direct response to

feeding, not only at the site of attack, but also throughout the plant, although the response declines with plant age. They are also constitutively produced in seeds and storage organs of many staple crops (Jongsma and Bolter 1997). The signalling cascade that is initiated by feeding damage and leads to proteinase inhibitor gene expression is described by Koiwa *et al.* (1997). Proteinase inhibitors work by binding directly to the active sites of the enzymes to form complexes, mimicking the normal substrates but effectively blocking the active sites. Digestion of plant protein is inhibited and the insects are effectively starved of amino acids and prone to amino acid deficiencies.

Soybean trypsin inhibitor was the first proteinase inhibitor shown to be toxic to insects, and the trypsin inhibitors are particularly well known, partly because trypsin is commonly used in screening procedures for proteinase inhibitors (Lawrence and Koundal 2002). Based on primary sequence data, there are at least eight families of serine proteinase inhibitors in plants (Koiwa *et al.* 1997). Cysteine proteinase inhibitors (cystatins) are best studied in rice and are effective against some Coleoptera, whereas serine proteinase inhibitors are most effective against Lepidoptera. Effects on performance are commonly evaluated in insects feeding either on artificial diets containing the proteins or on the transgenic crops (this provides an opportunity to evaluate the effect of plant allelochemicals in natural diets, by comparing herbivore performance with that on unmodified crops). Jongsma and Bolter (1997) present data from numerous studies investigating the effects of proteinase inhibitors on various fitness parameters of insects. Frequently, the results of feeding experiments have been disappointing in comparison to those from *in vitro* experiments with gut extracts and proteinase inhibitors. Moreover, inhibitory effects may be surprisingly poor in transgenic plants, as shown by tomato moth larvae, *Lacanobia oleracea* (Noctuidae) subjected to a soybean inhibitor in artificial diets and in transgenic tomato plants (Fig. 2.13) (Gatehouse *et al.* 1999). There are many factors that may be responsible for such discrepancies, such as expression levels in the plant tissue, inhibitor–enzyme affinity, diet quality,

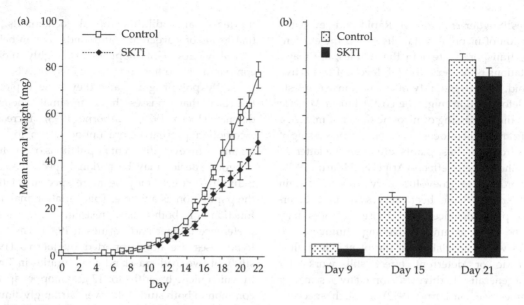

Figure 2.13 Effect of a soybean trypsin inhibitor (SKTI) on the mean mass (±SE) of surviving tomato moth larvae (*Lacanobia oleracea*) feeding on (a) artificial diet and (b) transgenic potato plants.

Note: The inhibitor was expressed at 2% of total protein in potato leaf-based diet and at 0.5% of total protein in potato plants. Reduction of growth was much more apparent for larvae feeding on artificial diet than for those on SKTI-expressing plants.

Source: Reprinted from *Journal of Insect Physiology*, **45**, Gatehouse *et al.*, 545–558. © 1999, with permission from Elsevier.

and rapid adaptation on the part of the insect pest. Insects can compensate for the loss of activity by hyperproduction of endogenous proteinases or by upregulation of new, inhibitor-insensitive proteinases, but both strategies are expensive in terms of amino acid utilization (Broadway and Duffey 1986; Jongsma and Bolter 1997). We might expect better adaptation in specialist herbivores, but the Colorado potato beetle, *Leptinotarsa decemlineata*, is only partially able to compensate for the effects of induced proteinase inhibitors in potato leaves (Bolter and Jongsma 1995).

Most proteinase inhibitors have little effect against phloem-feeding insects, whose diet is rich in free amino acids. However, the activity of lectins against homopteran pests is receiving considerable attention. Lectins are a diverse group of antinutrient proteins, often accumulated in plant storage tissues, which bind to carbohydrates (Peumans and Van Damme 1995). They have multiple binding sites and may bind directly to glycoproteins in the midgut epithelium, or may bind to and clog the peritrophic matrix. Snowdrop lectin, when expressed in transgenic potato crops, confers

partial resistance to aphids (Down *et al.* 1996). Because lectins bind to the gut epithelium and enter the haemolymph, they have the potential to act as carrier proteins for delivery of insect neuropeptides as insecticides, when oral administration of the peptides alone is ineffective (Fitches *et al.* 2002).

Coping with plant allelochemicals

Apart from behavioural avoidance, insects can deal with allelochemicals by detoxifying and excreting them. Polysubstrate monooxygenases (mixed-function oxidases) are non-specific detoxification enzymes, rapidly induced by the presence of toxins. The terminal component is cytochrome P-450, which catalyses the oxidation of toxins to produce more polar compounds that are excreted or further metabolized. Multiple cytochrome P-450 genes are typically expressed simultaneously, hence the wide range of chemically dissimilar toxins (including pesticides) on which they act. To take a familiar insect as example, the specialist tobacco feeder *M. sexta* absorbs ingested nicotine into the midgut cells and metabolizes it to cotinine-N-oxide, which is cleared from the haemolymph by the Malpighian tubules (Snyder *et al.* 1994). Rapid and reversible induction of nicotine metabolism, and the efficient active transport system in the tubules, are major adaptations of *M. sexta* to high levels of this active alkaloid. The reversibility of the response suggests that detoxification might be costly, but in *M. sexta* larvae the processing of nicotine does not impose a significant metabolic cost, nor does the processing of toxins from non-host plants, although the latter do have other adverse effects (Appel and Martin 1992).

According to coevolutionary theory, certain insect species have been successful in counteracting plant defences and those defences may then be used as unique feeding stimulants for species which specialize on the plant. Meanwhile the toxic or deterrent effect still works for other, generalist herbivores (for many fascinating examples see Harborne 1993). In such specialist feeders, allelochemicals may be sequestered for chemical protection, as in insects which sequester cardiac glycosides from milkweeds (Asclepiadaceae). Moreover, the allelochemicals become feeding attractants and oviposition stimulants, contributing to the evolution of close insect–plant associations and the enormous diversity of angiosperm feeders (Farrell 1998). Molecular phylogenies of *Blepharida* (Coleoptera, Chrysomelidae), many of which are monophagous, and *Bursera* species, which are rich in terpenes, show that host shifts have been strongly influenced by host plant chemistry (Becerra 1997). The role of chemistry in plant–animal interactions is beautifully illustrated by the cactus–microorganism–*Drosophila* system of the Sonoran desert (reviewed by Fogleman and Danielson 2001). Four species of *Drosophila* feed and reproduce in the necrotic tissue of five species of columnar cacti, each fly on a specific cactus, the specificity being due to the allelochemistry of the cacti (the presence or absence of certain sterols, alkaloids and terpenoids) and the volatile cues resulting from microbial action. Molecular ecological studies are now focusing on the evolution of the multiple cytochrome P-450 enzymes involved in these *Drosophila*–cactus relationships.

Host-plant specificity
There are two main benefits to feeding on a mixture of plants: selection of a suitable balance of nutrients, and dilution of allelochemicals, so that levels of particular compounds remain below critical values. Grasshoppers are highly mobile compared to other phytophagous insects and generally polyphagous, and they have exploited the fact that grasses have minimal chemical defence (Joern 1979; Harborne 1993). Increased locomotion presents more opportunities for diet mixing. However, different populations of an oligophagous species may be regional specialists, and individual insects may be more specialized than the population as a whole. This variation may be a function of both plant resistance and insect preference (Singer and Parmesan 1993). The latter hypotheses are not supported by Joern's (1979) study of two arid grassland communities in Texas, in which niche breadths for 12 grasshopper species common to both study sites were strikingly similar, or by close field observations of feeding in the polyphagous grasshopper *Taeniopoda eques*, in which single meals of individual females included up to 11 food items (Raubenheimer and Bernays 1993). The more phylogenetically derived insect orders (with more sedentary larvae) have tended towards diet specialization, which suggests greater efficiency than polyphagy, but there is not much supporting evidence for the idea that increased performance on one plant species is correlated with reduced performance on others, or that there is a physiological advantage to be gained from feeding specialization (Jaenike 1990). In a detailed comparison of 20 species of Lepidoptera larvae of varying degrees of specialization, the feeding generalists had slower consumption and growth rates, but this was mainly because they tended to be tree-feeders (Scriber and Feeny 1979). The advantages may be more ecological than physiological (as when a specialized insect becomes adapted to detoxify that plant's allelochemicals, and may even sequester and use the toxins for its own defence). For a readable account of the complexities of host plant specificity see Schoonhoven *et al.* (1998). It is also becoming clear that the study of tritrophic interactions may be necessary to explain host plant selection of some insect herbivores that select plants that are apparently suboptimal in nutritional quality (Singer and Stireman 2003).

2.5 Growth, development, and life history

The amount and quality of food consumed by an insect determines its performance; in the larval stage this is measured as growth rate, development time, body mass, and survival, and in the adult as fecundity, dispersal, and survival (Slansky and Scriber 1985). In the previous section we examined some of the variation in food quality experienced by insect herbivores; now we turn to variation in nutritional needs of insects during growth, development and reproduction. Much of this section concerns trade-offs between competing fitness functions, and these become apparent as a result of developmental plasticity (i.e. environmentally caused variation within a single genotype during development; see also Chapter 5). Insects provide excellent opportunities for experimentation on the nutritional basis of life history trade-offs, so it is not necessary to rely on correlations to show causality.

Nutritional factors are important in explaining the success of holometabolous development. Caterpillars have high protein requirements for rapid tissue growth, but are relatively sedentary, while cockroaches or grasshoppers need more carbohydrate to sustain higher activity levels, but their growth rates are lower (Bernays 1986b; Waldbauer and Friedman 1991). Caterpillars have double the consumption rate, double the gut capacity, and ECD values which are 50 per cent higher than do acridids. They also produce and maintain much lighter integuments: The cuticle of acridids is 10 times as heavy, and up to 50 per cent of total dry mass excluding the gut contents (Bernays 1986b).

2.5.1 Development time versus body size

Three traits central to life history theory are closely interrelated: adult size, development time, and growth rate. It is commonly accepted that there is a trade-off between short development time and large adult size (assuming constant growth rates), but an organism that grows at a high rate can achieve both (Arendt 1997; Nylin and Gotthard 1998). These negative associations between traits are exacerbated by stressful conditions, suggesting competition between different organismal demands

for limited resources. Physiological studies aimed at elucidating the mechanistic basis of life history trade-offs were reviewed recently by Zera and Harshman (2001).

Extending development time increases the risk of predation, but so can high growth rates which depend on increased feeding rates. Bernays (1997) clearly demonstrated the risks that caterpillars face from parasitoids and predators when they feed. Examination of feeding behaviour throughout the fourth and fifth instars of *Helicoverpa armigera* caterpillars shows that exponential growth is sustained more by increased ingestion rates than increased time spent feeding, especially during the late fifth instar which is most susceptible to bird predation (Barton Browne and Raubenheimer 2003). The fitness cost of high growth rate in terms of predation risk has been tested experimentally by using photoperiod to manipulate growth rate in the wood butterfly *Pararge aegeria* (Nymphalidae) (Gotthard 2000). Shorter day length induced slower growth, corresponding to late summer conditions when larvae enter diapause in the pupal stage, and this was accompanied by 30 per cent lower mortality due to a generalist predator, *Picromerus bidens* (Heteroptera, Pentatomidae), introduced to the cages. In seasonal environments development time is complicated by diapause, which only occurs in certain stages—this situation favours genotypes capable of plasticity in growth rate (Nylin and Gotthard 1998). Butterflies such as *P. aegeria* can either speed up their development to produce an additional generation before winter, or slow down and enter diapause. Gotthard (2000) distinguishes between the instantaneous mortality risk of the fast-growing caterpillars in his experiment, and their total mortality risk during the larval stage, which might actually be lower because of the shorter development time. Flexible growth strategies have recently been investigated in an alpine beetle, *Oreina elongata* (Chrysomelidae), and late season light conditions led to an increase in growth rate and shorter development time, but no change in prepupal weight (Margraf *et al.* 2003). The authors argue that if 'catch-up growth' occurs in this species, in which a short and unpredictable growing season might be expected to select for rapid growth, then it may be common in temperate insects.

Classic examples of the trade-off between body size and development time are seen in male butterflies which emerge first (protandry) and are consequently smaller than females (Lederhouse *et al.* 1982). However, the assumption that both sexes are growing at the same rate is not always true. Males of *Pieris napi* which develop directly instead of entering diapause are under severe time constraints and respond to selection for large size and protandry by increasing their growth rate (Wiklund *et al.* 1991).

Prolonged development can be viewed as a means of increasing food consumption on sub-optimal foods (Slansky 1993). If *M. sexta* are exposed to low dietary protein levels as early instars, low growth rates persist in the fifth instar even after transfer to a better diet (Woods 1999). Fig. 2.14 demonstrates how supernumerary moults by larvae of the African armyworm *Spodoptera exempta* (Noctuidae) enable them to reach the same final size when they are reared on poor quality grasses (Yarro 1985). Some female insects undergo an additional larval instar in order to store more nutrients for oogenesis (e.g Stockoff 1993). The seed beetle *Stator limbatus* (Bruchidae) varies by an order of magnitude in adult body size, due to resource competition between multiple larvae in a single seed, but egg sizes do not differ much and a longer

development time allows the initially smaller progeny of small females to pupate at the same size as those of large females (Fox 1997). This can be considered another example of 'catch-up growth'.

Laboratory selection experiments using *Drosophila* have been a powerful research tool for demonstrating life history trade-offs, such as that between extended longevity and early female reproduction. In laboratory-reared *D. melanogaster*, increased food quality or quantity is correlated with an increase in reproduction and decrease in longevity and starvation resistance. This suggests that the trade-off between reproduction and survival can be manipulated by diet. Simmons and Bradley (1997) used supplementary live yeast to explore the quantitative basis of this trade-off in long-lived (O) and control (B) populations of *D. melanogaster*. Diet enrichment caused both O and B females to produce more eggs, and both lines showed reduction in energy stores. However, the trade-off is not quantitative (Fig. 2.15): Decreasing somatic storage does not account for increasing egg production, and most of the additional energy allocated to

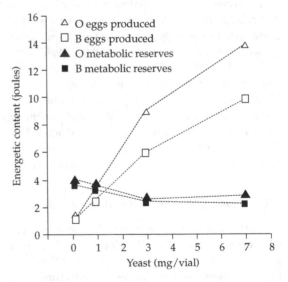

Figure 2.15 Changes in reproductive and somatic energy (joules) in female *Drosophila melanogaster* in response to supplementary yeast in the diet. Energy content was calculated for the eggs produced and the lipid and carbohydrate reserves in long-lived (O) and control (B) populations.

Source: Reprinted from *Journal of Insect Physiology*, **43**, Simmons and Bradley, 779–788. © 1997, with permission from Elsevier.

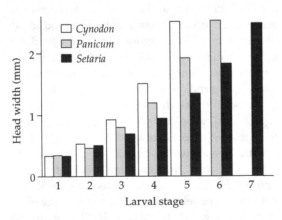

Figure 2.14 Effects of food quality on size and number of larval instars in *Spodoptera exempta* (Noctuidae). Super-numerary moults are used to achieve final body size (measured as head width) in larvae feeding on two nutritionally poor grasses, *Panicum* and *Setaria*.
Source: Data from Yarro (1985).

reproduction comes from the supplementary yeast. Divergence in life history characters of O and B flies during selection are discussed by Simmons and Bradley (1997), and the O flies appear to have been selected for the ability to acquire nutrients. Harshman and Hoffman (2000), among others, have drawn attention to possible artefacts with laboratory selection in *Drosophila*, such as strong directional selection, constraints on normal behaviour such as dispersal, and an overabundance of food. The latter tends to favour adaptive responses involving the storage of energy reserves (glycogen and triacylglycerol), which are commonly measured in laboratory selection experiments, rather than conservation of energy by lowering metabolic rate, which might be more evident in the wild. Section 4.5 deals with laboratory evolution and comparative studies in relation to water balance of *Drosophila*.

Developmental trade-offs may not be apparent if larvae are provided with discrete resources (van Noordwijk and de Jong 1986). In mass-provisioning solitary Hymenoptera, adult body size is controlled by maternally provided resources in a protected environment (Klostermeyer *et al.* 1973). An energy budget for reproduction is easily constructed by collecting either intact provisions, or fully grown larvae and their faeces, from sealed brood cells. The pollen–nectar mixture provided by female solitary bees is a high-quality food, and high assimilation efficiencies have been measured in carpenter bees *Xylocopa capitata* (Anthophoridae) (Louw and Nicolson 1983) and leafcutter bees *Megachile pacifica* (Megachilidae) (Wightman and Rogers 1978). Female offspring receive larger provisions than males and are larger than males, and this greater investment has implications for the sex ratio of the offspring, males being cheaper to produce (Bosch and Vicens 2002). These authors verified the use of body size as an estimate of production costs in bees by showing that weight loss of *Osmia cornuta* (Megachilidae) throughout the life cycle did not differ significantly between the sexes, in spite of differences in metabolic rate, water content, cocoon construction, and development time. Male size dimorphism in bees (involving different size classes) is also controlled by female provisioning decisions (Tomkins *et al.* 2001). Similar relationships

between provision weight and adult weight apply to solitary wasps, except that larvae are given prey such as paralysed spiders—nicely demonstrated by Marian *et al.* (1982) with energy budgets for a wasp nesting in the holes of electrical sockets. The digger wasp *Ammophila sabulosa* (Sphecidae) provisions cells with caterpillars of varying size, but offspring size is controlled by a flexible provisioning strategy which results in the same total weight of prey in each cell (Field 1992).

2.5.2 Developmental trade-offs between body parts

Metamorphosis allows resources to be redistributed among body compartments. Holometabolous insects are therefore ideal organisms for examining how resources accumulated during larval stages are allocated to reproductive or somatic tissues. Most of the growth of imaginal discs occurs in a closed system after the larva stops feeding. Removal of hind wing imaginal discs from larvae of a butterfly results in disproportionately large forewings, and use of juvenile hormone treatment to reduce horn development in male dung beetles, *Onthophagus taurus* (Scarabaeidae), leads to a compensatory increase in size of the compound eyes, which develop in close proximity to the horn (Nijhout and Emlen 1998). In these examples, competition between body parts is suggested by increased growth of one trait at the expense of growth in another, with no change in overall body size. Horn development is dimorphic in *O. taurus*, occurring only in males above a threshold size. The threshold is lower in populations subsisting on poor quality food (cow manure rather than horse dung) (Moczek 2002). In case-building caddis flies, larval resources can be manipulated by inducing them to build new cases and produce additional silk. Stevens *et al.* (2000) have demonstrated empirically that a short-lived caddis species preserves abdomen size (an index of reproductive allocation) at the expense of the thorax, while a long-lived species preserves thorax size in order to maintain longevity. This flexible trade-off between larval defence and adult body size again implies the partitioning of finite resources between parts of the body.

Adult feeding and reproduction in Lepidoptera

Butterflies provide excellent material for examining allocation of larval nutrients to reproduction or body building, which essentially means allocation to abdomen or thorax. In two species of Nymphalidae that feed only on nectar as adults, thorax mass decreased with age but abdomen mass decreased more, so flight ability was probably not impaired. In contrast, pollen feeding by a third nymphalid, *Heliconius hecale*, led to increases in both thorax and abdomen mass with age (Karlsson 1994). Pollen feeding among adult butterflies is unique to *Heliconius*, and is associated with long adult life and eggs laid singly over a prolonged period. Boggs (1981) predicted that the ratio of reproductive reserves to soma at eclosion would vary inversely with expected adult nutrient intake, and directly with expected reproductive output (assuming that organisms are equivalent in terms of larval nutrition). Her model was supported by data on nitrogen allocation in three species of closely related heliconiines, one of which does not collect pollen. Females obtain nitrogen from larval feeding, adult pollen feeding, and male spermatophores contributed during mating. Because the abdomen of a newly eclosed butterfly consists mainly of reserves stored in fat body, haemolymph, and developing oocytes, the ratio of abdomen total nitrogen to whole body total nitrogen can be used as an estimate of the allocation of larval resources to reproduction. Pollen feeding is unusual in butterflies, but predictions concerning larval and adult nutrient allocations to reproduction are supported by subsequent studies on other butterflies with different life histories (e.g. May 1992).

Nutrients acquired by adult foraging or from males during mating are renewable, while those accumulated during the larval stage are not. This leads to the distinction between capital breeders with non-feeding adults, and income breeders which accumulate resources for reproduction in both juvenile and adult stages. Nutrient allocation dynamics have recently been examined in more detail in nectar-feeding Lepidoptera in which proteins carried over from the larval stage play a major role in adult fecundity. Boggs (1997) used radiotracers to examine the use of glucose and amino acids acquired in larval and adult stages in two nymphalid butterflies. Glucose and amino acids labelled with ^{14}C and ^{3}H were painted on leaves to assess larval contributions, or included in nectar solutions to assess adult contributions. Because the adult diet is carbohydrate-rich, incoming glucose is used in preference to stored glucose (storing and then remobilizing nutrients incurs additional costs). By contrast, nitrogen is scarce in the adult diet, and juvenile reserves of amino acids are used throughout adult life. Male nutrient donations at mating, assessed by mating females with males which were labelled as larvae, are less predictably allocated, being immediately used in egg production (Boggs 1997). Stable isotopes have provided information on the dietary sources of amino acids used in egg manufacture by a day-flying hawkmoth, *Amphion floridensis* (O'Brien *et al.* 2000, 2002). These authors fed larvae on grape leaves (*Vitis*, C_3 species) and adults on sucrose purchased as either beet sugar or cane sugar (C_3 and C_4 plants, respectively). C_3 plants are substantially depleted in ^{13}C relative to C_4 plants, so the dietary sources of the carbons in specific egg amino acids can be identified. Essential amino acids originate entirely from the larval diet, whereas nonessential amino acids are synthesized from nectar sugar (Fig. 2.16). Amino acids in nectar contribute insignificantly to egg provisioning. After initial use of larval carbon sources, adult nectar meals provide 60 per cent of the carbon allocated to eggs, but the need for essential amino acids places an upper limit on their use in reproduction. Note that aphids also derive amino acids from dietary sucrose (Fig. 2.12), but their symbionts can synthesize the carbon skeletons of essential amino acids.

Re-allocation of larval nutritional resources can occur after metamorphosis. Flight muscle may be histolysed to provide amine groups for synthesizing non-essential amino acids in egg manufacture (Karlsson 1994). Alternatively, longevity may be favoured at the expense of reproduction, and oocytes are then resorbed when adult food is limited (Boggs and Ross 1993).

Larval performance is more important in Lepidoptera that do not feed as adults. Even as immatures, the sexes differ in food consumption and other performance criteria. Stockoff (1993)

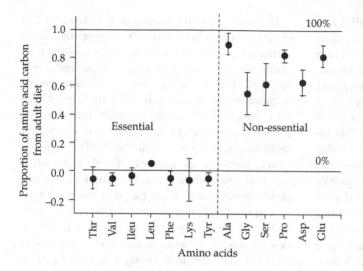

Figure 2.16 Proportion of amino acid carbon derived from the adult diet in eggs of the hawkmoth *Amphion floridensis*, measured using stable isotopes.

Note: Data are for eggs laid on day 12, when carbon isotopic composition has stabilized (see O'Brien *et al.* 2000). All essential amino acids are within measurement error of zero, indicating their exclusive origin in the larval diet.

Source: O'Brien *et al.* (2002). *Proceedings of the National Academy of Sciences of the USA* **99**, 4413–4418. © 2002 National Academy of Sciences, U.S.A.

reared gypsy moth larvae, *Lymantria dispar* (Lymantriidae) through several instars with differing access to two artificial diets, and found that an initial preference for high protein content shifted to one for high lipid content, especially in male larvae. Males need lipid for flight fuel, but female moths do not fly. Neither sex feeds as adults and there is no need to search for adult food plants. Moths eclose with mature eggs (1300 per female under laboratory conditions) which contain 50 per cent of the nitrogen assimilated by the larvae (Montgomery 1982). Incidentally, it is Lepidoptera with this type of life history that are most likely to reach outbreak densities in homogeneous food plant environments (Miller 1996). Spring-feeding forest Lepidoptera in the Geometridae and Lymantriidae exhibit a high incidence of wing reduction in females, although Hunter's (1995) phylogenetic analysis showed no statistically significant increase in fecundity.

Flight polymorphism

Polyphenisms or developmental switches are irreversible changes in phenotype in response to environmental information during a critical phase of development; as in the castes of social insects or the wet and dry seasonal phenotypes of some butterflies. Aphids exhibit a complex sequence of generations with different phenotypes, and it is commonly accepted that the development of winged aphids is a result of poor nutritional

quality of their host plants, although this hypothesis is only partially supported by the accumulated evidence (Müller *et al.* 2001). All insect polyphenisms are likely to be controlled by changes in endocrine physiology (Nijhout 1999), and the best evidence for endocrine control comes from wing polymorphism in crickets. This arises from a combination of genetic and environmental variation, and is controlled by elevated titres of juvenile hormone which block the development of wings and flight muscle during a critical period of development (Zera and Denno 1997). The term polymorphism is often used when there is a genetic component to the morphological differences. However, the terms polymorphism, polyphenism, and phenotypic plasticity are sometimes not carefully distinguished (see Section 5.2.1 for discussion of terminology).

Wing polymorphism is a commonly studied dispersal polymorphism because it involves discontinuous variation in flight capability, and wing morphs are easily recognized (reviewed by Zera and Denno 1997). The winged morph usually has fully developed, functional flight muscles and lipid stores for flight fuel, while the flightless morph has short wings, non-functional flight muscles and reduced lipid stores. In some species a second flightless morph is derived from long-winged adults by histolysis of flight muscles (although this morph was not distinguished from flight-capable morphs in earlier studies). Flightless females normally show increased fecundity, and histolysis of flight muscles coincides with ovarian growth in long-winged female crickets. These correlations have been interpreted as demonstrating a fitness trade-off between flight ability and reproduction, but are not sufficient to prove a causal relationship. However, recent nutritional studies provide unequivocal evidence for this fitness trade-off. Gravimetric feeding trials on *Gryllus* species have demonstrated that ovarian growth in flightless morphs (either natural or hormonally engineered) may be due either to increased food consumption or to relative allocation of the same quantity of absorbed nutrients, depending on the species. Only the latter situation represents a trade-off (Zera and Harshman 2001). Thus, reduced nutrient input can magnify a trade-off, whereas an increase in nutrients can eliminate it. This has been emphasized by measuring nutritional indices for three morphs of *G. firmus* (long-winged, short-winged, and flightless morphs with histolysed flight muscles). Values of ECD were significantly elevated in both types of flightless morph: compared to the flight-capable morph, flightless morphs converted a greater proportion of absorbed nutrients into body mass, mainly ovarian mass, and allocated a smaller proportion to respiration. Low-nutrient diets increased the discrepancy in ECD values, indicating that the trade-off between respiration and early reproduction was magnified (Zera and Brink 2000). Lipid accumulation in the first few days of adult life results in triglyceride reserves which are 30–40 per cent greater in the flight-capable morphs of *G. firmus*, and the magnitude of this difference suggests that limited space in the abdomen could also be a factor in the trade-off between lipid accumulation and ovarian growth (Zera and Larsen 2001). Newly emerged male and female dragonflies, *Plathemis lydia* (Libellulidae), are similar in mass and the mass of individual body parts, but the abdomens of females then increase fivefold in mass owing to ovarian development, and the thoraxes of males increase 2.5-fold in mass as a result of flight muscle growth (Marden 1989). Such a high investment in the thorax carries a cost in that gut mass and body fat content are minimal in territorial male dragonflies.

The widespread occurrence of flightlessness and wing polymorphism in insects suggests that flight carries substantial fitness costs in the construction, maintenance, or operation of wings and flight muscles. This is borne out by differences in respiration rate between the pink flight muscle of fully winged *G. firmus* and the white muscle of the short-winged adults or of long-winged adults after flight muscle histolysis (Zera *et al.* 1997). The flying insects surveyed by Reinhold (1999) had resting metabolic rates which were about three times those of non-flying insects. Wing polymorphism is a type of dispersal polymorphism, and Roff (1990), in reviewing the ecology and evolution of flightlessness in insects, concluded that secondary loss of wings is more frequent among females because of the trade-off with fecundity and is most likely in stable environments. Males remain mobile to find mates, and habitat fragmentation may select for rapid evolutionary change in flight-related morphology in both sexes, seen in an increased investment in the thorax as habitat area declines (Thomas *et al.* 1998).

2.6 Temperature and growth

2.6.1 Thermal effects on feeding and growth

In ectotherms, higher temperatures increase growth rates and decrease development times, and generally result in smaller adult body sizes (Atkinson 1994; Atkinson and Sibly 1997). In *Drosophila*, small body size resulting from development at high temperatures is due mainly to decreases in cell size (Partridge *et al.* 1994), although changes in cell number can also be involved (reviewed in Chown and Gaston 1999). The selective advantage of smaller size at higher temperatures, or larger size at lower temperatures, has long been considered obscure (Berrigan and Charnov 1994). Indeed, several authors have considered larger size at low temperatures an epiphenomenon of differences in the responses of growth and of differentiation to temperature, and therefore not adaptive at all (Chapter 7). Likewise, Frazier *et al.* (2001) have recently demonstrated interactive effects of hypoxia and high temperature on the development of *D. melanogaster*, and a discrepancy between oxygen delivery and oxygen demands is a possible explanation for smaller body size in ectotherms developing at high temperatures. Alternatively, other studies have suggested advantages to large size at lower temperatures, but not necessarily to small size at higher temperatures (e.g. McCabe and Partridge 1997; Fischer *et al.* 2003; see also Chapter 7). Given substantial variation in final body size relative to environmental temperature in the field (Chown and Gaston 1999), and that mechanisms of temperature-related size variation have yet to be resolved (Berrigan and Charnov 1994; Blanckenhorn and Hellriegel 2002), this field remains interesting and productive. Recently, stoichiometric work has started providing additional insights into the implications of temperature-related body size variation. Chemical analyses of *Drosophila* species show substantial variation in nitrogen and phosphorus contents, with the N and P contents of individual species being positively correlated with each other and with the N and P composition of the breeding sites (Jaenike and Markow 2003). Moreover, a variety of cold-acclimated organisms, including *Drosophila* species, exhibit substantially increased levels of N and P, as a result of both concentration changes and larger body sizes at low temperatures (Woods *et al.* 2003). It might, therefore, be useful for herbivores to forage on plants growing in cool microenvironments.

There is an extensive literature on the temperature dependence of larval growth rates in insects (Ayres 1993; Casey 1993; Stamp 1993). Most laboratory studies have been carried out under constant temperatures, but growth may be stimulated or inhibited in a fluctuating temperature regime, which better resembles field conditions (Ratte 1985). As an example of differing thermal response curves, Knapp and Casey (1986) compared the temperature sensitivity of growth rates in two caterpillar species. Eastern tent caterpillars *Malacosoma americanum* (Lasiocampidae), which hatch early and behaviourally thermoregulate, grow at rates that are highly dependent on temperature. By contrast, those of gypsy moth caterpillars *L. dispar* (Lymantriidae), which hatch later and are thermal conformers, are independent of temperature at ecologically relevant temperatures of 25–30°C. The

African armyworm *Spodoptera exempta* (Noctuidae) shows a density-dependent polymorphism resembling that of migratory locusts, and the *gregaria* phase develops much faster than the *solitaria* phase (Simmonds and Blaney 1986). Casey (1993) suggested that *solitaria* caterpillars are thermal conformers, switching to a thermoregulating strategy in the *gregaria* phase. However, Klok and Chown (1998a) found no evidence of behavioural thermoregulation in *gregaria* in the field and concluded that different development rates recorded in several studies most likely result from differences in utilization of food.

Behavioural thermoregulation (see Chapter 6) has physiological consequences for feeding, growth and reproduction, exemplified in field and laboratory studies of acridid grasshoppers. Populations of *Xanthippus corallipes* at different altitudes maintain similar metabolic rates, because smaller body mass at high elevations, which leads to relatively higher mass-specific metabolic rates, is offset by lower T_b in the field. Stabilization of field metabolic rate enhances total egg production, assessed by counting *corpora lutea* of females at the end of the reproductive season (Ashby 1998). During sunny days, *Melanoplus bivittatus* regulates its body temperature (T_b) between 32–38°C and has an essentially non-functional digestive system during the night (Harrison and Fewell 1995). The rate of digestive throughput is strongly limiting at low temperatures but this process has a high Q_{10} so growth is fast at high temperatures. Variables such as food consumption and faecal production are more temperature-sensitive (have a higher Q_{10}) than variables reflecting chewing and crop-filling rates. Within the range of preferred T_b, ingestion and processing rates are well matched. Faster growth at high temperatures is more an effect of faster consumption than of increased efficiency, and this is also true of caterpillars. In general, thermoregulating caterpillars and grasshoppers have higher Q_{10} values for feeding and growth than thermoconformers, which must be able to grow over wide temperature fluctuations in the field. Note that endothermic insects will also experience intermittent digestive benefits.

Social caterpillars in the family Lasiocampidae (tent caterpillars) show highly synchronized bouts of foraging activity, alternating with digestive phases when they rest in or on their tent, and these activity patterns can be electronically recorded under field conditions (Fitzgerald *et al.* 1988; Ruf and Fiedler 2002). The duration of both foraging bouts and digestive phases is inversely related to temperature in *Eriogaster lanestris*, which has an opportunistic feeding pattern in relation to thermal conditions. In contrast, another lasiocampid, the eastern tent caterpillar *M. americanum*, forages only three times a day, possibly because predation risk may outweigh the need for feeding efficiency. The effects of temperature on caterpillar foraging and growth are reviewed by Casey (1993) and Stamp (1993), but generalizations are difficult because such effects are intimately connected with the natural history of a species (which includes constraints due to predators and parasites).

The majority of caterpillar species are solitary, palatable, and cryptic and do not have the option of thermoregulation. Kingsolver (2000) measured peak consumption and growth rates around 35°C in fourth instar *Pieris brassicae* in the laboratory, and integrated these with the operative temperatures of physical models placed under leaves in a collard garden to demonstrate that infrequent high temperatures can make disproportionate contributions to caterpillar growth—because growth rates are so much faster at higher temperatures. In *M. sexta*, another thermoconformer (Casey 1976a), short-term consumption and growth rates show similar shallow thermal response curves, with Q_{10} values less than 2.0 for temperatures up to 34°C (Kingsolver and Woods 1997). When the thermal sensitivity of growth and its component processes in *M. sexta* was compared by measuring growth rate, consumption rate, protein digestion, methionine uptake, and respiration rate over the range 14–42°C, the thermal performance curve for growth rate was most similar to that for consumption rate, and declining growth above 38°C could not be ascribed to decreased digestion or absorption rates or increased respiration rates (Fig. 2.17). Reynolds and Nottingham (1985) similarly found that nutritional indices in *M. sexta* were unaffected by temperature, although chronic exposure resulted in smaller size at high temperatures.

2.6.2 Interactions with food quality

Caterpillars of *M. sexta* have again been used as a model system in examining nutritional interactions between temperature and dietary protein levels. Although low temperatures reduce rates of consumption and growth, low protein levels lead to increased consumption rates through compensatory feeding but may not affect growth rates. In short-term experiments (4 h), fifth-instar larvae of *M. sexta* failed to compensate for low protein levels at the most extreme temperatures of 14 and 42°C, but long-term experiments (duration of the fifth stadium) over a narrower temperature range showed little evidence of interactions between temperature and dietary protein (Kingsolver and Woods 1998). These authors suggest that compensatory feeding responses may be less effective in diurnally fluctuating temperatures than in the constant temperatures in which they are commonly examined. This study was extended by examining interactive effects on instars 1–3, 4, and 5 separately, and there were some striking differences between instars (Petersen *et al.* 2000). Mean growth rate was highest at 34°C during the first three instars but highest at 26°C during the fifth instar (the latter is the temperature at which laboratory colonies of *M. sexta* are maintained). Fifth instar caterpillars were surprisingly sensitive to high temperatures, which is significant in view of the fact that most studies of caterpillar nutritional ecology use this instar (Petersen *et al.* 2000). The gypsy moth is a spring-feeding forest insect which develops during a period of declining leaf nitrogen and increasing ambient temperatures. Lindroth *et al.* (1997) found interactive effects of temperature and dietary nitrogen on several performance parameters measured through the fourth instar, but there were no interactions over the larval period as a whole. The combination of poor quality food and low temperature can result in exceptionally slow growth and long life cycles, as in the Arctic caterpillar *Gynaephora groenlandica* (Lymantriidae) which has a development time of 7 years in spite of well developed basking behaviour (Kukal and Dawson 1989).

Temperature's effects become even more complex when allelochemicals or other trophic levels are

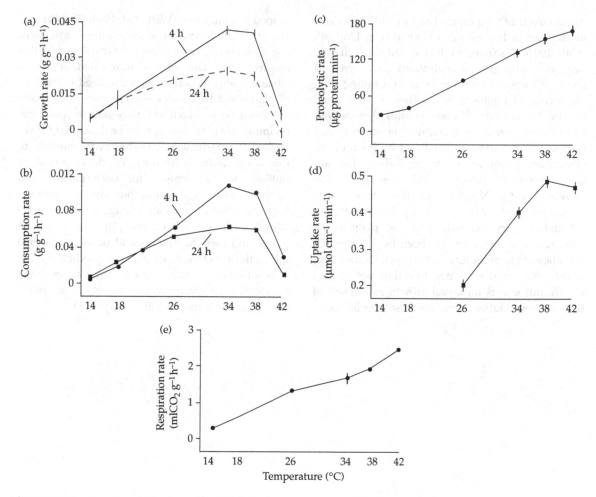

Figure 2.17 The thermal sensitivity of growth and feeding in *Manduca sexta* caterpillars, illustrated by performance curves measured over the temperature range 14–42°C for (a) mass-specific growth rate, (b) mass-specific consumption rate, (c) proteolytic digestion rate, (d) methionine absorption rate, and (e) mass-specific respiration rate.

Note: Growth and consumption rates (a and b) were measured over 4 and 24 h.

Source: *Physiological Zoology*, Kingsolver and Woods, **70**, 631–638. © 1997 by The University of Chicago. All rights reserved. 0031-935x/97/7006-96117$03.00.

involved. Stamp (1990) measured growth of caterpillars (*M. sexta*) as influenced by interactions between temperature, nutrients, and a common phenolic, rutin. Numerous interactions were found, the most important being that the effects of diet dilution and rutin on larval growth rate were a function of temperature. More time was spent moulting at low temperatures and when rutin was included in the diet, resulting in lower growth rates. In temperate regions, interactions of temperature, food quality, and natural enemies are most likely to affect gregarious spring-feeding caterpillars. In contrast, solitary behaviour reduces predation risk but forces caterpillars to forage in microhabitats that are suboptimal in terms of food quality and temperature (for review see Stamp 1993).

Finally, the interplay between temperature and nutrition is critical in the context of global climate change. Reduced foliar nitrogen is a consistent response of plants grown in enriched CO_2 atmospheres (Ayres 1993; Lincoln *et al.* 1993; Watt *et al.* 1995). Insect performance may be altered by the direct effects of temperature and by indirect effects of changes in leaf nitrogen content (e.g. Lindroth *et al.* 1997). Changes in leaf quality are due to decreased nitrogen concentrations and increased levels of carbohydrates (e.g. starch) resulting from the increased photosynthesis. There is contradictory evidence from different studies that carbon-based allelochemicals increase due to increased carbon accumulation, as predicted by Bryant *et al.* (1983), but phenolics are more likely to increase than terpenoids (Bezemer and Jones 1998). The increased leaf $C:N$ ratio is predicted to lead to a decline in performance of many herbivores, but the details vary depending on the plant–insect pair. As usual, research has been biased towards leaf-chewing Lepidoptera, mainly agricultural pests, which show generally negative (but not always significant) effects on larval growth as a result of the nitrogen dilution effect and resulting increases

in food consumption (Watt *et al.* 1995; Coviella and Trumble 1999). Many orders, including Orthoptera and Coleoptera, have been almost ignored in this research area. One of the anticipated effects of climate change is that the synchronized phenology of plants and herbivores, well known for temperate forest pests which are early-season specialists on immature plant tissue, may be disrupted (Ayres 1993). For a particular plant–herbivore interaction, the temperature sensitivity of development is unlikely to be identical for both parties, and temperature change will favour one or the other. Multiple effects of climate change are a powerful stimulus for research on the physiology of both plants and their herbivores and on the ecological interactions between them, but predictions are difficult when the effects of enhanced atmospheric CO_2 tend to be specific to each insect–plant system (Coviella and Trumble 1999; Chapter 7).

CHAPTER 3

Metabolism and gas exchange

The concept that insect respiration depends only on diffusion supplemented in larger species by ventilation is in need of an overhaul: the situation is much more complex.

Miller (1981)

Insects, like all living organisms, are far-from-equilibrium, dissipative structures. That is, they actively take up energy (and nutrients) and in doing so alter both themselves and their surrounding environment. Initially, the changes in both directions might appear insignificant, but on a longer time scale their impact can be profound (Brooks and Wiley 1988) (insect outbreaks are, perhaps, the best testimony to this). The dissipation of energy to the environment enables insects to undertake the enormous behavioural repertoire that has not only contributed to their success as a group, but which has also captivated human curiosity. This behaviour is usually the first response to a changing environment, and has been labelled the better part of regulatory valour (Bennett 1987). In response to unsuitable environmental conditions, insects might either leave by migrating or seek more suitable microclimates. If such behavioural regulation is insufficient to modulate the effects of change, insects can also avoid adverse conditions through physiological modification in the form of diapause, aestivation, or some form of dormancy (Danks 2000, 2002; Denlinger 2002). Active stages are also capable of considerable physiological responses to changing conditions.

This suite of responses to the environment, ranging from migratory behaviour to physiological regulation, is dependent on energy produced by the metabolism of substrates. Moreover, metabolism enables insects to acquire additional resources, to grow and reproduce, and to participate in the manifold interactions that characterize any given community. Although metabolism has both anabolic and catabolic components, in this chapter we will be concerned mostly, but not exclusively (see Wieser 1994), with catabolism. That is, we are concerned with the largely oxidative metabolism of substrates for energy provision, and the ways in which oxygen required for this process is transported to the tissues and carbon dioxide removed from them (water balance is dealt with mostly in Chapter 4). Although the rate of the entire process is often termed metabolic rate, especially where oxygen uptake and CO_2 production rates are concerned, it is important to make a distinction between oxidative catabolism as a cellular level process (respiration), and gas exchange as the physical transfer of gases between the atmosphere and the tissues/haemolymph (Buck 1962; Lighton 1994).

In insects, metabolism during both rest and activity (especially flight) is generally aerobic. However, occasionally ATP provision can be via anaerobic metabolism (Gäde 1985), although this is most common in special situations (Conradi-Larson and Sømme 1973). During more normal pedestrian activity, anaerobic metabolism only accounts for a small proportion of ATP production, and it is virtually non-existent during flight. The demand for ATP is not consistent, but varies widely both over the short and longer terms. The most pronounced, virtually instantaneous rates of change are those associated with the transition from the alert quiescent state to flight, although endothermic heat generation is also responsible for substantial increases in metabolic rate. Metabolic rate is likewise dependent on body size, environmental temperature, and physiological status of the insect (such as dehydration and feeding), and it is responsive to both short- and longer-term changes in the partial pressures of oxygen and carbon dioxide. These physiological changes are not simply passive responses to the changing environment, but include active modulation of metabolic rate, sometimes via phenotypic plasticity, and often via evolutionary change.

3.1 Method and measurement

Over the last several decades it has become increasingly clear that the circumstances and methods used to assess a given physiological variable have a considerable influence on the outcome of the experiment or trial in which they are used (Baust and Rojas 1985; Harrison 2001). This is certainly the case when it comes to the measurement of gas exchange rates. Although closed system (constant pressure and constant volume) methods are intrinsically accurate for the measurement of gas exchange (Sláma 1984), in all but the most quiescent of stages (such as diapausing pupae) it is difficult to distinguish activity metabolism from standard metabolic rates (SMR) owing mostly to the poor temporal resolution and the integrative nature of the technique (Lighton and Fielden 1995). This means that comparisons of SMR, which are essential for determining whether insects modulate metabolic rates in response to different environmental circumstances (Chown and Gaston 1999), are unlikely to be possible. Comparisons between species and populations will be confounded by an experimental technique that fails to consider activity, especially because changes in levels of activity appear to be one of the responses of insects to a changing environment. For example, in laboratory-selected *Drosophila melanogaster*, desiccation-selected flies are far less active than control flies. Desert-dwelling flies appear to show a similar response compared to their more mesic congeners (Gibbs 2002*a*).

Fortunately, modern flow-through gas analysis (Lighton 1991*a*), continuously recording calorimetric (Acar *et al.* 2001), and activity detection (Lighton 1988*a*) methods allow periods of activity to be separated from those where the animal is at rest, so facilitating accurate estimation of SMR (because insects do not have a thermoneutral zone these rates cannot be referred to as basal). The difference between these open system methods and the closed system techniques previously used is usually substantial (Lighton 1991*b*). Lighton and Wehner (1993) provided the first quantitative data on beetles and ants highlighting this problem, pointing out that $\dot{V}O_2$ is not necessarily reduced in xeric species. In a later, more detailed analysis, Lighton and Fielden (1995) showed that previous estimates for insects based on closed system methods may have errors of 134 per cent of the true value for an insect weighing 0.001 g, and 18 per cent for a 10 g individual. These errors are a consequence of different estimates of both the coefficient and exponent in the scaling relationship between metabolic rate and mass. The general overestimation of SMR was borne out in a later comparative analysis (Addo-Bediako *et al.* 2002), but the exponent of the relationship between mass and metabolic rate remains contentious (Section 3.4.5).

The influence of technique on the outcome of experimental work has also been highlighted recently in three other areas of investigation. These are the measurement of oxygen partial pressures in tissues, the importance of the coelopulse system for gas exchange and its regulation, and the 'postural effect' that might be responsible for the lack of congruence between resting metabolic rate estimates for insects running on a treadmill and those measured during quiescence. In the first instance, oxygen partial pressures in tissues are often measured using oxygen microelectrodes inserted into the tissue (Komai 1998). Although care is usually taken to assess the effects of this invasive technique, it has recently been demonstrated that tissue damage might be responsible for underestimates of pO_2. The use of implanted paramagnetic crystals of lithium phthalocyanine and *in vivo* electron paramagnetic resonance oximetry has shown that pO_2 might be considerably higher than estimated by oxygen microelectrodes, owing to acute mechanical damage caused by the latter, which probably stimulates oxygen consumption in the fluid surrounding the electrode (Timmins *et al.* 1999). Implantation of paramagnetic crystals and a subsequent period allowed for tissue repair mean that the effects of tissue damage could be substantially reduced.

The second example concerns the debate on the importance of extracardiac pulsations and the coelopulse system for gas exchange and its regulation, respectively (Sláma 1999) (Section 3.4.1). These extracardiac pulsations are visually imperceptible abdominal segmental movements (caused by contraction of abdominal intersegmental muscles), but with an apparently significant effect on haemocoelic pressure, and therefore on gas exchange (Sláma 1988, 1999). Later investigations by Tartes and his colleagues (review in Tartes *et al.* 2002) of several species and stages identical to those used by Sláma (1999) have failed to confirm the importance of these pulsations for the regulation of gas exchange. Indeed, both Tartes *et al.* (2002) and Wasserthal (1996) ascribe many of the abdominal movements recorded in earlier reasonably invasive experiments (e.g. connection of the abdominal tip to a position sensor) to artefacts of the experimental design. Furthermore, Tartes *et al.* (2002) concluded that these miniscule abdominal movements are functionally similar to large abdominal movements, though reduced in magnitude, with only a small influence on ventilation, and then only in a few species.

The effects of measurement technique are also obvious in the calculation of the costs of transport. It has long been known that in both vertebrates and invertebrates the *y*-intercept of the relationship between running speed and metabolic rate, usually obtained from data gathered using animals on a treadmill, provides an estimate of resting metabolic rate that is higher than empirical measurements of resting metabolic rate. This elevated metabolic rate was ascribed to a 'postural effect' by Schmidt-Nielsen (1972). In subsequent investigations of insects, this explanation, and several others that have been mooted to explain the elevated *y*-intercept, were questioned (Herreid *et al.* 1981). It was subsequently demonstrated that the elevated *y*-intercept in insects is a consequence of the use of treadmills. In ants running voluntarily in a respirometer tube, there is no *y*-intercept elevation (Lighton and Feener 1989), and this has also recently

been confirmed for ants in the field (Lighton and Duncan 2002).

3.2 Metabolism

The fat body is the principal site of synthesis and storage of carbohydrates, lipids, amino acids, and proteins. It is the major location of trehalose and glycogen synthesis, respectively the main haemolymph and storage carbohydrates in insects, the principal site of proline synthesis from alanine and acetyl CoA, and the principal site of lipid synthesis and storage (mostly as di- or triacylglycerol) (Friedman 1985). When either flight muscles, or other muscle groups, place a demand on the system for fuel, initially this fuel is provided from stores in the muscles themselves, but shortly thereafter energy-providing macromolecules from the haemolymph and eventually from the fat body must be used. Small peptide hormones produced by the *corpora cardiaca* are responsible for the control of fuel mobilization, and many members of this adipokinetic (AKH) family of hormones, which are responsible for lipid, carbohydrate and proline mobilization, are now known from insects (Gäde 1991; Gäde *et al.* 1997; Gäde and Auerswald 2002).

The bulk of the energy provided by carbohydrates, lipids, and proline is made available by aerobic metabolism. Indeed, insect flight, one of the most energetically demanding activities known in animals, is wholly aerobic. In contrast, pedestrian locomotion can be fuelled partly by anaerobic metabolism, and anaerobic metabolism serves to provide energy during periods of environmentally induced hypoxia.

3.2.1 Aerobic pathways

Because flight is such an energetically demanding activity, and because so many insect species use flight as a major method of locomotion, much of the work done on substrate catabolism has been concerned with the aerobic provision of ATP for flight. Nonetheless, it is worth noting that only adult insects are capable of flight, that in ants and termites only the alates fly and then usually for just a brief period, and that in many species the adults are flightless (Roff 1990), although, amazingly, some stick insects have secondarily regained their powers of flight (Whiting et al. 2003). Even among those adult insects that can fly, some, such as dung beetles, parasitic wasps, flies, leafhoppers, and grasshoppers, spend much of their time walking, running, or hopping (Chown et al. 1995; Gilchrist 1996; Krolikowski and Harrison 1996).

Carbohydrates are among the most widely known flight fuels in insects. Small amounts of trehalose and glycogen (the most common carbohydrate stores) can be found in the muscles, but the former is usually found in large quantities in the haemolymph, and the latter is stored in the fat body. Carbohydrates are utilized by most flies and hymenopterans, although they are also used in species, such as locusts, that shift from one kind of fuel to another or that alter their reliance on different fuels depending on food availability (Joos 1987; O'Brien 1999; Dudley 2000). The biochemical pathways involved in the use of glycogen and trehalose have recently been reviewed in detail (Nation 2002).

Lipids (fatty acids) are also widely known as fuels for insect flight. In locusts they provide a major source of energy in flights that last for longer than approximately 30 min, as they do for non-feeding or starved hawk moths (Ziegler 1985; O'Brien 1999; Dudley 2000). Adipokinetic hormone mobilizes triacylglycerol stored in fat body cells as diacylglycerol (via triacylglycerol lipase), and this is subsequently transported as a lipoprotein complex (the apolipoprotein is apoLp-III, and the complex known as lipophorin) through the haemolymph to the flight muscle cells, where it is utilized via β-oxidation (Ryan and van der Horst 2000).

Although proline was first thought to be important only in tsetse flies and Leptinotarsa decemlineata (Coleoptera, Chrysomelidae), or as a primer of the Krebs cycle in some species (such as the blowfly, Phormia regina) it has now been shown to be an important flight fuel especially in beetles (Gäde and Auerswald 2002). In species that make use of both proline and carbohydrates for energy provision, proline varies in its contribution to flight metabolism from approximately 14 per cent in the meloid beetle Decapotoma lunata, to about

50 per cent in the cetoniine scarab beetle, Pachnoda sinuata. As much as half of this energy may come from haemolymph stores in beetles with large haemolymph volumes, and although measurable levels of lipids are present in the haemolymph, they are not used for flight metabolism (Gäde and Auerswald 2002). In P. sinuata, warm-up prior to flight is powered entirely by proline (Auerswald et al. 1998), and in those true dung beetles (Scarabaeinae) that have been examined, it appears that flight (or walking in flightless species) is powered entirely by proline (Gäde and Auerswald 2002). The pathways involved in proline utilization are discussed in Nation (2002), and regulation of its use is reviewed by Gäde and Auerswald (2002).

3.2.2 Anaerobic pathways and environmental hypoxia

Insects generally do not have well developed anaerobic metabolic capabilities. Nonetheless, the majority of species show a remarkable ability to recover from either hypoxia or anoxia, and at least some taxa are capable of surviving, being active, and reproducing under conditions of profound hypoxia. While high altitude insects (such as those found in the high Himalayas) experience prolonged hypoxia and hypobaria, hypoxia or anoxia is also characteristic of several other environments. These include hypoxic water and anoxic mud in aquatic systems (Gäde 1985), and hypoxic burrows, dung, carrion, and decomposing wood, and anoxic flooded and ice-covered ground in the terrestrial environment (Conradi-Larson and Sømme 1973; Hoback and Stanley 2001). For example, larvae of Orthosoma brunneum (Coleoptera, Cerambycidae) inhabit wood where oxygen levels are as low as 2 per cent and CO_2 levels reach 15 per cent (Paim and Beckel 1963). Moreover, adult females seek CO_2 concentrations of 90 per cent in which to lay their eggs (Paim and Beckel 1964). Likewise, several species of insects (mostly beetles) that inhabit wet dung pats in Denmark encounter and can tolerate, over the short term, O_2 levels as low as 1–2 per cent and CO_2 levels as high as 25–30 per cent, although generally they seek out regions of the dung pats that have conditions closer to ambient levels (Holter 1991; Holter and Spangenberg 1997).

Functional anaerobiosis in insects, such as that induced by exhaustive activity, does not contribute substantially to the total energy budget, with lactate production accounting for as little as 7 per cent of total ATP production in grasshoppers (Harrison *et al.* 1991). By contrast, under conditions of environmental hypoxia or anoxia, anaerobic metabolism can be of considerable importance as a source of ATP. For example, under low oxygen conditions and after their haemoglobin O_2 supply is exhausted, aquatic *Chironomus* midge larvae switch to anaerobic metabolism with glycogen as the primary substrate and ethanol as the end product (Gäde 1985). Likewise, *Chaoborus* phantom midge larvae use anaerobic metabolism for energy production (primarily from malate with succinate and alanine as the end products) when they rest in anoxic mud during the day.

Most terrestrial insects are metabolic regulators. With declining pO_2, oxygen consumption remains constant until a critical oxygen tension of about 5–10 kPa (generally lower in adults) (Loudon 1988). Thereafter, metabolic rate declines precipitously, and ATP levels are generally not defended while concentrations of ADP, AMP, and IMP increase (Hoback and Stanley 2001; Kölsch *et al.* 2002). At least some energy is made available via anaerobic metabolism with lactate and alanine forming the major end-products. In species that are regularly exposed to anoxia, such as the tiger beetle *Cicindela togata*, metabolism is rapidly downregulated and ATP levels are defended for at least 24 h. Thereafter anaerobic metabolism is responsible for energy provision (Hoback *et al.* 2000). Survival of long-term anoxia in the Arctic carabid beetle *Pelophila borealis* (Carabidae) is associated with metabolic downregulation and with provision of small amounts of energy via a lactate pathway (Conradi-Larsen and Sømme 1973).

Metabolic downregulation or arrest plays a major role in the response of insects to pronounced hypoxia, and downregulation of ion channel activity is thought to be an important component of the response allowing survival of low oxygen conditions. In *Locusta migratoria* (Orthoptera, Acrididae), anoxic conditions result in a reduction in outward K^+ currents in neurons, and possibly also an inhibition of Na^+ currents (Wu *et al.* 2002). Thus, ATP demand is reduced and membrane-

related changes that lead to cellular damage are presumably prevented (Zhou *et al.* 2001). These responses are in keeping with some of those found in vertebrates (Hochachka *et al.* 1996).

Short-term experiments indicating that insects can remain active and maintain virtually unchanged metabolic rates down to even fairly low pO_2 levels (e.g. Holter and Spangenberg 1997) suggest that insects are largely unaffected by hypoxia. By contrast, longer-term rearing under hypoxic conditions reveals a rather different picture. Under these conditions, initial metabolic rate declines are often reversed (depending on pO_2), and growth and development can take place. However, especially at low pO_2, growth and development is reduced, leading to longer development times, to the addition of instars (in species where this is possible), and to a reduction in final body size (Fig. 3.1), especially if instar number is constrained (Loudon 1988; Frazier *et al.* 2001; Zhou *et al.* 2001). Reductions in body size often have a direct effect on fecundity, whereas increases in larval duration expose animals to greater risk of predation. Moreover, at pO_2 levels which seem to have little short-term effect, mortality can be very high, even if exposures are not permanent, but take place on a regular basis (Loudon 1988).

Hypoxia is also likely to have a pronounced effect on the short-term performance of active

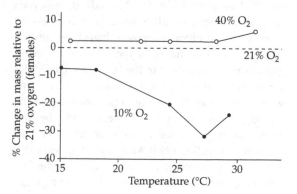

Figure 3.1 Change in body mass of *Drosophila melanogaster* females reared at a variety of temperatures in either 10% or 40% O_2 relative to individuals reared under normoxic conditions.

Note: Although size declines with temperature under normoxic conditions this is not shown.

Source: *Physiological and Biochemical Zoology*, Frazier *et al.*, **74**, 641–650. © 2001 by The University of Chicago. All rights reserved. 1522-2152/2001/7405-01011$03.00.

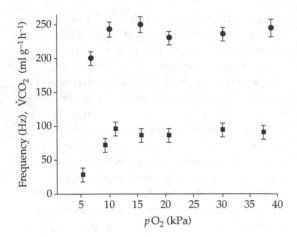

Figure 3.2 Means and 95% confidence intervals of metabolic rate (circles) and wing-stroke frequency (squares) as a function of ambient pO_2 in flying honeybees.

Source: *Physiological Zoology*. Joos *et al.*, **70**, 167–174.
© 1997 by The University of Chicago. All rights reserved.
0031-935X/97/7002-9637$02.00

insects. In both honeybees (Joos *et al.* 1997) and dragonflies (Harrison and Lighton 1998), flight metabolism is sensitive to ambient oxygen partial pressures below 10 kPa. In honeybees, this is thought to limit their altitudinal distribution to below 3000 m, because above this altitude convective oxygen delivery limits metabolic rates to levels insufficient to meet the aerodynamic power requirements of hovering flight (Fig. 3.2). Indeed, these data show that oxygen supply is not always in excess for insects, as is often supposed based on resting metabolism, and might serve to explain insect gigantism characteristic of the late Paleozoic (Miller 1966). Ambient oxygen concentrations at this time were as much as 35 per cent higher than at present, and gigantism among flying insects (and other terrestrial arthropods) was a characteristic of the Carboniferous (Dudley 1998). Moreover, it appears that during environmental hyperoxia, of both the late Paleozoic and late Mesozoic periods, flapping flight in insects and pterosaurs and birds and bats, respectively, may have evolved (Dudley 1998).

3.3 Gas exchange structures and principles

In many small arthropods (such as some springtails and mites) and in insect eggs, gas exchange takes place by diffusion through the cuticle, and aeropyle or chorionic airspaces, respectively. However, in most other arthropods there are specialized gas exchange structures (Fig. 3.3), and none is more widely known than the tracheal system of insects. This internal system of anastomozing tubes includes the main tracheae, which are often enlarged, especially in adult insects, to form air sacs, and which convey air from the spiracles into the system and through much of the body. These tracheal tubes divide up into finer branches, with conservation of cross-sectional area in some cases, but apparently not in others (Buck 1962; Locke 2001), eventually forming the tracheoles. These fine tubes end blindly, are usually liquid filled (though this apparently varies with oxygen partial pressure) (Wigglesworth 1935), are highly responsive to oxygen demand (Jarecki *et al.* 1999), and are intimately associated with the tissues. In the case of the flight muscles the tracheoles indent the plasma membrane and penetrate deeply into the muscle fibre (see Chapman 1998 for review).

Throughout the following sections, we focus almost entirely on gas exchange in terrestrial insects, mostly because of their predominance in terms of both species richness and absolute abundance. However, gas exchange in aquatic insects is no less intriguing and is based on the same principles, especially in those insects that use tracheal gills, physical (or gas) gills, or plastron respiration. The behavioural, morphological, and physiological adaptations for gas exchange in aquatic insects have been reviewed several times (Nation 1985, 2002; Chapman 1998), and the principles underlying this gas exchange are dealt with extensively by Kestler (1985).

Before proceeding with a discussion of gas exchange, it is necessary to clarify our use of the terms diffusion, convection, and ventilation. In the literature, these terms are often used either to refer to the exchange of gases more broadly, or in the case of ventilation and convection, are used interchangeably. Here, diffusion is used in the standard, physical sense; convection means mass movement of the medium; and ventilation refers to convection that is assisted by some form of pumping activity, usually associated with muscular contraction and relaxation.

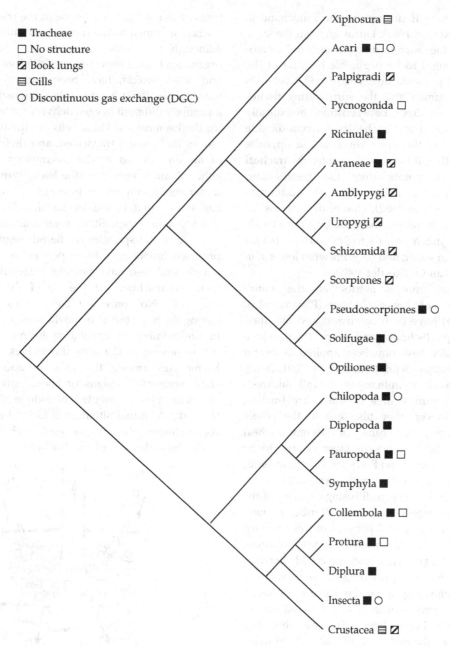

Figure 3.3 Gas exchange structures in the arthropods, also indicating those taxa which have at least some species characterized by discontinuous gas exchange.

3.3.1 Gas exchange and transport in insects

Gas exchange via the tracheae

The general morphology and functioning of the tracheal system and the spiracles has long been known and has now been well described for many species (Snodgrass 1935; Miller 1974; Nation 1985). Typically, gas exchange is thought to proceed as follows. Oxygen moves along a pathway from the spiracles through the main tracheal tubes (via convection or diffusion—see Section 3.3.2) to the tracheoles where it diffuses to the mitochondria (Buck 1962; Kestler 1985). Diffusion from the large tracheae to the surrounding tissues or haemolymph is thought to be negligible because of the small partial pressure difference (ΔpO_2) between the tracheal lumen and the surrounding tissues compared to the ΔpO_2 between the metabolically active tissues and the tracheoles. Carbon dioxide does not follow the same route in the opposite direction. Rather, it is thought to enter the tracheal system at all points from the tissues and haemolymph where it is buffered as bicarbonate (Bridges and Scheid 1982). One of the reasons for the difference in routes followed by O_2 and CO_2 might be the much larger Krogh's constant (K) for CO_2 than O_2 in water, and the somewhat lower K in air for CO_2 than O_2 (Kestler 1985).

Stereological morphometrics of the entire tracheal system of *Carausius morosus* (Phasmatodea, Lonchodidae) have both confirmed and modified this paradigm (Schmitz and Perry 1999). These methods, which have now been applied to several arthropod groups (Schmitz and Perry 2001, 2002), allow systematic examinations of wall thickness and tracheal surface area of the entire tracheal system. Moreover, they also enable the whole tracheal volume, the volumes of different tracheal classes and the volumes of other organs to be determined (Schmitz and Perry 1999). In *C. morosus*, class I tracheae (or the tracheoles) account for 70 per cent of the oxygen diffusing capacity of the system, while larger classes II and III tubes account, respectively, for 17 and 7 per cent of this capacity, and may thus be important for gas exchange. Diffusion through larger tracheae could therefore be an effective way of providing all tissues with O_2, especially following spiracular closure, although this clearly depends on metabolic demand (ΔpO_2) (Schmitz and Perry 1999). The morphometric analyses also showed that the lateral diffusing capacity for the tracheal walls is much higher for CO_2 than for O_2. Thus, tracheal classes I and II contribute 85 per cent to diffusing capacity, and classes III and IV the remaining 15 per cent, indicating that the entire tracheal system plays a large role in the elimination of CO_2.

Recently, it has also become clear that gas exchange does not take place only via direct dif-

fusion from the gaseous phase in the tracheal and tracheolar lumen to the tissues and mitochondria. Although the aeriferous tracheae, which serve organs such as the ovaries in *Calliphora* blowflies and other insects, have been known for some time, Locke (1998) has recently demonstrated that an entirely different oxygen delivery system is used for the haemocytes. These cells are the only insect tissues that are not tracheated, and their tolerance of anoxia has led to the assumption that they either obtain oxygen from the haemolymph or live under anoxic conditions. In contrast, Locke (1998) has shown that in the larvae of *Calpodes ethlius* (Lepidoptera, Hesperiidae) some tracheae leading away from the spiracles of the 8th segment form profusely branching tufts suspended in the haemolymph, and also have branches extending to the tokus compartment at the tip of the abdomen (Fig. 3.4). Not only are haemocytes abundant among the tufts, but also anoxia causes an increase in the number of circulating haemocytes, with many moving to the tufts themselves. Moreover, haemocytes among the tufts in anoxic larvae show none of the signs of anoxic stress typical of haemocytes distributed elsewhere throughout the body. A similar situation is found in the tokus compartment, where haemocytes attach themselves to the basal lamina of the tracheae. The tracheal

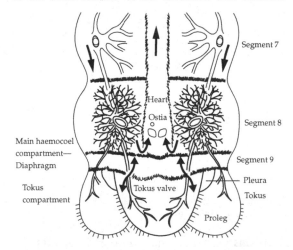

Figure 3.4 Haemolymph flow among the aerating tracheae and tokus compartment of the larvae of *Calpodes ethlius*. These tracheae serve as a lung for circulating haemocytes.

Source: Reprinted from *Journal of Insect Physiology*, **44**, Locke, 1–20, © 1998, with permission from Elsevier.

tufts (or aerating tracheae) and tokus compartment, therefore, form a lung for haemocytes in this caterpillar, and a similar arrangement has been found in several other species from a variety of lepidopteran families (Locke 1998).

Respiratory pigments

Owing to the presence of the tracheal system, it has also been assumed that, with a few well known exceptions (*Chironomus* (Diptera, Chironomidae) and *Gasterophilus* (Diptera, Gasterophilidae) larvae, and adult Notonectidae (Hemiptera) in the genera *Anisops* and *Buenoa*), respiratory pigments (specifically haemoglobin, Hb) are absent in insects (Weber and Vinogradov 2001; Nation 2002). However, Burmester and Hankeln (1999) recently identified the presence and expression of a haemoglobin-like gene in *D. melanogaster* by comparing *Chironomus* globin protein sequences with anonymous cDNA sequences of the former species. Further investigation revealed a haemoglobin similar to that of other insect globins in terms of its ligand binding properties and expression patterns (Hankeln *et al.* 2002). As is the case in other insects with more pronounced Hb expression (Weber and Vinogradov 2001), the *Drosophila* Hb is synthesized in the fat body and tracheal system. Similarity between the oxygen affinity and tissue distribution of the *Drosophila* Hb and those in other insects suggests that its role is the transport and storage of oxygen even under normoxic conditions. Hb in the tracheoles might facilitate the movement of oxygen from the tracheolar space into the tissues, and might be associated with storage of oxygen in the fat body surrounding metabolically active tissues such as the brain (Hankeln *et al.* 2002). Haemoglobin may also function to enhance tolerance of anoxia or might serve as an oxygen sensor involved in the regulation of tracheal growth. In the latter case, Hb could serve as a source of oxygen for nitric oxide synthase needed for nitric oxide synthesis, which is in turn involved in a signalling pathway that is thought to stimulate tracheal growth under hypoxia (Wingrove and O'Farrell 1999). Alternatively, it might also be involved in oxygen signalling of tracheal growth via the Branchless Fibroblast Growth Factor (Jarecki *et al.* 1999). Whatever the actual role of *Drosophila* Hb, its

discovery suggests that oxygen supply in insects is likely to be more complex than previously thought, and may depend on some form of haemoglobin-mediation or storage. Nonetheless, the restriction of Hb to the tracheoles and terminal cells of the tracheal system, and to fat body cells, suggests that the supply of oxygen to the tissues is still largely fulfilled by the tracheae. Recently, a haemocyanin has been found in the stonefly *Perla marginata* (Hagner-Holler *et al.* 2004).

3.3.2 Gas exchange principles

Although the physical principles underlying gas exchange are set out in a variety of texts and reviews (Piiper *et al.* 1971; Farhi and Tenney 1987), in terms of insect gas exchange the most thorough treatment is provided by Kestler (1985). Among the 47 equations presented to facilitate theoretical investigations of gas exchange in insects, and on which Kestler (1985) bases his arguments regarding the relative contributions of diffusion and convection to gas exchange, and the scaling of water loss, several can be considered most essential in the broader context of insect responses to the environment.

Gas exchange and water loss

The molar rates of transport $\dot{M}_x$ (mmol s^{-1}), of gas x in a tube containing the medium m, are given, in the case of diffusion, by

$$\dot{M}_x = (A/L) \cdot D'_{x,m} \cdot \beta_{x,m} \cdot \Delta p_{x,m}, \tag{1}$$

where A is the cross-sectional area of the tube (cm^2), L its length (cm), $D'_{x,m}$ the effective diffusion coefficient (cm^2 sec^{-1}), and $\beta_{x,m}$ the capacitance coefficient (mmol cm^{-3} kPa^{-1}) and $\Delta p_{x,m}$ the partial pressure difference (kPa). In the case of convection, the equation is

$$\dot{M}_x = \dot{V}_m \cdot \beta_{x,m} \cdot \Delta p_{x,m}, \tag{2}$$

where $\dot{V}_m$ is the rate of volume flow (cm^3 s^{-1}), which is the product of the frequency of volume changes (f, s^{-1}) and the volume change (ΔV_m, cm^3).

These equations represent the situation for pure diffusion and pure convection, respectively, and apply equally to N$_2$, O$_2$, CO$_2$, and water vapour (Kestler 1985). Given that $D'_{x,m}$ and $\beta_{x,m}$ remain the same in a given medium for a given gas, it is clear that $\dot{M}_x$ can be altered in a pure diffusion system by changes in A, L and/or $\Delta p_{x,m}$, and in a pure convection system by changes in f, ΔV_m and/or $\Delta p_{x,m}$. This has considerable implications for gas exchange in insects, especially for a tracheal system that functions to exchange O_2 and CO_2, while conserving water (Kaars 1981).

For example, because the partial pressure difference for water is independent of metabolic rate (tissues remain saturated irrespective of metabolic rate), alteration of metabolic rate will not effect a change in the rate of water loss. However, if the partial pressure gradient for CO_2 can be enhanced, then the same molar rate of CO_2 transport can be maintained with a reduction in cross-sectional area of the spiracles. The latter often forms the major resistance to water loss (O_2 is not considered here because partial pressure gradients are generally higher for O_2 than for CO_2, as is K (Kestler 1985; Lighton 1994; Wasserthal 1996)). In other words, changes in metabolic rate as a means to reduce or enhance water loss are only effective in a pure diffusion system in as much as they alter the area term in equation (1). In a pure convection system, any alterations to f or to ΔV_m will have profound effects on water loss rates.

Based on similar reasoning, several null models for the likely extent of water loss in small and large terrestrial insects utilizing either pure convective or pure diffusive gas exchange have been developed (Kestler 1985). If metabolic rate scales with an exponent lying somewhere between $m^{0.67}$ and $m^{1.0}$ (Section 3.4.5), tracheal cross-sectional area scales as $m^{0.67}$, and tracheal length as $m^{0.33}$, in a pure diffusion system respiratory water loss should scale as $m^{0.33}$. In a convective system, ventilation frequency does not scale with mass (unlike the situation in vertebrates—see Lighton 1991a; Davis et al. 1999), and ventilation volume scales as $m^{1.0}$ (Kestler 1985; Lighton 1991a; Davis et al. 1999). Therefore, respiratory water loss scales as $m^{1.0}$. To adjust both modes of oxygen flux to the same intensity, ventilation frequency must equal diffusive oxygen uptake, and net convective respiratory water loss should, therefore, be 80 per cent of diffusive loss (Kestler 1985). Notwithstanding this consistent difference, it is clear that according to the null model, respiratory water loss by diffusion will always be larger than convective water loss at small body sizes (Fig. 3.5). Therefore, from a purely theoretical perspective, selection should favour a system dominated by convective gas exchange in small insects.

However, water balance is not just a function of respiratory water loss. It is also dependent on the rate of cuticular transpiration, which scales as $m^{0.67}$ (Chown and Davis 2003). When this transpiratory route is set to contribute half of the total water loss, then the difference between diffusion and convection remains pronounced, with diffusion resulting in greater water loss at small body sizes and convection showing the converse. However, if cuticular water loss makes up a large proportion of total transpiration ($c.85-90$ per cent), then the difference between rates of water lost by diffusion or convection are small, especially for small-bodied insects (see Chapter 4). Moreover, the theoretical analysis is not especially sensitive to alteration of

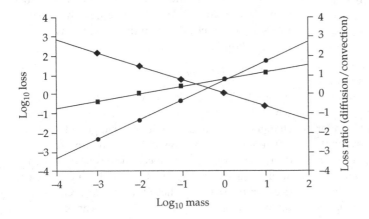

Figure 3.5 Null model for the relationship between mass and water loss in a pure diffusion system (squares) and a pure convection system (circles). Variation in the ratio between the two relative to mass is given by the third line (diamonds). This shows that water loss in a diffusion system, relative to a convection system, increases as body mass declines. The loss units are arbitrary, whereas mass varies between 0.1 mg and 100 g.

the scaling exponents to fit the allometric scaling laws proposed by West and his colleagues (West *et al.* 1997, 2002) (see also Section 3.4.5). They assume that tracheal tubes are space filling, that their cross-sectional area remains constant with divisions in the tubes, that the finest tubes are scale invariant, and that energy is minimized. Based on these and other assumptions they predict and show that metabolic rate should scale as $m^{0.75}$, and that changes in cross-sectional area should scale as a quarter power of body mass too (West *et al.* 1997). These values are considerably different from some of those proposed by Kestler (1985), but all that is required for the null models to hold is that the scaling exponent for convection should be larger than that for diffusion.

Perhaps of more significance than the change in the relative contributions of gas exchange by diffusion and convection, is the body mass at which this change might take place. Kestler (1985) suggested that this changeover should take place at a body mass of approximately 1 g. This is likely to be the case if net convective respiratory water loss is always *c*.80 per cent of diffusive loss. However, if it is not, then it is not clear at what body size the changeover should take place. Despite the elapse of more than 15 years since Kestler (1985) published his analysis, little empirical work has been undertaken to investigate these theoretical predictions (but see Harrison 1997). Although comparisons of the relative proportions of cuticular and respiratory transpiration are commonly made in an effort to understand the significance of respiratory water loss, they are unlikely to reveal the way in which water balance has been modified by selection. Relative proportions might vary by as little as a few per cent despite major differences in the gas exchange mode and/or differences in cuticular versus repiratory transpiration (Chown 2002). Rather, an understanding of the way in which changes in respiratory mode might contribute to the conservation (or elimination) of water requires a comparative analysis of the absolute and relative contributions of each of the major routes of water loss across a suite of closely related taxa in the context of this theory. Doing this is relatively straightforward in insects that show discontinuous gas exchange cycles (DGCs) (see Section 3.4.2), and

a comparative study of dung beetle water loss has illustrated the benefits of using a rigorous, analytical approach to separate cuticular and respiratory water loss (Chown and Davis 2003). The analytical methods developed by Gibbs and Johnson (2004), which allow calculation of cuticular water loss as the intercept of the intra-individual regression of water loss on metabolic rate irrespective of respiratory pattern, should make such comparisons more straightforward in insects that do not show discontinuous gas exchange.

Diffusive–convective gas exchange

Based on the demonstration that diffusive gas exchange alone is unlikely in insect tracheal systems, Kestler (1985) developed a see-saw model for gas exchange, arguing that a pure diffusion-based system is unlikely to occur in insects, and that diffusive–convective or pure convective gas exchange is much more likely. He further developed Buck's theory of flow-diffusion and corrected the analysis, showing that during suction ventilation in insects (Section 3.4.2) oxygen shows cocurrent diffusive–convective gas exchange at the spiracles, described by the following equation:

$$\dot{M}_x^{S-COCU} = -A^S \cdot v \cdot \beta_x \cdot (\Delta p_x^S \\ \times 1/(1 - \exp(v \cdot L^S/D_x')) - p_x^A), \quad (3)$$

where A^S is spiracular cross-sectional area, v the flow velocity, β_x is the capacitance coefficient of gas x, Δp^S is spiracular partial pressure gradient for gas x, L^S is the length of the spiracular tube, D' is the effective diffusion coefficient, and p^A is the partial pressure of gas x at the opening of the tubular valve.

CO_2, water vapour, and nitrogen show anticurrent diffusive–convective gas exchange, described as

$$\dot{M}_y^{S-ANCU} = A^S \cdot v \cdot \beta_y \cdot \left(\Delta p_y^S \\ \times \exp\left(-v \cdot L^S/D_y'\right) \\ /(1 - \exp(-v \cdot L^S/D_y')) - p_y^A\right). \quad (4)$$

Kestler (1985) then examined these principles in the light of experimental work undertaken by himself on *Periplaneta americana* (Blattaria, Blattidae) and by others (mostly Schneiderman and his colleagues—see below) on the silkmoth

Hyalophora cecropia (Lepidoptera, Saturniidae), providing support for his theoretical analyses. Despite these analyses, the extent to which gas exchange might take place in insects by diffusion only remains contentious (Lighton 1996, 1998; Sláma 1999). Nonetheless, Kestler's (1985) equations provide a foundation for examining this issue, and will make further exploration thereof more straightforward than otherwise might have been the case.

3.4 Gas exchange and metabolic rate at rest

Following the theoretical analysis by Krogh, and subsequent confirmation thereof by Weis-Fogh (review in Miller 1974), which showed that diffusion could serve the gas exchange requirements of small insects and most large insects at rest, diffusion has regularly been assumed to be the only means by which insects exchange gases at rest. Indeed, this idea continues to permeate modern discussions (West *et al.* 1997), despite the fact that ventilation, either by abdominal pumping or by movements of other sclerites, has been suspected since 1645, and has been confirmed by observation and using modern experimental methods at least since the 1960s (Schneiderman and Williams 1955; Miller 1981). Even ventilation by means of tracheal expansion and contraction, which was recently hailed as a major new discovery (Westneat *et al.* 2003) has been known since the 1930s (Herford 1938). Gas exchange patterns and mechanisms in insects have been reviewed several times (Miller 1974, 1981; Kaars 1981; Wasserthal 1996), and the importance of both diffusion and convection (by ventilation in many cases) for gas exchange at rest is now well established.

Within a particular species, metabolic rates at rest vary over the course of development (Clarke 1957), during diapause (Denlinger *et al.* 1972), over the course of a day (Crozier 1979; Takahashi-Del-Bianco *et al.* 1992), between seasons (Davis *et al.* 2000), and as a consequence of changing temperature, water availability, and size. The latter variation is most interesting from an ecological perspective, and has consequently been the source of most controversy.

3.4.1 Gas exchange patterns

At rest, insects show a large variety of gas exchange patterns. These range from continuous gas exchange, by diffusion–convection or convection (including active ventilation), with the spiracles held open (Fig. 3.6), to DGCs (Fig. 3.7) where gas exchange is either of the diffusive–convective kind, or where, in the case of active ventilation, ventilation and spiracular opening are highly coordinated (Kaars 1981; Miller 1981; Lighton 1996). Between these extremes, insects show a continuum of patterns, ranging from barely cyclic through to strongly cyclic gas exchange that differs only marginally from DGC (Hadley and Quinlan 1993; Williams and Bradley 1998; Shelton and Appel 2001*a*) (Fig. 3.8).

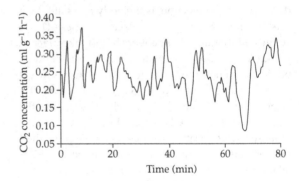

Figure 3.6 Continuous gas exchange in the aphodiine dung beetle, *Aphodius fossor*.

Source: Chown and Holter (2000).

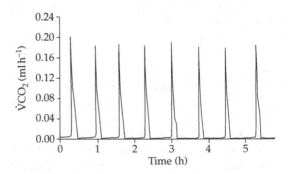

Figure 3.7 Discontinuous gas exchange cycles in *Omorgus radula* (Coleoptera, Trogidae) recorded at 24°C.

Source: Bosch *et al.* (2000). *Physiological Entomology* **25**, 309–314, Blackwell Publishing.

Ventilation at rest

In those species where active ventilation accompanies gas exchange at rest, ventilatory movements can include active abdominal pumping (by means of dorso-ventral and telescoping movements), head protraction and retraction, and prothoracic pumping (Miller 1981; Harrison 1997). The proportion of tracheal air estimated to be exchanged with each stroke is in the region of 5–20 per cent for *Schistocerca gregaria* (Orthoptera, Acrididae), although some recent estimates (Westneat *et al.* 2003), have suggested a much higher value for other insects. Although some small insects, such as ants, appear not to use ventilation for gas exchange (Lighton 1996), others ventilate vigorously (Miller 1981).

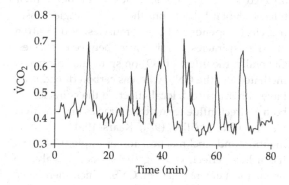

Figure 3.8 Cyclic gas exchange in pseudergate of *Incisitermes tabogae* (Isoptera, Kalotermitidae).

Source: *Comparative Biochemistry and Physiology A*, **129**, Shelton and Appel, 681–693, © 2001, with permission from Elsevier.

Haemolymph circulation also plays a considerable role in tracheal ventilation, especially in adult insects. This is largely as a consequence of the presence of many large air sacs and the compartmentalization of the haemocoel into an anterior section comprising largely the thorax and often the first abdominal segment, and a posterior section, consisting of the abdominal segments (Wasserthal 1996; Hertel and Pass 2002). This compartmentalization is especially distinct in the aculeate Hymenoptera owing to their petiole, the narrow segment that separates the first abdominal segment from the remainder of the abdomen. In flies such as *Calliphora*, haemolymph flow reversal, associated with heartbeat reversal, interacts with the air sacs to produce inspiration via the abdominal spiracles during forward heartbeat, and collapse of the abdominal air sac and expiration via the abdominal spiracles during retrograde heartbeat (Fig. 3.9). Similar interactions between circulation and ventilation have been documented in many other adult insects (see review for the holometabolous insects provided by Wasserthal 1996). In bumblebees, moths, and beetles these interactions are important for thermoregulation (Chapter 6). Accessory pulsatile organs also play a significant role in circulation (insects have more 'hearts' than most other organisms), and the interaction between circulation and gas exchange (Hertel and Pass 2002).

In several insect species, extracardiac pulsations associated with contraction of abdominal, intersegmental muscles are thought to be important for gas exchange as a consequence of their

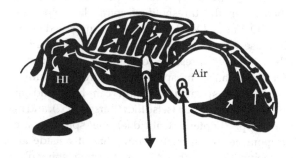
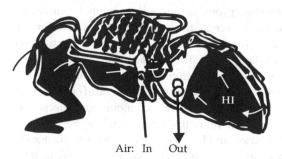

Figure 3.9 Gas exchange dynamics in *Calliphora vicina* (Diptera, Calliphoridae) resulting from haemolymph oscillation between the anterior and posterior body.

Source: Reprinted from *Advances in Insect Physiology*, **26**, Wasserthal, 297–351, © 1996, with permission from Elsevier.

pronounced effect on haemocoelic pressure (Sláma 1988, 1999). Although heartbeat reversal (cardiac pulsation) is an important contributor to gas exchange in lepidopteran pupae (Hetz *et al.* 1999), and abdominal movements are likewise important for gas exchange in some beetle species (Tartes *et al.* 2002), the significance of extracardiac pulsations remains moot. While Sláma (1994, 1999) has vigorously defended both the importance of extracardiac pulsations and the significance of the coelopulse system for regulation of respiration, it appears that they play only a minor role in gas exchange and the regulation of respiration, respectively (Lighton 1994; Wasserthal 1996; Tartes *et al.* 2002).

Ventilatory movements are largely controlled by the central nervous system (CNS) via the action of the abdominal ganglia and a central pacemaker located in the fused abdominal or metathoracic ganglia, or in abdominal ganglia (Ramirez and Pearson 1989), as well as by afferent feedback from abdominal ganglia and proprioceptors. Manipulations of pO_2 and pCO_2 have a pronounced effect on ventilation rate (Miller 1981), suggesting that there is central chemoreception of oxygen and carbon dioxide in the CNS (Miller 1974; Harrison 1997). Recently, Bustami *et al.* (2002) provided the first clear evidence for such central chemoreception, showing similar ventilatory responses to changes in pO_2 and pCO_2 of both the CNS *in vitro* and whole animals. Moreover, they also demonstrated a biphasic response to hypoxia, characterized by an initial extreme increase in ventilation rate, followed by complete cessation of activity, and argued that such a response, which is more typical of vertebrates, provides an indication that ventilatory networks across the animal kingdom share many basic functions. Other experimental work on grasshoppers such as *Romalea guttata* and *Schistocerca americana* has demonstrated that in alert, but quiescent individuals, manipulation of endotracheal pO_2 and pCO_2 has pronounced effects on abdominal pumping rate, as might be expected (Gulinson and Harrison 1996). However, changes in pH have little effect on abdominal pumping rates (except through pCO_2 changes associated with injection of HCO_3^-). Indeed, it appears that in grasshoppers and possibly in other insects, the ventilatory system participates indirectly in acid–base regulation by maintaining a relatively constant tracheal pCO_2, while the excretory system (gut and Malpighian tubules) is responsible for responding to pH changes and for regulating non-volatile acid–base equivalents (Gulinson and Harrison 1996; Harrison 2001).

Spiracular control

Gas exchange pattern and efficacy are markedly affected by the degree of synchronization between the spiracles and abdominal pumping (Miller 1973). In many insects there is close central coordination of spiracular and ventilatory movements (Harrison 1997), which often results in unidirectional, retrograde airflow at rest. As is the case with ventilatory movements, spiracular opening and closing are affected by both pO_2 and pCO_2, although their effects depend both on the type (single closer muscle, or opener and closer muscles) and location of the spiracles, and vary between species. Generally, the effect of pO_2 on spiracular control is mediated via the CNS, whereas carbon dioxide can have either a direct, local effect on the spiracles, or both a direct influence on the spiracles and an effect via the CNS (Miller 1974; Kaars 1981). In some mantids and cockroaches, for example, CO_2 first has a local effect, causing the spiracular valve to open, and later reaches the CNS, which then has a direct effect on the muscle, allowing further spiracular opening.

Spiracular movements are also affected by the hydration state of the organism and the humidity of the surrounding environment. In various dragonfly species, partial dehydration causes an increase in the threshold of spiracular responses to CO_2 and O_2. Miller (1964) suggested that increases in concentration of certain ions, rather than osmotic pressure overall, are responsible for the change in spiracular control. A similar increase in spiracular control can be found in partially dehydrated tsetse flies (Bursell 1957), and in *Aedes* sp. (Diptera, Culicidae) the spiracles are responsive to the relative humidity of outside air (Krafsur 1971). Other factors affecting spiracular control include injury, starvation, and temperature (Section 3.4.2), while pH has little effect either directly or via the CNS.

Although spiracular control is often synchronized with ventilatory movements to produce a unidirectional airflow, there is much variation in the degree of spiracular coordination, ventilatory movements, and airflow, among species, among individuals, and even within individuals. Thus, gas exchange may take place through either one or all of the spiracles, spiracular coordination can either be tightly regulated or completely unsynchronized, and airflow may be unidirectional (in either direction and reversible), tidal, or in the form of cross-currents (Buck 1962; Kaars 1981; Miller 1981; Sláma 1988). In flightless beetles, it has long been maintained that airflow is unidirectional, from the thoracic spiracles to the abdominal spiracles, thus ensuring expiration into a subelytral chamber that would restrict water loss (Hadley 1994a,b). However, investigations of a large, flightless scarab beetle, *Circellium bacchus*, have raised doubts concerning this conventional interpretation of the role of the subelytral chamber in beetle water economy (Duncan and Byrne 2002; Byrne and Duncan 2003). Rather than being characterized by retrograde airflow at rest, airflow in this species is either tidal or anteriograde, with a clear division of labour between the thoracic and abdominal spiracles, and the entire respiratory demand often being served by a single mesothoracic spiracle. A similar division of labour between the thoracic and abdominal spiracles has been found in the desert-dwelling ant *Cataglyphis bicolor* (Lighton *et al.* 1993a), and might reflect a means of restricting water loss in insects, particularly because the abdominal spiracles are generally small and play an important role in oxygen uptake.

A large proportion of the variation in gas exchange patterns within and among individuals is clearly a function of external factors (e.g. temperature, water availability) or the physiological status of the individual (e.g. levels of dehydration, starvation, or activity) (Lighton and Lovegrove 1990; Quinlan and Hadley 1993; Chappell and Rogowitz 2000). However, insects at rest also show much variability in gas exchange patterns that cannot be ascribed to these factors. For example, Miller (1973, 1981) reported considerable among- and within-individual variation in ventilatory patterns of *Blaberus* sp. (Blattaria, Blaberidae), although he

thought that at least some of this variability might have been due to activity of the insects. In the case of lepidopteran pupae, movement is unlikely to be a major source of variation, yet, in some species the pupae show substantial variation in gas exchange patterns both within and between individuals of the same age. Recognizing the significance of this variation in the context of understanding gas exchange patterns, Buck and Keister (1955) undertook an analysis of variance to demonstrate that most of the variability was among rather than within individuals. This early formal analysis of variance in gas exchange patterns, and later detailed investigations (though with only qualitative analyses) of variability in spiracular movements and ventilatory patterns have subsequently largely been overlooked in favour of a focus on the average characteristics of particular gas exchange patterns (Lighton 1998). This focus on the average pattern is also reflected in something of a dissociation between work on the neurophysiology of rhythmic respiratory behaviour and that on gas exchange patterns identified by flow-through respirometry (though see Harrison *et al.* 1995; Bustami and Hustert 2000). While at first this dissociation might appear surprising, it is, perhaps, understandable given that initial investigations of gas exchange patterns and mechanisms in adult insects were concerned with the extent to which they might also show the discontinuous gas exchange patterns so characteristic of diapausing pupae. Recently, however, attention has once again been drawn to variability in gas exchange patterns, and the reasons for this variability, especially to cast light on the growing debate on the adaptive value of the DGC (Lighton 1998; Chown 2001; Marais and Chown 2003).

3.4.2 Discontinuous gas exchange cycles

Discontinuous gas exchange cycles (DGCs) are one of the most striking gas exchange patterns shown by insects. Discontinuous gas exchange was originally described in adult insects (Punt *et al.* 1957; Wilkins 1960), but it was the investigation of DGCs in diapausing saturniid pupae by Schneiderman and his colleagues (e.g. Schneiderman 1960; Levy and Schneiderman 1966a,b; Schneiderman and

Schechter 1966) that resulted in a comprehensive understanding both of the pattern and the mechanisms underlying it. Subsequently, discontinuous gas exchange has been documented in a wide variety of both adult and pupal insects, although the patterns and control thereof show considerable variation among species. To date, discontinuous gas exchange has been found in cockroaches (Kestler 1985; Marais and Chown 2003), grasshoppers (Harrison 1997; Rourke 2000), hemipterans (Punt 1950), beetles of several families (Lighton 1991*a*; Davis *et al.* 1999; Bosch *et al.* 2000; Chappell and Rogowitz 2000), lepidopterans (Levy and Schneiderman 1966*b*), robber flies (Lighton 1998), wasps (Lighton 1998), and ants (Lighton 1996; Vogt and Appel 2000). Similar patterns have been found in a variety of other insects including thysanurans, termites (Shelton and Appel 2001*a,b*) and *Drosophila* (Williams *et al.* 1997), but these patterns are not considered DGCs because they deviate somewhat from the standard, three-period cycles. Nonetheless, the considerable variation that is characteristic of DGCs means that the distinction between this gas exchange pattern and other forms of cyclic gas exchange is often difficult to make. Moreover, convergence in pattern does not necessarily mean that the same underlying mechanism is responsible for it (Lighton 1998; Lighton and Joos 2002).

Discontinuous gas exchange cycles have variously been referred to as discontinuous ventilation, discontinuous ventilatory cycles, and discontinuous respiration. Obviously, cellular respiration is not discontinuous, making that title inappropriate. Likewise, gas exchange is not always by ventilation, making that term inappropriate as a general descriptor too. In consequence, discontinuous gas exchange remains the most appropriate broad descriptive name for the gas exchange pattern described below, and if the pattern is repeated then the outcome can be referred to as 'discontinuous gas cycles'. Discontinuous gas exchange cycles are characterized by the repetition of three major periods, during which the exchange of CO_2 and O_2 is partially separated (Fig. 3.10). In terms of spiracular behaviour, these periods are an open (O) period when the spiracles are fully open, a closed (C) period during which the spiracles are tightly shut, and a

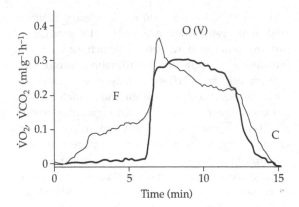

Figure 3.10 Discontinuous gas exchange in *Psammodes striatus* (Coleoptera, Tenebrionidae), indicating the Closed (C), Flutter (F), and Open (O) periods.

Note: Oxygen (thin line) and carbon dioxide (thick line) are partially separated in time.

Source: Reprinted from *Journal of Insect Physiology*, **34**, Lighton, 361–367, © 1988, with permission from Elsevier.

flutter (F) period during which the spiracles open and shut by means of reduced fluttering movements. Although this spiracular behaviour has been confirmed for both lepidopteran pupae (Schneiderman 1960), and adult ants (Lighton *et al.* 1993*a*), it is generally assumed to be characteristic of other insects showing DGCs on the basis of the gas exchange (especially CO_2 output) trace recorded during an investigation. Likewise, the majority of the work on the control of the DGC has been conducted on lepidopteran (usually saturniid) pupae, and mostly assumed to apply more generally to adult insects. However, there is considerable variation in control of the DGC, and it cannot be assumed that it is necessarily the same across all insects, irrespective of the taxon or stage (Harrison *et al.* 1995). Nonetheless, the patterns and control mechanisms are sufficiently similar in those insect species that have been investigated for them to be dealt with as a whole here.

During the closed period, no gas exchange takes place through the tightly shut spiracles. Oxygen in the endotracheal space is depleted by respiration, but CO_2 is buffered in the haemolymph, leading to a slow decline in endotracheal pressure, and to a steady decline in pO_2, to *c*.5 kPa. A CNS-mediated O_2 setpoint (see Section 3.4.1) then causes the spiracles to partially open and close in rapid succession, and the flutter period is initiated. At least in saturniid pupae, *P. americana*, and the ant *C. bicolor*, the negative endotracheal pressure is responsible for convective movement of air into the tracheae (Levy and Schneiderman 1966c; Kestler 1985; Lighton *et al.* 1993a), and this exchange of gases by means of inward convection has been termed passive suction ventilation (Miller 1974). Convective gas transport restricts outward movement of both CO_2 and H_2O and is thought to be one of the major ways in which the DGC is responsible for water conservation. However, in some species of ants and probably in the tenebrionid beetle *Psammodes striatus*, it appears that gas exchange during the F-period is primarily by means of diffusion (Lighton 1988b; Lighton and Garrigan 1995), thus reducing its water saving potential (see below).

During the F-period in saturniid pupae, endotracheal pressure increases rapidly towards ambient in a series of steps caused by spiracular opening and closing. During the open periods, gas exchange takes place predominantly by diffusion, whereas when the spiracles are more constricted convective gas exchange predominates (Levy and Schneiderman 1966b,c; Brockway and Schneiderman 1967, see also Lighton 1988b). According to Brockway and Schneiderman (1967) spiracle filters in saturniid pupae impede inward air flow and prolong the period during which convective gas exchange takes place, thus enhancing water conservation. A similar role has been ascribed to sieve plates in scarab beetles (Duncan and Byrne 2002), although these plates might also serve to exclude parasites which can have a substantial effect on performance (Brockway and Schneiderman 1967; Miller 1974; Harrison *et al.* 2001). In saturniid pupae, when endotracheal pressure reaches ambient values, the F-period continues, with microcycles of decreasing and increasing pressure associated with spiracular fluttering. During this time, pO_2 remains at c.5 kPa and the partial pressure gradient is sufficient to ensure that tissue oxygen demand is met. In both *P. americana* and *P. striatus*, oxygen uptake during the F-period is modulated to match cellular demand. Although the F-period in saturniid pupae comprises two readily distinguishable parts, this

division is not discernible in *P. americana* (Kestler 1985), and has not been sought in other species.

Carbon dioxide continues to accumulate during the F-period until pCO_2 reaches an endotracheal value of approximately 3–6 kPa, depending on the species (Brockway and Schneiderman 1967; Burkett and Schneiderman 1974; Harrison *et al.* 1995; Lighton 1996). At this partial pressure, CO_2 affects the spiracles both directly and indirectly (see Section 3.4.1), causing them to open widely, with concomitant egress of CO_2 and H_2O, and ingress of oxygen. During this O-period, gas exchange may take place either predominantly by diffusion (Levy and Schneiderman 1966a; Lighton 1994) or by ventilation as a consequence of active ventilatory movements (Lighton 1988b; Lighton and Lovegrove 1990; Harrison 1997). In *P. americana*, both forms of gas exchange can occur in the O-period, and Kestler (1985) referred to the former as CFO type and the latter as CFV type. Over the course of the O-period, pO_2 increases to approximately 18 kPa, and pCO_2 declines to initial values (approximately 3 kPa in saturniid pupae). In some species, such as ants and trogid beetles, the spiracles close shortly after pCO_2 has declined to initial values, so producing a single spike representing the O-period (Fig. 3.7). However, in lepidopteran pupae (Brockway and Schneiderman 1967; Hetz *et al.* 1999), and in scarab and carabid beetles (Davis *et al.* 1999; Duncan and Byrne 2000; Duncan and Dickman 2001), the O-period is characterized by an initial rapid release of CO_2 and then by several smaller bursts owing to repeated opening and restriction of the spiracles. These smaller bursts decline in amplitude until there is total constriction of the spiracles (Fig. 3.11), and in the beetle species these bursts are accompanied by abdominal pumping. It has been suggested that this alternative opening and closing of the spiracles during the O-period might be a mechanism to reduce water loss (Duncan and Byrne 2000; Duncan and Dickman 2001), although evidence for this idea is wanting. It also appears likely that in at least some cases these patterns might be a consequence of washout associated with slow flow rates in the experimental system (see Bartholomew *et al.* 1981; Lighton 1991b), although this is unlikely to account entirely for them.

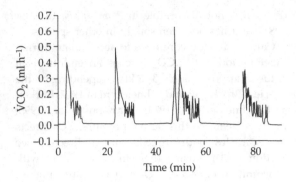

Figure 3.11 Discontinuous gas exchange in *Scarabaeus flavicornis* (Coleoptera, Scarabaeidae) showing smaller bursts during the Open period.

Source: Duncan and Byrne (2000). *Oecologia* **122**, 452-458, Fig. 3. © Springer.

3.4.3 Variation in discontinuous gas exchange cycles

Each of the periods in a single DGC can be characterized by several parameters, including duration, gas exchange rate, and volume of gas exchanged. Moreover, the entire cycle can be characterized by its frequency, mean peak height of the O-periods, proportional contributions (volume, duration) of each of the periods, and by the overall respiratory quotient ($RQ = CO_2/O_2$) if both gases are measured. Because CO_2 is the most straightforward of the gases to measure for small insects in a flow-through system (Lighton 1991*b*), rates and volumes of each of the periods often refer solely to CO_2. However, where both CO_2 and O_2 are measured (Lighton 1988*b*), rates and volumes can refer to either of the gases, and short term RQs (i.e. for a single period, and also known as respiratory exchange ratios, RER) can be calculated. Measurements of CO_2, O_2, and H_2O are also sometimes made (Hadley and Quinlan 1993), although combined measurements of CO_2 and H_2O are more common (Lighton 1992; Lighton *et al.* 1993*b*; Quinlan and Lighton 1999; Chown and Davis 2003). In the latter case, molar ratios of CO_2 to H_2O loss can also be calculated for each of the periods.

Because characterization of DGCs in adult insects, and the effects of mass, temperature, and $\dot{V}CO_2$ on them, have been the focus of much of the recent literature, mean values for the parameters characteristic of each period and of the DGC as a whole (for one or more temperatures) are generally provided in each investigation. Whether these summary statistics are valuable for understanding the reasons for interspecific differences in DGC (such as the apparent utility of an extended F-period for conserving water—see Lighton 1990; Davis *et al.* 1999; Duncan and Byrne 2000; Chown and Davis 2003), depends fundamentally on the extent of the variation of the DGC both within and among individuals, compared to that among species. Some recent reports have suggested considerable variation both within and among individuals (Lighton 1998; Chown 2001), indicating that within- and among-individual variation might be almost as large as, or sometimes larger than, that found between species. Clearly, there is much merit in investigating the partitioning of variance in gas exchange parameters, from the within-individual to among-species level, in several monophyletic groups (see Chown 2001 for further discussion). To date, only a single investigation has provided a careful quantification of the contribution of within- and among-individual variation to total variation in the parameters of the DGC (Marais and Chown 2003). In that study it was shown that repeatability (= among-individual variation) in a cockroach species characterized by at least four different gas exchange patterns (see Fig. 1.2) is high and generally significant, and that size-correction prior to estimates of repeatability should not be undertaken.

The work by Marais and Chown (2003) has also provided an indication of the correct level at which statistical analyses of the variation in gas exchange parameters should be undertaken. In several studies, investigations of relationships between various characteristics of the DGC are undertaken using measurements for each cycle of a gas exchange trace for a given individual. Thus, if seven individuals are investigated, and the parameters of four cycles are measured in each individual, the sample size is given as 28. If within-individual variation in DGC parameters was much greater than that between individuals, then this procedure might not provide any cause for concern. However, Marais and Chown (2003) demonstrated that variation within individuals is quite low, and this also seems to be most commonplace in the published literature. Therefore, statistical analyses undertaken using single cycles as independent data points are flawed. Not only will the degrees of freedom for the analysis be overestimated, leading to an increase in the likelihood of Type I statistical errors, and hence erroneous conclusions, but also estimates of the variability of gas exchange cycles will be confounded. Therefore, either mean values for each of the parameters for each individual animal should be calculated and used in analyses (Lighton and Wehner 1993; Lighton and Garrigan 1995), or variation within and among individuals should be fully explored.

Temperature

Temperature, body mass, and metabolic rate are the three most important factors influencing the DGC and each of its periods, although metabolic rate is clearly also influenced by temperature and body mass. An increase in temperature usually results in an increase in metabolic rate (Section 3.4.6). In all of the species examined to date, this temperature-related increase in metabolic rate is accompanied by an increase in DGC frequency. However, in some species, such as adult *Camponotus vicinus* (Hymenoptera, Formicidae) (Lighton 1988*a*), several species of *Pogonomyrmex* harvester ants (Quinlan and Lighton 1999), the fire ant *Solenopsis invicta* (Vogt and Appel 2000), adult carabid beetles (Duncan and Dickman 2001), and several lepidopteran pupae (Buck and Keister 1955; Schneiderman and Williams 1955), O-period volume declines with an increase in temperature. In others, such as the ant *C. bicolor* (Lighton and Wehner 1993), several species of dung beetles (Davis *et al.* 1999), two species of *Phoracantha* (Cerambycidae) beetles (Chappell and Rogowitz 2000), and termites (Shelton and Appel 2001*b*), O-period volume does not change (Fig. 3.12).

At present, it is not clear what mechanisms are responsible for these changes in frequency and volume (Lighton 1988*a*, 1996). An increase in temperature results both in a decline in CO_2 solubility and a decline in haemolymph pH, so affecting haemolymph buffering capacity (Lighton 1996). Thus, the effect of temperature is not restricted to a change in metabolic rate. The change in buffering capacity and increased metabolic demand presumably lead to an increase in DGC frequency and a decline in O-period volume. However, the

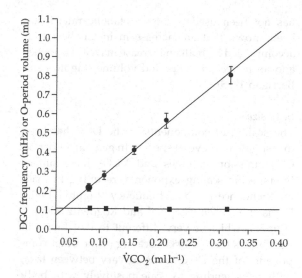

Figure 3.12 Relationships between mean $\dot{V}CO_2$ and DGC frequency (mean ± SE, circles) and O-period emission volume (mean ± SE, squares) in *Scarabaeus westwoodi* (Coleoptera, Scarabaeidae).

Source: Physiological and Biochemical Zoology, Davis *et al.*, **72**, 555–565. © 1999 by The University of Chicago. All rights reserved. 1522-2152/1999/7205-98157$03.00

maintenance of constant volumes with changing temperature is more difficult to explain. Presumably, species in which volumes remain constant must either be capable of modulating the O-period CO_2 setpoint, or of offsetting temperature-related changes in haemolymph buffering capacity. Davis *et al.* (1999) suggested that the former is the case in the dung beetles they investigated, although their conclusion was not experimentally verified. It has been suggested that there might be an adaptive advantage to the temperature-independence of O-period emission volumes because DGC frequency might not increase at as high a rate as might otherwise be expected, and therefore the exposure of the tracheal system to dry air might be reduced (Lighton 1994). However, Davis *et al.* (1999) found no interspecific differences in temperature-related modulation of the DGC despite the fact that the species were collected across a broad range of habitats (mesic to xeric). At present, the limited number of investigations of modulation of frequency of the DGC and O-period volume with changes in metabolic rate make any general conclusions unwarranted. Moreover, where temperature has not been used to alter metabolic rate, it has been shown that an increase in metabolic rate is accompanied by both an increase in DGC frequency and an increase in O-period volume (Lighton and Berrigan 1995).

Body size

The scaling of components of the DGC has been investigated in several species. In general, O-period CO_2 emission volumes and $\dot{V}CO_2$ have similar interspecific scaling exponents, resulting in mass independence of DGC frequency (Lighton 1991a; Davis *et al.* 1999; Chappell and Rogowitz 2000), a situation which is very different to that found in vertebrates (Peters 1983). Scaling of the other components of the DGC tends to vary between taxa, with rates tending to scale positively with body size, durations showing no relationship with size, and the scaling of F and C-period emission volumes being variable in their significance. Fewer investigations of intraspecific scaling have been undertaken. In those species where body mass variation is low, relationships tend to be insignificant as might be expected (Bosch *et al.* 2000), whereas in other species, where there is a reasonable range in size, both DGC frequency and O-period emission volume increase with mass (Lighton and Berrigan 1995). Insufficient attention has been given to the relationships between intra- and interspecific scaling exponents of the DGC and its characteristics though there are some taxa where this could be readily done, such as size-dimorphic scarab beetles (Emlen 1997) and size-variable harvester ant workers (Lighton *et al.* 1994). Likewise, to date, only a single study has investigated the effects of $\dot{V}CO_2$ on the characteristics of the DGC while accounting for the effects of both mass and temperature. Chappell and Rogowitz (2000) found that DGC frequency, O-period emission volume, and O-period peak rate increased with increasing $\dot{V}CO_2$, whereas the combined CF period duration declined. The former findings are in keeping with those of Lighton and Berrigan (1995).

Proportional duration of the periods

Several studies have suggested that differences in the proportional durations of the DGC periods (especially the F-period), and the ways in which these proportional durations change with temperature might reflect adaptations of the DGC to various environmental conditions, particularly water availability (Lighton 1990; Davis *et al.* 1999; Bosch *et al.* 2000; Duncan *et al.* 2002a, but see also Lighton *et al.* 1993a). These studies have all relied on calculating the relative durations of each of the DGC periods. Lighton (1990) argued that because the sum of the constants of the regression relationships between total DGC duration and durations of each of the periods is likely to be close to zero, these proportional durations can be calculated as the exponents of the least squares linear regressions of the durations of the C, F, and O-periods, on DGC duration, respectively. These ventilation phase coefficients have subsequently been used in several investigations (Lighton 1991a; Duncan and Lighton 1997), although it has now been demonstrated that they might give rise to considerable errors in estimating the proportions of the DGC occupied by each of the phases (Davis *et al.* 1999). This latter finding applies to any investigation in which the slope of a regression equation is used to determine a proportion. The slope of the regression equation provides an estimate of the change in the dependent variable with a given change in the independent variable. This change can be thought of as a proportion only if the intercept of the relationship is zero. If the intercept is a non-zero value or if the relationship is non-linear, this proportion changes systematically with a change in the independent variable (as has long been appreciated for Q_{10} values—see Cossins and Bowler 1987).

3.4.4 Origin and adaptive value of the DGC

Discontinuous gas exchange cycles have long been thought to represent a water-saving adaptation in insects (Chapter 4). The main reasons for this idea are that spiracles are kept closed for a portion of the DGC, thus reducing respiratory water loss to zero, and a largely convective F-period restricts outward movement of water (Kestler 1985; Lighton 1996). If the spiracles are artificially held open, water loss increases considerably (Loveridge 1968), and by calculation it can also be shown that a continuous O-period would lead to large increases in overall water loss (Lighton 1990; Lighton *et al.* 1993a,b).

Indeed, the ideas that the DGC is an adaptation to conserve water, and that changes in various characteristics reflect additional selection for water economy permeate the modern literature (Sláma and Coquillaud 1992; Lighton 1996; Duncan *et al.* 2002*a*). Clearly, an invaginated gas exchange surface does lead to a reduction in water loss (Kestler 1985). This can, perhaps, be seen most clearly in the pronounced trend in isopods towards an internalized lung system with spiracles in highly terrestrial species, from the original surface-borne lung system on the pleopodite exopods, which is found in species occupying moister habitats (Schmidt and Wägele 2001). However, whether the DGC itself and modifications thereof represent adaptations for restricting water loss is more controversial, although recent work has once again come out in favour of this idea (Chown and Davis 2003).

F-period gas exchange
One of the more controversial aspects of the water-saving (or hygric genesis) hypothesis for the origin and subsequent maintenance of the DGC is the nature of F-period gas exchange (Lighton 1996). If gas exchange is predominantly by convection, then the outward diffusion of water will be minimized and both a DGC, and increases in the duration of the F-period, will lead to enhanced water conservation (Kestler 1985). On the other hand, if gas exchange during the F-period takes place predominantly by diffusion, then there is unlikely to be any benefit to water economy in either possession of such a period, or regulation thereof. It is clear that at least in lepidopteran pupae (Brockway and Schneiderman 1967), the ant *C. bicolor* (Lighton *et al.* 1993*a*), and *P. americana* (Kestler 1985), the F-period is characterized by largely convective gas exchange. However, diffusion seems to predominate in the tenebrionid beetle *P. striatus* (Lighton 1988*b*) and in *C. vicinus* (Hymenoptera, Formicidae).

In experiments to determine the nature of gas exchange in the latter species, Lighton and Garrigan (1995) manipulated external oxygen concentrations to investigate the nature of F-period gas exchange, and made several predictions. Because ambient oxygen concentration determines the amount of

oxygen available at the start of the C-period, a reduction in ambient oxygen concentration should result in a decline in C-period duration. Declining ambient oxygen concentrations should also prevent the generation of a substantial negative endotracheal pressure. Thus, if F-period gas exchange is predominantly by convection, F-period duration should decline owing to reduced gas exchange capabilities due to the lower Δp (equation 2). Alternatively, if F-period gas exchange is dominated by diffusion, then metabolic requirements during declining ambient oxygen concentrations could be met by increasing spiracular opening (the area term in equation 1). If this is done, the time taken to reach the hypercapnic setpoint that usually triggers O-period spiracular opening will increase, resulting in an increase in F-period duration. In *C. vicinus*, F-period increased with declining oxygen concentration suggesting that gas exchange takes place mostly by diffusion (it cannot be solely by diffusion—see Kestler 1985). By contrast, in *H. cecropia*, declining external pO_2 results in a reduction in F-period duration (Schneiderman 1960), as might be expected from a species in which changes in endotracheal pressure in the direction predicted by convective gas exchange take place (Brockway and Schneiderman 1967). A similar response to declining external pO_2 was found in the dung beetle *Aphodius fossor* (Fig. 3.13), suggesting that in this species the F-phase is also predominantly convective (Chown and Holter 2000).

Based on the partial pressure gradients for both CO_2 and O_2, it has also been suggested that respiratory exchange ratios should be in the region of 0.2 during the F-period if gas exchange takes place primarily by diffusion. Calculated F-period RERs are not significantly different from this value in *P. striatus* (Lighton 1988*b*), and in *C. vicinus* (Lighton and Garrigan 1995), which has led Lighton (1996) to conclude that diffusion is the predominant form of gas exchange during the F-period in many insect species. However, RERs can be notoriously difficult to calculate owing to limitations of flow-through oxygen analysers (Kestler 1985).

Because only a small number of species have been investigated it is not possible to determine whether diffusion or convection is the predominant form of F-period gas exchange in insects. Moreover,

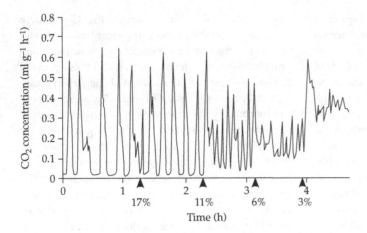

Figure 3.13 Gas exchange in the dung beetle *Aphodius fossor* showing a decline in F-period duration with declining ambient pO_2, and the absence of a C-period at the lowest oxygen concentrations. *Source*: Chown and Holter (2000).

in at least some species of desert-dwelling tenebrionid beetles the F-period has been modified such that gas exchange is reduced to a number of small bursts which are accompanied, at least in some species, by ventilatory movements, suggesting that the F-period might be modified in several other ways (Lighton 1991*a*; Duncan *et al.* 2002*a*). In addition, in *C. bicolor*, there is temporal partitioning of the F-period such that F-period gas exchange initially takes place via the small, abdominal spiracles and then via the larger thoracic spiracles. Lighton *et al.* (1993*a*) suggested that the limited diffusive capability of the abdominal spiracles might assist water retention during the initial part of the F-period when the pressure gradient is large, but that as this pressure gradient diminishes the larger thoracic spiracles supply O_2 via diffusion owing to their larger cross-sectional area (see also Byrne and Duncan 2003). Therefore, a definitive conclusion regarding the role of the F-phase in water economy and the likely adaptive nature of the DGC from this perspective cannot be provided for the insects as a whole.

Alternative hypotheses

It is not only the nature of F-period gas exchange that has raised doubts regarding the contribution of the DGC to water balance in insects (see Chapter 4). The small contribution of respiratory transpiration to total water loss has also raised doubts about the extent to which modulation of gas exchange patterns might affect water balance (see

Hadley and Quinlan 1993; Shelton and Appel 2001*a*, and review in Chown 2002). Moreover, many insects seem to abandon the DGC just when water economy is required most—when desiccated or at high temperatures (Hadley and Quinlan 1993; Chappell and Rogowitz 2000), presumably because haemolymph buffering of CO_2 cannot take place (Lighton 1998). For example, worker ants, which spend much time above ground exposed to hot, dry conditions, should show DGCs, whereas female alates, which spend the larger part of their lives underground in a humid environment should not. In *Messor* harvester ants exactly the opposite pattern is found, and this has lead Lighton and Berrigan (1995) to propose the chthonic (= underworld) genesis hypothesis for the origin of the DGC. They argued that female alates are likely to spend much time in environments that are profoundly hypoxic and hypercapnic, especially during their claustral phase when the burrow entrance is sealed. Under these conditions, in insects that rely predominantly on diffusion (as ants apparently do), gas exchange can be promoted either by increasing the area term (spiracular opening) or by increasing the partial pressure difference between the endotracheal space and the environment (equation 1). Opening the spiracles fully, as would be required in the former case, would result in elevated water loss, and presumably fairly rapid dehydration unless the surrounding air were completely saturated. However, alteration of the partial pressure gradients would not incur a water loss penalty, but would require that endotracheal gas concentrations are significantly hypoxic and hypercapnic relative to the environment. One way of ensuring this is would be to adopt discontinuous gas exchange, during which pCO_2 is enhanced and pO_2 depleted relative to the environment. In worker ants, this extreme form of gas exchange, which carries penalties in terms of internal homeostasis, would not be required because of the large amount of time they spend above ground (Lighton and Berrigan 1995).

Based on a qualitative review of gas exchange patterns in various species, Lighton (1996, 1998) has suggested that the chthonic genesis hypothesis better explains the absence of the DGC in some species (largely above-ground dwellers) and its presence in others (psammophilous and subterranean species), and is therefore more likely an explanation for the origin of this gas exchange pattern than the pure hygric genesis hypothesis. Since then, two direct tests of this hypothesis have been undertaken. In the first, it was found that the dung beetle *A. fossor*, which encounters profoundly hypoxic and hypercapnic environmental conditions in wet dung (Holter 1991), abandons the DGC as conditions become more hypoxic (Chown and Holter 2000) (Fig. 3.13). This suggests that DGCs, which are highly characteristic of the species at rest, do not serve to enhance gas exchange under hypoxic and hypercapnic conditions. Rather, under hypoxic conditions, this species increases spiracular conductance to meet its largely unchanging metabolic requirements, which is unlikely to carry a biologically significant water loss penalty because cuticular transpiration in this species is high and because it lives in a habitat where water is plentiful. In the second test, Gibbs and Johnson (2004) found that in the harvester ant *Pogonomyrmex barbatus*, which shows several different gas exchange patterns, the molar ratio of water loss to CO_2 excretion does not vary with gas exchange pattern. Therefore, they concluded that the DGC does not serve to improve gas exchange while reducing water loss. Several other authors have argued that the chthonic genesis hypothesis cannot explain the origin and maintenance of the DGC, based mostly on the grounds of habitat selection and the presence or absence of DGCs in the species they have

investigated (Vogt and Appel 1999; Duncan *et al.* 2002*a*). In particular, Duncan *et al.* (2002*a*) pointed out that subsurface conditions in sand are unlikely to become severely hypoxic and hypercapnic (Louw *et al.* 1986), highlighting the need for additional information on environmental gas concentrations (Section 3.2.2).

A second, alternative adaptive explanation for the origin of the DGC is the oxidative damage hypothesis proposed by Bradley (2000). He suggested that because oxygen is toxic to cells, DGCs might serve to reduce the supply to cells when metabolic demand is low. Although this is an interesting alternative hypothesis, it does not explain the absence of DGCs in many insects, nor does it account for the fact that oxidative damage might be limited more readily by alteration of fluid levels in the tracheoles (Kestler 1985).

In contrast to the three adaptive hypotheses (hygric genesis, chthonic genesis, oxidative damage), Chown and Holter (2000) proposed that DGCs are not adaptive at all. They suggested that the cycles might arise as an inherent consequence of interactions between the neuronal systems that control spiracular opening and ventilation (see also Miller 1973). Under conditions of minimal demand, interacting feedback systems can show a variety of behaviours ranging from a single steady-state to ordered cyclic behaviour (May 1986; Kauffman 1993), and the periodic nature of several physiological phenomena is thought to be the consequence of such interactions (Glass and Mackey 1988). In the case of cyclical gas exchange patterns there are several reasons for considering this to be likely. First, the isolated CNS is capable of producing rhythms very similar to those of the DGC (Bustami and Hustert 2000; Bustami *et al.* 2002), and is characterized by neuronal feedback (Ramirez and Pearson 1989). Second, changes in pO_2 have a marked influence on CO_2 setpoints and vice versa, suggesting that considerable interaction is characteristic of the system (Levy and Schneiderman 1966*a*; Burkett and Schneiderman 1974). Third, although much of the recent literature suggests that DGCs are rather invariant in the species that have them, there is much evidence showing that gas exchange patterns vary considerably within species and within individuals (Miller 1973, 1981;

Lighton 1998; Chown 2001; Marais and Chown 2003; Gibbs and Johnson 2004). The latter is reminiscent of the kinds of variability characteristic of interacting feedback systems.

A consensus view?

By the mid-1970s the DGC story appeared to have been told: the cycles are characteristic especially of diapausing pupae in which there is likely to have been strong selection for reductions in water loss, resulting in a water-saving convective F-period and a C-period when the spiracles are entirely closed. Thirty years later, the situation is less clear. DGCs are characteristic of many adult and pupal insects (living in a variety of environments), show considerable variability, and are sometimes abandoned just when they seem to be needed most. None of the alternative adaptive hypotheses proposed to explain the origin and maintenance of the DGC appear to enjoy unequivocal support either, and recent work has once again highlighted the fact that metabolic rate and DGC patterns (specifically O-period and F-period durations) can be modulated to save water (Chown and Davis 2003). Moreover, it has also recently been suggested that DGCs might represent convergent patterns that have evolved along two entirely different routes. That is, DGCs might include CFO and CFV patterns of the kind that we have been most concerned with, but might also have convergently arisen from intermittent–convective gas exchange (Lighton 1998). Although there is little evidence for this idea, it does highlight the fact that similar patterns may not necessarily be the consequence of identical mechanisms.

The likely independent evolution of DGCs several times within the arthropods lends additional support to this idea. Discontinuous gas exchange cycles are also found in soliphuges (Lighton and Fielden 1996), ticks (Lighton *et al.* 1993c), pseudoscorpions (Lighton and Joos 2002) and centipedes (Klok *et al.* 2002), (but not in mites (Lighton and Duncan 1995) or harvestmen (Lighton 2002)) and have probably evolved independently at least four times in the arthropods (Fig. 3.3). Moreover, it appears that in pseudoscorpions (or at least the one that has been investigated) the O-period is triggered by hypoxia rather than by hypercapnia as is the case in the

insects that have been investigated to date. Thus, there might be considerable variation in both the origins of and mechanisms underlying DGCs in arthropods, and given the patchy distribution of DGCs and their considerable variety among the insects, it seems likely that in this group they have also evolved independently several times.

Whether natural selection has been responsible for this evolution is a more difficult question to resolve. Arguably, there is a variety of circumstances that could impose selection for discontinuous gas exchange. These include drought, hypoxia, hypercapnia, and perhaps also the need for insects to reduce metabolic rates during dormancy (Lighton 1998), especially in long-lived adults that have to cope with unpredictably unfavourable environments (Chown 2002). In addition, these circumstances are probably not important in active insects that have ready access to food and water, and which can generally tolerate starvation and desiccation for periods that are much longer than those routinely encountered during the peak activity season (Chapter 4). Thus, the various environmental factors that promote the DGC are most likely to be important during dormancy, and probably also act in concert to promote both DGCs and low metabolic rates (to overcome the threat of starvation), making it difficult to distinguish between the major adaptive hypotheses solely by using experimental manipulations. One solution to this problem might be to combine experimental work with broad comparative studies within a strong inference framework (Huey *et al.* 1999). Another is to examine more broadly the extent to which variation in the DGC is partitioned within and among individuals. At the moment it appears that even in highly variable species most of the variation in the DGC and its characteristics is partitioned among individuals, and much less within individuals, so providing natural selection with much to work on (Marais and Chown 2003). If this is the case in most species then the adaptive hypotheses will be much more difficult to dismiss (Chown 2001), particularly because metabolic characteristics also respond to selection (Gibbs 1999).

Thus, it would seem prudent to conclude, based on the evidence of the current literature, that DGCs have evolved as adaptations to cope with unfavourable environmental conditions, and particularly limited water availability. However, the alternative, non-adaptive hypothesis—that cyclic gas exchange is a basal characteristic of all arthropods that have occludible spiracles, which results from interactions between the regulatory systems responsible for spiracular control and ventilation—has not been sufficiently well explored for it to be rejected. Indeed, cyclic gas exchange may well have had a non-adaptive origin, but subsequently found itself pressed into other forms of service.

3.4.5 Metabolic rate variation: size

Discontinuous gas exchange in ticks is thought to be one of the ways in which these animals maintain the very low metabolic rates required by their sit-and-wait strategy, which includes long periods of fasting (Lighton and Fielden 1995). Scorpions are also thought to have uncharacteristically low metabolic rates, and this has prompted considerable speculation regarding the benefits of low metabolic rates in both groups (Lighton *et al.* 2001). In turn, this speculation has raised the question of what a 'characteristic' metabolic rate is for arthropods, including insects, of a given size. In other words, what values should the coefficient (*c*) and exponent (*z*) assume in the scaling relationship

$$B = cM^z, \tag{5}$$

where *B* is metabolic rate (usually expressed in µW) and *M* is body mass (usually expressed in g). This question has long occupied physiologists and ecologists, and can indeed be considered one of the most contentious, yet basic issues in environmental physiology. The controversy concerns both the empirical value of *z* (but also of *c*, see Heusner 1991), and the theoretical reasons why a particular value of *z* should be expected. In addition, theoretical investigations have focused mostly on endotherms, while empirical work often includes unicells, ectotherms (vertebrate and invertebrate) and vertebrate endotherms (Robinson *et al.* 1983; West *et al.* 2002).

From a theoretical perspective, for endotherms, it was originally suggested (as far back as 1883 by Rubner) on the basis of a simple dimensional analysis that $z = 0.67$, the same as the scaling

relationship for surface area. Heusner (1991) subsequently argued that this exponent reflects an underlying dimensional relationship between mass and power and is, therefore, of less interest than the coefficient (*c*) of the relationship. The early surface area arguments appeared to be at odds with the empirically derived data (see Dodds *et al.* 2001 for discussion), spawning many more theoretical investigations. These have been based, *inter alia*, on

(1) the temperature dependence of metabolic rate (Gray 1981);

(2) considerations of organisms in terms of heterogeneous catalysis of enzyme reactors, fractal geometry, and nonlinear relationships for transport processes (Sernetz *et al.* 1985);

(3) the contribution of non-metabolizing and metabolizing tissues to total mass (Spaargaren 1992);

(4) nutrient supply networks (West *et al.* 1997, 1999);

(5) dimensional analyses and four-dimensional biology (Dodds *et al.* 2001); and

(6) multiple control sites in metabolic pathways (Darveau *et al.* 2002).

Of these analyses, those based on nutrient supply networks have gained the most attention recently. This is predominantly because it has been argued that the nutrient supply network model provides a mechanistic basis for understanding the primary role of body size in all aspects of biology, and that the quarter power scaling (i.e. $z = 0.75$) is the 'single most pervasive theme underlying all biological diversity' (West *et al.* 1997, 1999). These analyses assume that the rate at which energy is dissipated is minimized, transport systems have a space-filling fractal-like branching pattern, and the final branch of the network is size invariant. Based on these assumptions, and several others, such as branching patterns that preserve cross-sectional area, a scaling exponent of 3/4 is derived, which is apparently in keeping with previous empirical estimates of this value (see also Gillooly *et al.* 2001). Subsequently, Dodds *et al.* (2001) have argued that there are several inconsistencies in West *et al.*'s (1997, 1999) mathematical arguments, suggesting that the exponent need not be 0.75, as the latter have claimed. For example, Dodds *et al.* (2001) argued that under a pulsatile flow system, and using the arguments of West et al. (1997), the scaling exponent should be 6/7, rather than 3/4. Since then, several, often critical, discussions have appeared of the assumptions of the nutrient supply network models, the empirical data that are used to support them, and their implications (see *Functional Ecology* 2004, Volume 18:2).

By contrast, Darveau et al. (2002) have argued (for endothermic vertebrates) that searching for a single rate-limiting step, which enforces scaling of metabolism, is simplistic, even though many previous models are based on such an assumption. Metabolic control is vested in both energy demand and in energy supply pathways and at multiple sites. Thus, there must be multiple causes of metabolic rate allometry. They argue that this relationship is best expressed as

$$MR = a \sum c_i M^{b_i}, \qquad (6)$$

where M is body mass, a is the intercept, b_i is the scaling exponent of process i, and c_i the control coefficient of that process. At rest, all steps in the O_2 delivery system display large excess capacities (arguably for any organisms that have factorial aerobic scopes above 1), therefore it is energy demand that is likely to set the scaling exponent. The two most likely energy sinks are protein turnover (25 per cent of total ATP demand) and the Na^+/K^+-ATPase (25 per cent), and using the scaling exponents and maximum and minimum values for the coefficients of these processes Darveau et al. (2002) find that the scaling exponent for basal metabolic rate lies between 0.76 and 0.79. On the other hand, maximal metabolic rate is likely to be limited by O_2 supply because it is known that, at maximum metabolic rate, supply shows almost no reserve capacity, whereas actomyosin and the Ca^{2+} pump (in muscles which are using 90 per cent of the energy during activity) have considerable reserve capacity. In consequence, the scaling exponent during maximal metabolic demand lies between 0.82 and 0.92. Thus, in this multi-site control model, b_i and c_i for all major control sites in ATP-turnover pathways determine the overall scaling of whole-organism bioenergetics, depending on the relative importance of limitations in supply versus demand. Any environmental

influences on scaling (e.g. Lovegrove 2000) must consequently have an effect through these proximal causes.

Subsequently, it has been shown that equation (6) is technically flawed because it violates dimensional homogeneity, and the utility of this approach for understanding scaling has been questioned (Banavar et al. 2003; West et al. 2003). Darveau et al. (2003) claim that the technical problems can be overcome by modifying equation (6) to

$$BMR = MR_0 \sum c_i (M/M_0)^{b_i}, \qquad (7)$$

where MR_0 is the metabolic rate of an organism of characteristic body mass M_0. However, this begs the question of what a characteristic mass might be of, for example, mammals that show eight orders of magnitude variation in mass, and indeed the choice of M_0 must be arbitrary (Banavar et al. 2003). Notwithstanding these difficulties, the analysis by Darveau et al. (2002) draws attention to the importance of multiple control pathways in metabolism.

What these theoretical analyses imply for the scaling of insect metabolic rate is not clear. West et al. (1997) argue that their models apply as much to insects as to other organisms. However, they also assume that gas exchange in insects takes place solely by diffusion, and that tracheal cross-sectional area remains constant during branching of the system. Clearly there are problems with both of these assumptions (Sections 3.3, 3.4.1). Moreover, empirical studies suggest that in insects SMR scales neither as $M^{0.75}$, nor as $M^{0.67}$. Although several early studies provided estimates of the scaling relationship for SMRs in insects (e.g. Bartholomew and Casey 1977; Lighton and Wehner 1993), the first consensus scaling relationship across several taxa, that took into account the likely effects of activity (Section 3.1), was the one provided by Lighton and Fielden (1995), and subsequently modified by Lighton et al. (2001). They argued that all non-tick, non-scorpion arthropods share a single allometric relation:

$$B = 973M^{0.856}, \qquad (8)$$

where mass is in g and metabolic rate in µW, at 25°C. The scaling exponent in this case lies midway between earlier assessments suggesting that the exponent is approximately 0.75 and others suggesting that it is closer to 1 (e.g. von Bertalanffy 1957). It is also statistically indistinguishable from the 6/7 exponent predicted by Dodds *et al.* (2001). Although tracheae are clearly capable of some flexibility (Herford 1938; Westneat *et al.* 2003), thus tempting speculation regarding the similarity of the empirical and theoretically derived values, pulsatile flow of the kind seen in mammals seems unlikely in insects. Moreover, it seems even more unlikely in scorpions, which have a similar exponent (Lighton *et al.* 2001), but lack tracheae.

Other empirical investigations have revealed a wide range of interspecific scaling exponents for metabolic rate in insects (0.5–1.0), although in many cases only a small number of species was examined (Hack 1997; Davis *et al.* 1999; Duncan *et al.* 2002*a*), or data obtained using a wide variety of methods were included in the analysis (Addo-Bediako *et al.* 2002). These results have led several authors to suggest that carefully collected data from a wider variety of species are required before a consensus scaling relationship for insect SMRs is adopted, especially because Lighton and Fielden's (1995) analysis was based on a limited range of taxa (mostly tenebrionid beetles and ants) (Duncan *et al.* 2002*a*). There can be little doubt that additional investigations of insect metabolic rate, which control carefully for both activity and feeding status (see Section 3.5.3), would be useful for elucidating the consensus scaling equation for insect metabolic rate. They might reveal that there is no consensus relationship, but that the relationship varies with size, taxonomic group, and geography, as seems to be the case for endothermic vertebrates (Heusner 1991; Lovegrove 2000; Dodds *et al.* 2001). Moreover, such a comparative analysis, including estimations based on phylogenetic independent contrasts (Harvey and Pagel 1991) (something that is rarely done for insects), might also reveal a very different scaling relationship to those that have been found to date, although even here analyses are likely to be problematic (see Symonds and Elgar 2002). Nonetheless, determining what form the scaling relationship for metabolic rate takes in insects, and whether it is consistent

at the intra- and interspecific levels is of considerable importance. The available data and analyses suggest that in insects the intraspecific and interspecific scaling relationships are quite different, with intraspecific relationships being indistinguishable from the 0.67 value predicted solely on geometric considerations (Bartholomew *et al.* 1988; Lighton 1989), and the interspecific slope being much higher. These results contradict West *et al.*'s (1997, 2002) models which suggest that the scaling relationship at the intra- and interspecific levels should be identical, and support previous suggestions that there is no reason why interspecific and intraspecific scaling relationships should be the same (see Chown and Gaston 1999 for review). From an ecological perspective, the scaling of metabolic rate provides insight into the likely energy use of insects of different sizes, which has a host of implications. These include those for the evolution of body size frequency distributions (Kozłowski and Gawelczyk 2002), for understanding energy use and abundance in local communities (Blackburn and Gaston 1999), and for understanding patterns in the change of insect body sizes across latitude (Chown and Gaston 1999).

3.4.6 Metabolic rate variation: temperature and water availability

Temperature and water availability are both thought to influence metabolic rate, especially over the longer term, resulting in adaptations that apparently reflect the need either for water conservation or starvation resistance, or the response to low environmental temperatures (Chown and Gaston 1999). The influence of temperature on metabolic rate over short timescales has been called the most overconfirmed fact in insect physiology (Keister and Buck 1964), and acute modifications of metabolic rate by temperature are certainly widely known for insects, with many modern studies continuing to document them. The short-term influence of humidity on metabolic rates has also been documented in several species, though with the advent of flow-through respirometry these effects are often not investigated, largely because rate measurements are made in dry air for technical reasons. In at least some instances, increases in metabolic rate with declining humidity may be the result of increased activity, rather than any other fundamental alteration in metabolism or gas exchange (Lighton and Bartholomew 1988).

Temperature

Much of the discussion of temperature effects on metabolic rate has been undertaken under the rubric of Q_{10}, or the change in the rate of a process with a 10°C change in temperature. Whether this concept (and the Arrhenius equation, see below) should be applied to the whole organism, rather than to specific reactions is an argument that dates back at least to Krogh (Keister and Buck 1961), and continues to generate discussion. Nonetheless, it is now widely known that whole-animal Q_{10} varies with the temperature range over which it is measured (Keister and Buck 1964; Cossins and Bowler 1987). In consequence, Q_{10} cannot strictly be called a parameter and should rather be termed the apparent Q_{10} when calculated in the normal fashion (Chaui-Berlinck et al. 2001). Moreover, the apparent Q_{10} is also supposedly inappropriate for deriving conclusions regarding metabolic control (i.e. up- or down-regulation of the response to temperature), and only if a Q_{10} is assumed in advance of an analysis can conclusions regarding Q_{10} be made (Chaui-Berlinck et al. 2001). However, there are substantial problems inherent in the solutions proposed by Chaui-Berlinck et al. (2001) to the 'apparent Q_{10}' problem.

The temperature dependence of Q_{10} was also recently rediscovered by Gillooly et al. (2001), who suggested that their 'universal temperature dependence' (UTD) of biological processes should be used in the place of Q_{10} to describe rate–temperature relationships. This UTD amounts to little more than a rediscovery of the Arrhenius equation, originally proposed at the turn of the last century, and extensively discussed by Cossins and Bowler (1987). Here, the relationship between rates at two temperatures is given by

$$k_1 = k_2 e^{\mu/R((T_1-T_2)/T_1 T_2)}, \qquad (9)$$

where, k_1 is the rate at absolute temperature T_1, k_2 is the rate at absolute temperature T_2, μ is a constant (the Arrhenius activation energy or critical thermal increment), and R is the universal gas constant ($8.314 \, \mathrm{J\,K^{-1}\,mol^{-1}}$).

The novelty of the Gillooly et al. (2001) approach is that it takes a concept initially applied within organisms (and species) and applies it to a cross-species problem. However, this application has generated some concern (Clarke 2004), and its validity, especially given reasonably constant metabolic rates of insects across latitude (Addo-Bediako et al. 2002), remains an issue open for discussion.

Although the Arrhenius equation does resolve the problems associated with temperature dependence of Q_{10}, Cossins and Bowler (1987) have noted that over the range of biologically relevant temperatures (0–40°C), the two measures scarcely differ. Given the concerns raised by Chaui-Berlinck et al. (2001) it would seem most appropriate to make use of the universal temperature dependence (or Arrhenius activation energy), which can readily be calculated from most data obtained for investigation of Q_{10} effects. In this context it is also worth noting that the use of analysis of variance (ANOVA) for assessing the temperature dependence of metabolic rate, as is sometimes done (Duncan and Dickman 2001), is inappropriate. ANOVAs are generally much less sensitive than regression techniques for determining the relationship between two variables (see Somerfield et al. 2002), and in consequence give spurious conclusions. In this context, ANOVAs might lead authors to conclude that there are no effects of temperature on metabolic rate, when in fact these effects are profound.

Long-term responses of insect metabolic rates to temperature have also been documented in many species and have generally been discussed in the context of the conservation of the rate of a temperature-dependent physiological process, in the face of temperature change. This temperature conservation, or metabolic cold adaptation (MCA), has been the subject of considerable controversy both in ectotherms in general (Clarke 1993), and in insects (Addo-Bediako et al. 2002; Chown et al. 2003; Hodkinson 2003).

Metabolic cold adaptation

Elevated metabolic rates of species (or populations) from cold climates relative to those from warm climates (at the same trial temperature) are thought to allow the species showing them to meet the elevated ATP costs of growth and development (Wieser 1994) necessary for completion of life cycles in the relatively short, cool growing seasons that characterize cold (high latitude or altitude) regions (Chown and Gaston 1999). Such high growth rates and their elevated metabolic costs are not maintained in species from all environments because of the likely fitness costs of rapid growth (Gotthard *et al*. 2000). Thus, MCA represents not only a physiological response to environmental temperature, but also a significant component of the life history of an organism, which allows it to respond, via internal alterations, to its environment. Metabolic cold adaptation in insects has been discussed extensively by Chown and Gaston (1999), and they have highlighted the importance of distinguishing between intra- and interspecific levels when addressing MCA. While the outcomes of intraspecific studies are varied, though often finding support for MCA, the ways in which interspecific studies should be undertaken and the outcome of these investigations are much more controversial.

At the intraspecific level, only a few investigations have been undertaken across latitudinal gradients, with some finding no evidence for MCA (Nylund 1991), while others have found variation in line with predictions of the MCA hypothesis (Berrigan and Partridge 1997). Presumably for reasons of species geographic range sizes (most tend to be small), studies across elevational gradients are more commonplace. In general, these have found evidence in favour of metabolic cold adaptation (Chown and Gaston 1999), although Ashby (1997) has pointed out that simultaneous changes of body size across elevational gradients (Blackburn *et al.* 1999), and preferred temperatures of the individuals at each site, complicate interpretation of the evidence. In her study, mass-specific metabolic rates of *Xanthippus corallipes* (Orthoptera, Acrididae) increased with elevation, though body mass declined, leading to conservation of rate across the gradient. In another grasshopper, *Melanoplus sanguinipes*, body size varied in the opposite direction to that found in *X. corallipes*, suggesting that size effects cannot account entirely for their higher metabolic rates at high elevations. Nonetheless, cooler activity temperatures meant that

compensation was not complete (Rourke 2000) (Fig. 3.14). These studies highlight the need to take both size and temperature into account in investigations of MCA, something that, until recently, has not been routinely done. Body size can be particularly problematic in this regard because it might either increase or decline with elevation (or latitude), even within closely related taxa (see

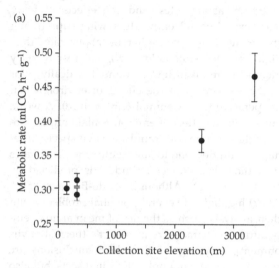

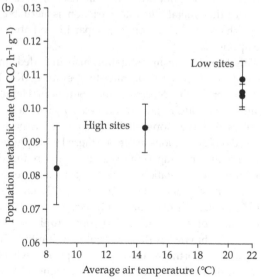

Figure 3.14 Mean (± SE) metabolic rate of different *Melanoplus sanguinipes* (Orthoptera, Acrididae) populations plotted as a function of (a) altitude (rates measured at 35°C), and (b) mean air temperature corrected for July air temperature.

Source: Rourke (2000).

Chapter 7 and Chown and Klok 2003a), and this change is not independent of the metabolic and developmental responses of individuals to their environments (Chown and Gaston 1999).

At the interspecific level, there have been many investigations of MCA (Chown and Gaston 1999), of which the most comprehensive latitudinal study was undertaken by Addo-Bediako et al. (2002). This investigation was based on a large compilation of insect metabolic rates, and the effects of body mass, experimental temperature, wing status (flying species tend to have higher metabolic rates than flightless ones—Reinhold 1999), and respirometry method were taken into account statistically prior to the assessment of the effects of environmental temperature (mean annual temperature). A weak, but significant effect of environmental temperature on metabolic rate was found such that species from higher latitudes tend to have higher metabolic rates than those from lower latitudes, as predicted by MCA (Fig. 1.5). Although Addo-Bediako et al. (2002) highlighted several potential problems with their analysis (such as the use of mean annual temperature to characterize an insect's thermal environment), they argued that their conclusions are robust, reflecting not only MCA in insects, but also the fact that metabolic rate variation is heritable, given that much variation in it is partitioned above the species level.

This broad-scale investigation also included an analysis of the slope of rate–temperature (R–T) curves across the Northern and Southern Hemispheres. Although data are limited for the latter, it appears that the slope of the R–T curve increases towards colder regions in the Northern Hemisphere, but remains unchanged in the south. Thus, reduced sensitivity of metabolic rate to temperature is likely to be characteristic of southern, cold-climate species, where climates are likely to be permanently cool and often cloudy, and opportunities for effective behavioural thermoregulation limited. By contrast, in northern, cold-climate species, where hot, sunny periods may be more frequent (Danks 1999), thus facilitating behavioural thermoregulation, greater R–T sensitivity would be expected. Hence, the metabolic benefits of R–T sensitivity that would accrue to species experiencing regularly sunny conditions in the Northern Hemisphere, are

likely to be outweighed by the costs associated with a permanently depressed metabolism under cloudy conditions in the Southern Hemisphere. Such hemisphere-associated asymmetry in responses is characteristic of other physiological variables (Chapters 5 and 7) and may well contribute to large-scale differences in the biodiversity of the south and the north.

Water availability

For those organisms that are able to occupy stressful environments, reductions in metabolic rate (for an organism of a particular size at a particular temperature compared to similar species or populations) are thought to be important adaptations to these conditions (Hoffmann and Parsons 1991). In the case of food and water stress a reduction in metabolic rate should mean a reduction both in ATP use and, in insects which exchange gases predominantly by convection, a reduction in water loss. Indeed, many authors have claimed that the reduced metabolic rates they find in insects they have investigated are a response either to dry conditions or to low food availability (see Chown and Gaston 1999 for review). Although the importance of reduced metabolic rates as a means to conserve water (Chapter 4), or as a general stress response is controversial, it is clear that, especially during dormancy, insects lower metabolic rates to overcome both a reduced energy intake and lack of water (Lighton 1991a; Davis et al. 2000; Chown 2002; Chown and Davis 2003). Moreover, in a few species, depression of metabolic rates appears to be actively promoted by means of behavioural aggregation, which promotes hypoxia, and consequently reduces metabolic rates (Tanaka et al. 1988; Van Nerum and Buelens 1997).

The controversy surrounding the idea of low metabolic rates as an adaptation to dry conditions in species inhabiting xeric environments is a consequence of three rather different issues. First, it is often stated that the contribution of respiratory transpiration to overall water loss is generally small, and therefore modification of metabolic rates is unlikely to contribute significantly to water conservation. Although this argument is now widely used, the data in support of it are contradictory (see especially Chown and Davis 2003) and the argument remains difficult to resolve (Chapter 4). Second, xeric environments are often characterized as much by patchy (in space and time) food resources as by patchy water availability. Therefore, it is difficult to determine which is likely to be the more important in selecting for low metabolic rates. Although responses to desiccation and starvation can be rather different in some species (Gibbs and Matzkin 2001; Gibbs 2002*a*), these two environmental factors might also act in concert to lower metabolic rates (Chown 2002). Third, because low temperature environments might select for higher metabolic rates and arid environments for low metabolic rates, and because temperature and water availability often show negative covariation (hot environments are often dry), it might be difficult to distinguish which of these factors is responsible for variation in metabolic rates (Davis *et al.* 2000).

These problems could be addressed in three ways. First, work could be conducted only on those animals for which a fully resolved phylogeny is available, along with information on both habitat temperature and water availability (or any other variables thought to be of significance). However, if covariation in the abiotic factors and metabolic rate is in the same direction across the phylogeny this method is unlikely to result in adequate resolution of the problem. Second, closely related species could be selected from environments that are largely similar with regard to most of their abiotic variables, but differ with regard to the variable of interest (although the problem of underlying causal variables may remain). Third, wide-ranging studies on a large variety of insects from many different habitats could be undertaken and their relationship to abiotic factors could be explored using general linear modelling techniques to obtain some idea of the most important abiotic variables affecting metabolic rate. Such a macrophysiological approach is likely to be confounded by problems associated with inadequate investigations of acclimation, as well as by relatively weak power to detect trends owing to variation associated with other factors. Nonetheless, it can provide a useful assessment of the factors likely to be influencing metabolic rate (Chown and Davis 2003). Owing to the paucity of these kinds of investigations, it seems reasonable to conclude that both temperature and water availability have considerable effects on metabolic rate, but their adaptive significance remains to be fully explored across a range of species.

3.5 Gas exchange and metabolic rate during activity

Insects show tremendous scope for the increase of metabolic rates above resting levels. This is due mainly to the energetic demands of flight, which can be responsible for 100-fold increases in metabolic rate. Although many insects, such as ants and honeybees, might never experience levels of metabolism as low as SMRs measured under solitary conditions (sometimes including anaesthesia), and therefore have factorial aerobic scopes more in keeping with those of vertebrates (4–10) (Harrison and Fewell 2002), large moths such as those examined by Bartholomew and Casey (1978) are often quiescent. Therefore, factorial aerobic scopes of more than 100 (as high as 170) are likely to be accurate reflections of the increases in aerobic metabolism that are possible in insects. Such high aerobic scopes are of considerable interest not only because insects often make a rapid transition from alert rest to flight, but also because flight metabolic rate might also constrain resting levels of metabolism, and in so doing affect both abundance and fecundity (Marden 1995*a*; Reinhold 1999).

Flight is not the only reason for elevated metabolic rates. Other forms of locomotion, including both pedestrian locomotion and swimming are responsible for increases in metabolic rate, as are calling (Lighton 1987; Reinhold 1999), sexual activity (Giesel *et al.* 1989; Woods and Stevenson 1996), thermoregulation (both for flight and in the absence of flight), colony homeostasis (Heinrich 1993), and feeding. For example, *Megasoma elephas* (Coleoptera, Scarabaeidae) responds to reduced ambient temperatures by endothermic heat production, in the absence of any form of activity (Fig. 3.15) (Morgan and Bartholomew 1982), while in another group of scarabaeids, the winter active rain beetles, males maintain elevated thoracic temperatures and metabolic rates while searching for females (Morgan 1987). In the leaf-cutting ant, *Atta sexdens*, leaf-cutting activity results in high metabolic rates with factorial aerobic scopes of

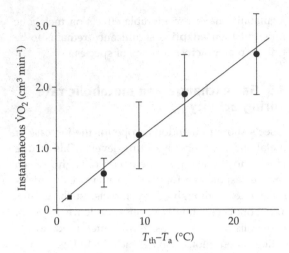

Figure 3.15 The relationship between the temperature excess (the difference between thoracic and ambient temperature) and mean (± 2SE) metabolic rate in *Megasoma elephas* (Coleoptera, Scarabaeidae) at rest.

Source: Reprinted with permission from Morgan and Bartholomew. 1982. *Science* **216**, 1409–1410. © 1982, with permission from AAAS.

approximately 30 (Roces and Lighton 1995). This metabolic activity is the consequence of high mass-specific metabolic rates of the mandibular muscles (194 W kg^{-1}, insect flight muscle varies from 300 to 3000 W kg^{-1}), and a high mandibular muscle mass (>25 per cent of total body mass). Despite the wide variety of reasons for variation in metabolic rates, we will deal in detail with just three of these, viz. flight, pedestrian locomotion, and feeding.

3.5.1 Flight

Metabolism during flight is strictly aerobic. In small insects, such as *Drosophila*, oxygen requirements are met almost exclusively by diffusion, and spiracular opening can be regulated to meet varying aerodynamic power requirements, thus conserving water (Lehmann 2001). In large insects, aerobic demands are met by ventilation of the tracheal system in one of four ways. In some insects, such as bees and wasps, abdominal pumping, usually with unidirectional flow, is thought to be responsible for most or all of the bulk oxygen flow during flight (Weis-Fogh 1967a; Miller 1974). By contrast, in several other groups of insects, such as locusts, dragonflies, beetles, cockroaches,

and moths, ventilation as a consequence of the flight muscle movements, or autoventilation, meets the gas exchange requirements of the flight musculature (Miller 1974), and has been examined in most detail in *S. gregaria*. During flight, tidal autoventilation, through spiracles 2 and 3, accounts for 250 l kg^{-1} h^{-1}, and serves the flight musculature, whilst abdominal pumping accounts for a further 144 l kg^{-1} h^{-1}, and is largely unidirectional through the dorsal orifice of spiracle 1 and out through spiracles 5–10 (Miller 1960; Weis-Fogh 1967a). Autoventilation meets the oxygen demands of many large beetles, such as *Goliathus regius* (Scarabaeidae) one of the largest known flying insects. However, in several species, but particularly cerambycid beetles in the genus *Pterognatha*, there is Bernoulli entrainment of air within the tracheal system. Air is pulled through the primary tracheae because of a pressure gradient owing to higher external airflow velocities near the posterior spiracles. This draught ventilation is supplemented by autoventilation of the secondary tracheae. Although draught ventilation is known only from a few beetles, it is thought to be more widespread, particularly in fast flying insects (Dudley 2000). Finally, in the hawkmoth, *M. sexta*, it has been shown that airflow during flight is unidirectional as a consequence of the position of the posterior thoracic spiracle in the subalar cleft and the functioning of the flight apparatus (Wasserthal 2001). During the downstroke, the posterior thoracic spiracles are closed automatically within the subalar cleft and thoracic deformation results in an increase in the volume of the thoracic air sacs. Air, therefore, enters through the anterior thoracic spiracle. During the upstroke, air sac volume declines and air is expired through the posterior spiracle (Fig. 3.16).

The type of ventilation used by insects could have profound implications for their response to atmospheric oxygen levels (Harrison and Lighton 1998). Insects using autoventilation can increase airflow by increasing wingstroke frequency or amplitude, but this also increases metabolic rate. In consequence, they might not be able to meet the increase in ratio of oxygen supply to demand that is required to compensate for hypoxia. By contrast, insects relying on abdominal ventilation might not face the same difficulties, and in honeybees facing

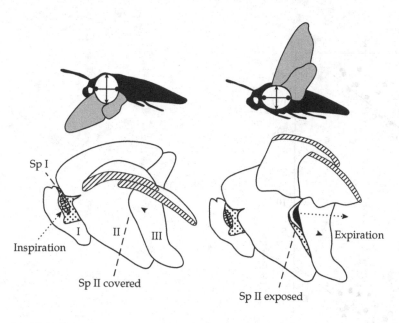

Figure 3.16 Schematic lateral view of the thorax of *Manduca sexta* (Lepidoptera, Sphingidae) during flight showing the deformation of the air sacs and the direction of airflow during the downstroke and upstroke.

Note: Thoracic segments I–III are labelled as are spiracles one (Sp I) and two (Sp II).

Source: Redrawn from Wasserthal (2001).

hypoxia it is clear that abdominal pumping rate is elevated to meet the demands associated maintenance of a constant metabolic rate under hypoxia (Joos *et al.* 1997).

Power production, like many features of muscle physiology, is temperature dependent, and tends to be greatest at the temperature characteristic of free flight of an insect (Dudley 2000). In many small insects this temperature is reached either very rapidly or is equivalent to that of ambient air temperature, while in larger species several minutes of preflight warm-up are required before flight can take place (Chapter 6). From an energetic perspective, flight at or marginally above ambient temperatures, such as that found in small species is least expensive. In general, metabolic rate varies positively with body size such that the smallest species have the lowest metabolic rates (e.g. Bartholomew and Casey 1978) (Fig. 3.17). However, there is considerable variation about this relationship, the slope of which also appears to vary between taxa (from *c.* 0.80 to 1.06) (Harrison and Roberts 2000). Wingstroke frequency and wing loading are also of considerable importance, explaining much of the variation in metabolic rate both within and between higher taxa (Casey *et al.* 1985; Casey 1989; Harrison and Roberts 2000). In general, metabolic costs of flight are lowest in species with low wingstroke frequencies and low

wing loading, and lower cost may be traded off against reduced performance. For example, the gypsy moth is a large species (0.1 g) with a large wing area (5.39 cm^2), and low wingstroke frequency. By contrast, the tent caterpillar moth is only marginally lighter (0.09 g), but has a much smaller wing surface area (2.36 cm^2). In consequence, the latter species has both a higher wingstroke frequency (27 s^{-1} versus 58 s^{-1}) and a higher metabolic rate (22.6 versus 60.0 mW) (Casey 1981). In the winter-flying geometrid moth, *Operophtera bruceata*, induced power output required for flight of an 11.7 mg male is approximately 27.9 µW, whereas in *Manduca sexta* males weighing approximately 1473 mg, induced power output required for flight is in the region of 12960 µW. Coupled with a low wing loading, and compensation of muscle performance for low temperatures, this low induced power output enables the geometrid males to remain active at temperatures between −3 and 25°C, whereas preflight warm-up, and consequently considerable energy expenditure, is required for flight at thoracic temperatures exceeding 30°C in *M. sexta* (Marden 1995a).

In endothermic species (or more correctly heterothermic species given that at least some part of the day is spent at ambient temperatures), both preflight warm-up and flight are energetically demanding. Endothermic warm-up results in an

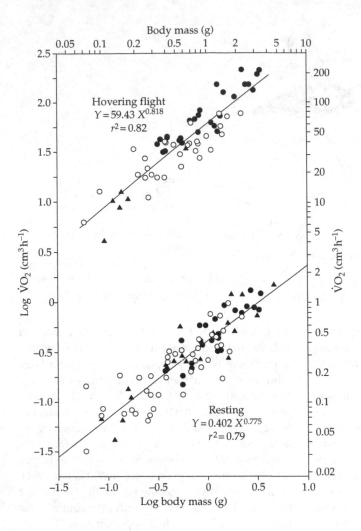

Figure 3.17 The relationship between body mass and oxygen consumption rates in resting and hovering heterothermic moths.

Source: Bartholomew and Casey (1978).

instantaneous increase in $\dot{V}O_2$ (Fig. 3.18), which is generally higher than that required for maintenance of flight temperature (Bartholomew *et al.* 1981; Nicolson and Louw 1982). The metabolic costs of warm-up also scale positively with body mass at the interspecific level (Bartholomew and Casey 1978), although the slope of the regression (0.9–1.0) varies with thoracic temperature in dragonflies (May 1979*a*). In endothermic species that closely regulate thoracic temperature, regulation is largely due to manipulation of convective, or sometimes evaporative, heat loss (Dudley 2000, Chapter 6). However, it is now clear that in several bee species thermoregulation can take place by variation of metabolic heat production (Chapter 6).

During flight, metabolic rate is independent of speed, at least between forward speeds of 0 (hovering) and 4 m s^{-1}, which are within the normal range of free-flying individuals (maximum speed of 5 m s^{-1}), in the only species examined to date, the bumblebee *Bombus lucorum* (Ellington *et al.* 1990). In consequence, estimates of metabolic costs for forward flight can simply be estimated from those measured during hovering (a more straightforward procedure—see Joos *et al.* 1997; Roberts and Harrison 1999). It does seem likely though that metabolic costs of flight will increase at very high airspeeds (Harrison and Roberts 2000), resulting in a J-shaped power curve, owing to increases in parasite power requirements (the power required to move the body through the air—Casey 1989). By contrast, in honeybees metabolic rate increases with an increase in load carried (which can equal body mass in undertaker honeybees) (Wolf *et al.* 1989; Feuerbacher *et al.* 2003), and

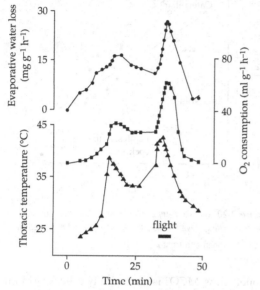

Figure 3.18 Fluctuations in thoracic temperature and oxygen consumption during warm-up and flight in the carpenter bee, *Xylocopa capitata* (Hymenoptera, Anthophoridae).

Source: Nicolson and Louw. *Journal of Experimental Zoology* **222**. © 1982. Reprinted by permission of Wiley-Liss, Inc., a subsidiary of John Wiley & Sons, Inc.

also varies with age, reward rate (Moffatt and Núñez 1997), allele frequencies of malate dehydrogenase, type of load carried and subspecies (Harrison and Fewell 2002; Feuerbacher *et al.* 2003) (Fig. 3.19).

Constructing, maintaining and fuelling the flight apparatus are clearly all energetically expensive. In consequence, it should come as no surprise that there is an environmentally dependent trade-off between maintaining flight machinery (and wings) and survival, development rate and fecundity (Roff 1990; Zera and Denno 1997, Chapter 2). This trade-off arises because of the energetic costs of flight and the flight apparatus, which can be diverted to other fitness components, and probably also because of architectural costs of maintaining a flight apparatus (Marden 1995). Nonetheless, flight clearly has many ecological benefits, to which the diversity of the insects (Harrison and Roberts 2000), and redevelopment of flight following its loss in some stick insects (Whiting *et al.* 2003) clearly testify.

3.5.2 Crawling, running, carrying

Although insects are best known as a group of flying arthropods, they spend much of their time running, hopping, or crawling (especially the

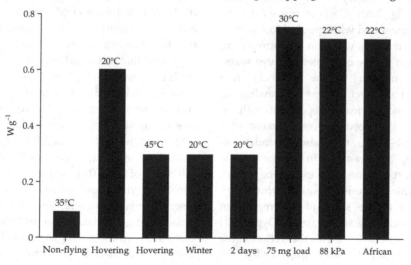

Figure 3.19 Metabolic rates during flight in honeybees under various conditions.

Note: European honeybees (*A. mellifera ligustica*): non-flying, walking and grooming, hovering at 20 and 45°C, winter bees at 20°C, two-day-old bees at 20°C, hovering with a 75 mg nectar load, flying in 88 kPa air, and Africanized honeybee (*A. m. scutellata*) flying at 22°C.

Source: Reprinted from *Comparative Biochemistry and Physiology A*, **133**, Harrison and Fewell, 323–333, © 2002, with permission from Elsevier.

juveniles of holometabolus species). Many adult insects such as all worker and soldier ants, and all flightless species have little other choice. Ballooning is open to small caterpillars, and passive dispersal by water or by wind is occasionally possible for small, desiccation-resistant, or hydrophobic species (Coulson *et al.* 2002). For the rest, pedestrian locomotion is the only alternative to flight. It is perhaps surprising then that pedestrian locomotion and its costs have been investigated in such a small range of species and stages: worker ants (Lighton and Duncan 2002 and references therein), the caterpillars of a moth species (Casey 1991), larvae and adults of a calliphorid fly (Berrigan and Lighton 1993, 1994), the adults of a large tropical fly (Bartholomew and Lighton 1986), and a few cockroach and beetle species (Herreid *et al.* 1981; Bartholomew and Lighton 1985; Lighton 1985; Kram 1996; Rogowitz and Chappell 2000). Moreover, at least some of these studies have been undertaken using treadmills and 'motivational hardware' (Lighton *et al.* 1987) that might result in inaccurate estimates of transport costs (Section 3.1).

The costs of pedestrian locomotion are much lower than those of flight (Berrigan and Lighton 1994; Roces and Lighton 1995), but can still be very high in limbless stages and those that rely on a hydraulic skeleton (Casey 1991; Berrigan and Lighton 1993) (Fig. 3.20). Costs of transport vary both with body mass and with speed of locomotion although the form of this variation is controversial. Costs of transport can be calculated in three ways. Gross cost of transport is the metabolic rate (hereafter MR) of an insect walking (or crawling) at a given speed and is clearly dependent both on SMR and locomotion speed. Net cost of transport is equivalent to MR–SMR, and is also dependent on speed. However, Taylor *et al.* (1970) showed that the costs of transport decline with increasing speed, eventually reaching a constant value, which they argued is equal to the slope of the regression relating mass-specific metabolic rate (ml $O_2 g^{-1} h^{-1}$) and running speed (km h^{-1}). This theoretical value is the minimum cost of transport (M_{run} or MCOT) and is independent of SMR and speed, at least at the highest running speeds. Whether these speeds are actually reached by insects is not clear, but seems unlikely in cockroaches (Herreid *et al.* 1981).

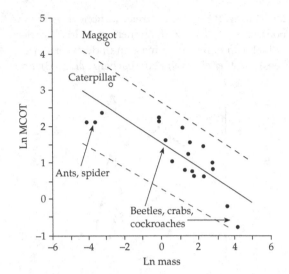

Figure 3.20 The allometry of minimum cost of transport in insects, showing elevated costs of transport in limbless species. *Source:* Berrigan and Lighton (1993).

Nonetheless, MCOT is now widely used for examining the costs of transport in insects, as it is in vertebrates, and is commonly used when the scaling of these costs is investigated.

In endothermic vertebrates, MCOT scales negatively with body mass either as $M_{run} = 8.46 M^{-0.40}$ (Taylor *et al.* 1970), or $M_{run} = 3.89 M^{-0.28}$ (Fedak and Seeherman 1979). Whether the minimum costs of transport of insects conform to this relationship has not been resolved. Early studies suggested that this is indeed the case (Herreid *et al.* 1981; Lighton 1985; Lighton *et al.* 1987), while later work has indicated that, at least in ants, minimum costs of transport are much lower than the vertebrate relationship would suggest (Lighton and Feener 1989; Lighton *et al.* 1993*d*; Lighton and Duncan 2002). Indeed, Lighton and Duncan (2002) conclude that estimates of MCOT based on vertebrate scaling equations yield several-fold overestimations for insects. Thus, at present, and in the absence of a broader range of data, it seems safest to conclude that minimum costs of transport are lower in insects than in endothermic vertebrates, but that these costs vary significantly with the form of locomotion adopted.

In ants, load carriage generally results in a reduction in running speed and an elevation of the costs of transport (Lighton *et al.* 1993*d*; Fewell *et al.* 1996; Nielsen 2001). In some species, the costs of load carriage are equivalent to carrying an extra amount of body mass (i.e. internal and external loads carry the same energetic price) (Bartholomew *et al.* 1988; Nielsen 2001), whereas in others external loads are less costly (Lighton *et al.* 1993*d*). The ways in which loads are carried also have an effect on both the cost and efficiency of carriage. This is nicely illustrated in a comparison of army ants and leafcutter ants. Workers of the former species carry light, energy-rich loads slung between their legs and load costs are low relative to those incurred by leafcutter workers, suggesting that this load carriage method is mechanically more effective. However, as the load ratio increases, the costs of carriage relative to SMR increase more rapidly in the army ant workers than in the leafcutters, making heavier loads relatively less expensive for the latter, whose long legs help them to balance bulky loads (Feener *et al.* 1988). Thus, the way in which loads are carried is likely to have a marked influence on relative leg length. Indeed, this relationship might be much more important in determining the relationship between body size and leg length (Nielsen *et al.* 1982), than the costs of long leg length for movement through rugose environments as predicted by the size-grain hypothesis

(Kaspari and Wieser 1999). To date, there has been no examination of costs of movement by different sized ants through planar and rugose environments, and field tests of ant body lengths in environments differing in rugosity are equivocal (Yanoviak and Kaspari 2000; Parr *et al.* 2003).

The costs of foraging, in relation to the benefits, are central to foraging theory, and range from negligible in some ants such as seed harvesters to values that are comparable to the costs of high energy input foragers such as bees (Fewell *et al.* 1996). These benefit to cost (*B/C*) ratios have strong effects on foraging strategy. Ants with low *B/C* ratios are generally sensitive to foraging costs and vary their behaviour in response to foraging costs, whereas those with high *B/C* ratios choose strategies that maximize net foraging gains per unit time. These strategies can be thought of as those which maximize efficiency and those that maximize power. They are closely linked to habitat conditions and resource type, and have a considerable influence on reproductive strategy (Fewell *et al.* 1996).

3.5.3 Feeding

As is the case in other animals, metabolic rate increases during and/or shortly after feeding in

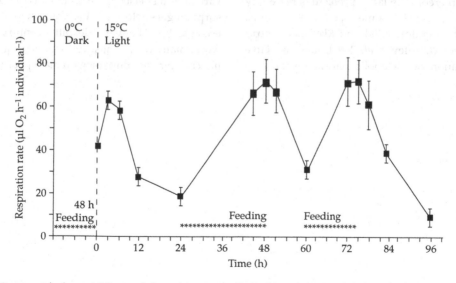

Figure 3.21 Oxygen uptake (mean ± SE) rates of *Gynaephora groenlandica* (Lepidoptera, Lymantriidae) during feeding and starvation. *Source*: Bennett *et al.* (1999).

insects. This metabolic cost of food processing is known as specific dynamic action, and it has been demonstrated in several species, including lepidopteran caterpillars (Aidley 1976; McEvoy 1984; Bennett *et al.* 1999), tsetse flies (Taylor 1977), ants (Lighton 1989), locusts (Gouveia *et al.* 2000), whiteflies (Salvucci and Crafts-Brandner 2000), beetles (Duncan *et al.* 2002*b*), and bugs (Bradley *et al.* 2003). In general, metabolic rates increase between two and fourfold (but can be as high as 15-fold—Bradley *et al.* 2003), and the time course of the increase and subsequent decline in metabolic rate varies substantially between species. The increase and the decline in metabolic rate can be either rapid or prolonged (Fig. 3.21), and, as might be expected, in those species showing discontinuous gas exchange at rest, the elevation in metabolic rate necessitates a change to a more continuous form of gas exchange (Duncan *et al.* 2002*b*).

The costs of feeding are associated with a fixed cost of food processing (Taylor 1977; Willmer *et al.* 2000), behavioural arousal (Gouveia *et al.* 2000), and perhaps, in a few species, such as leafcutter ants (Roces and Lighton 1995), the costs of chewing. Failure to take the costs of digestion into account sometimes gives rise to energy budgets that will not balance (Schmidt and Reese 1986), and also has a marked influence on estimates of SMR and the scaling thereof. The latter problem is not easily resolved, because in some species of insects, such as ants (Lighton 1989) and *Rhodnius prolixus* assassin bugs (Bradley *et al.* 2003), there is active downregulation of metabolic rate in response to starvation, making the term 'post-absorptive' difficult to apply (see also Chapter 2).

3.6 Metabolic rate and ecology

The investigation of metabolic rate variation, and the gas exchange patterns that support it, are among the most venerable pursuits in environmental physiology. Yet, at least in insects, there is much that remains controversial. In particular, the scaling exponents of metabolic rate during rest, pedestrian locomotion, and flight are not well resolved. It is not known whether the intraspecific and interspecific exponents differ consistently, nor whether phylogenetic non-independence has a large effect on the latter. Likewise, it is not entirely clear why insects at rest show discontinuous gas exchange, nor whether characteristics of the DGC, or for that matter metabolic rates, always show high repeatability. What is clear, though, is that a sound comprehension of metabolic rate variation is essential for understanding the ecology and evolution of insects. Metabolic rate variation not only influences the abundance and distribution of insects (Chown and Gaston 1999), but it probably also determines their likely success in colonizing new environments (Vermeij and Dudley 2000), and might also be instrumental in determining global variation in species richness (Allen *et al.* 2002). The surprising conclusion, therefore, is that from an ecological point of view, despite a century of careful work, many of the most fundamental questions in insect respiratory physiology remain poorly resolved.

CHAPTER 4

Water balance physiology

Most of the recent work has been concerned with the exploration of particular mechanisms, and there is now both a need and an opportunity to take a wider view and attempt to synthesise such knowledge into an understanding of water balance in whole animals in natural situations.

Edney (1977)

Water availability and temperature are the two most important abiotic variables influencing the distribution and abundance of insects, although they are continuously modified by two other major climatic factors, solar radiation and wind. The physiological effects of climate are usually related to water and thermal balance, but these effects occur on a fine scale. Small size is commonly accepted as disadvantageous for insects in terrestrial environments because of the small storage capacity for water and relatively large surface area for losing it; however, it also enables them to escape to microhabitats and take advantage of favourable microclimates (Willmer 1982).

Much water balance work has been undertaken in desert regions (Addo-Bediako *et al.* 2001) and has been concerned with demonstrating superior desiccation resistance in species from arid environments, so it is not surprising that the emphasis has been predominantly on insects conserving water. However, high water turnover can occur for various reasons, some of which may be combined in one species: diet (animal or plant fluids), flight (especially when associated with large body size and endothermy), metamorphosis (especially water loss at adult eclosion) and aquatic habitats. The emphasis on adult insects observed by Edney (1977) is still largely true: when Addo-Bediako *et al.* (2001) surveyed the anglophone literature on insect water balance as far back as 1928, they found that >70 per cent of the studies concerned adult insects. However, eggs and early instar larvae can be

expected to have the worst surface area:volume ratio problems (Woods and Singer 2001), and eggs and pupae are frequently the dormant stages in which dehydration is likely to be prolonged and access to water is extremely limited (Yoder and Denlinger 1991; Danks 2000).

The physiological mechanisms involved in water balance and osmoregulation of individual insects were thoroughly reviewed in Edney's (1977) monograph and its successor by Hadley (1994*a*), and also by Wharton (1985) with a rather different emphasis, on the kinetics of water exchange. In the standard approach to water balance physiology, the overall flux of water through an insect is partitioned into different avenues of water loss and recovery, which vary greatly in both absolute and relative terms. A recent study on water-stressed caterpillars (Woods and Harrison 2001) illustrates this hierarchical nature of organismal physiology: fitness-related performance traits represent the aggregate outcome of numerous, more mechanistic physiological traits. The performance trait in question is the growth rate of the caterpillars, which is depressed during water shortage, and the dominant mechanistic traits in this particular case involve modulation of faecal and evaporative water losses.

4.1 Water loss

The use of water activity for comparing aqueous and vapour phases is recommended (Edney 1977;

Wharton 1985). Relative humidity RH is 100 times a_v, the activity of water in air. The term a_v can be directly compared to the haemolymph concentration as a_w, and the difference $a_v - a_w$ between the air and the insect determines whether the insect will gain water from the surrounding air or lose water to it. Insect haemolymph has a_w values in the 0.995 range, while atmospheric a_v is almost always much lower, so the gradient for water movement is almost invariably outwards. Note that doubling of the haemolymph osmolality from 300 to 600 $mOsmol\,kg^{-1}$ would decrease a_w from 0.995 to 0.99 and thus can have little effect on the water activity gradient between air and insect (Willmer 1980).

The water content of insects varies widely, from 40 to 90 per cent of wet mass (Edney 1977; Hadley 1994a). It is lower in insects with high fat reserves or heavy cuticles, and varies between species, instars, individuals, and over time in the same individual (Wharton 1985). Because of the inverse relationship between water content and lipid content, water content does not necessarily give a good indication of the state of hydration of an insect. Dormant insects tend to have lower water contents because of fat accumulation (Danks 2000). Water content is best expressed as $mg\,H_2O\,mg^{-1}$ dry mass rather than the commonly used percent water. The relationship between percent water content and the absolute water content is an exponential function (Coutchié and Crowe 1979a). For example, a small change in the percent water, from 66 to 80 per cent, represents a doubling of the absolute water content of the animal (from 2 to 4 $mg\,H_2O\,mg^{-1}$ dry mass).

4.1.1 Cuticle

Measurement of cuticular water loss
This has never been a simple matter (Loveridge 1980; Noble-Nesbitt 1991; Hadley 1994a). Essentially, there are three ways of measuring cuticular water loss—gravimetric, isotopic, and electronic—which have been applied to whole insects, portions of their cuticles, or *in vitro* cuticle preparations. Early gravimetric measurements on living insects inevitably included respiratory water loss. However, spiracular closure during continuous and

sensitive recording of body mass enables periodic measurement of minimum or cuticular water loss; provided the insect in question exhibits discontinuous gas exchange (Kestler 1985; Machin *et al.* 1991). Isotopic methods allow unidirectional flux to be measured (while gravimetric methods measure net flux). Other advantages of isotopic methods are sensitivity, suitability for use at high a_v (gravimetric measurements are usually made in dry air), and direct comparison with aquatic animals (Croghan *et al.* 1995). Nicolson *et al.* (1984) developed a technique using ventilated capsules and tritiated water to measure cuticular transpiration of desert tenebrionid beetles, and Croghan *et al.* (1995) modified this technique to measure simultaneous water and CO_2 loss from cockroaches. Electronic moisture sensing in flow-through systems, first developed by Hadley *et al.* (1982) and Nicolson and Louw (1982), is now widely used in concurrent measurement of water loss and metabolic rate by means of CO_2/H_2O analysers, but the emphasis is usually on respiratory water loss (see below). For cuticular water loss only, the ventilated capsule technique of Nicolson *et al.* (1984) and later variations thereof (for diagrams see Hadley 1994a: 72) allow measurement of transpiration across small areas of cuticle *in vivo*, and the same can be achieved with experiments using excised discs of cuticle *in vitro*. Both techniques can be used to investigate regional differences in permeability, which are averaged in whole-insect studies (for review see Hadley 1994a). Avoidance of damage during handling is critical with isolated cuticle preparations. The units used to express cuticular water loss data have also led to considerable confusion in the literature, discussed by Noble-Nesbitt (1991): commonly, the concentration gradient is expressed in terms of saturation deficit, so that transpiration rates in $\mu g\,cm^{-2}\,h^{-1}$ can be converted to permeability in $\mu g\,cm^{-2}\,h^{-1}\,Torr^{-1}$ (Loveridge 1980). Potential errors in estimating surface area are discussed by Loveridge (1980).

The possibility of hormonal control of cuticle permeability was suggested on the basis of increased mass loss after decapitation in *Periplaneta americana* and its reversal by injection of brain or corpora cardiaca homogenates (Treherne and Willmer 1975). There is still insufficient experimental evidence to confirm this (for different opinions see Noble-Nesbitt 1991; Hadley 1994*a*). Moreover, the adaptive value of such increased cuticle permeability has been questioned when other avenues of water loss in cockroaches vary greatly (Machin *et al.* 1991). *Periplaneta americana* has been the traditional subject of research on cuticular water loss, yet methodological differences have led to enormous variation in the data obtained for this one species. Machin and Lampert (1987) demonstrated that all previous permeability estimates for cockroaches were too high because of damage to the cuticle of individuals in crowded cultures (although this damage is continually repaired), and further damage incurred during experimental manipulation. When cuticular and respiratory water losses of quiescent *P. americana* were measured by continuous weighing, water loss rates were up to an order of magnitude less than those obtained during intermittent weighing. This was attributed to ventilatory water loss during the periodic disturbances required by intermittent weighing (Machin *et al.* 1991). Regardless of the difficulties involved in accurate measurement of cuticle permeability, the lipid component of the cuticle is accepted as the major barrier to water movement.

The lipid barrier

Lipids, by definition, do not interact with water: thus, it is not surprising that they have major waterproofing functions in plants as well as in arthropods (Hadley 1989). Insect epicuticular lipids include a diverse array of nonpolar, hydrophobic molecules, mainly long-chain saturated hydrocarbons, which differ in their physical properties. The composition of these lipids has been extensively studied (for a recent review see Gibbs 1998) and found to differ widely at various levels of organization. Individuals of a species vary in hydrocarbon composition and rates of cuticular water loss (Toolson 1984), while variation between species has been used as as a taxonomic character (Lockey 1991). Epicuticular lipids in the grasshopper *Melanoplus sanguinipes* vary in composition and biophysical properties between individuals, families, and populations, and with thermal acclimation (Gibbs *et al.* 1991; Gibbs and Mousseau 1994; Rourke 2000). There has long been broad consensus

among insect physiologists that cuticular water loss is related to habitat water availability, and that there is a strong correlation between cuticular water loss and the quantity of epicuticular lipids (Edney 1977; Hadley 1989, 1994*b*). Waterproofing properties are also affected by structural differences in cuticle lipids; waterproofing increases with hydrocarbon chain length, but decreases with methyl-branching and unsaturation, both of which prevent close packing of molecules (Gibbs 1998).

Extreme development of surface lipids occurs in certain insects. Examples are the filamentous wax blooms which cover Namib Desert tenebrionid beetles in response to desiccating conditions (McClain *et al.* 1985), and the particulate wax deposits of whiteflies (Homoptera, Aleyrodidae) (Byrne and Hadley 1988).

Rates of cuticular water loss increase with temperature, gradually at first and then rapidly above a critical or transition temperature. This discontinuity is thought to signify a change in molecular organization of the lipid barrier, when surface lipids begin to melt and their resistance to water movement is impaired. The transition temperature is species-specific and may be ecologically relevant or may exceed the lethal temperature of the species concerned. There has long been controversy about the relationship between lipid melting and increased cuticular permeability, and indeed about the thermodynamics of water movement through cuticle, specifically whether vapour pressure deficit is the appropriate driving force (for review see Noble-Nesbitt 1991). Gilby (1980), for example, emphasized the complexity of the system and the need to measure cuticle as well as air temperatures during transpiration experiments: he concluded that transition temperatures were artefacts. A kinetic approach based on Arrhenius plots has been advocated by some authors (Wharton 1985; Noble-Nesbitt 1991). Arrhenius plots of ln [permeability] against the reciprocal of the absolute temperature give the same estimate for transition temperature whether or not the permeability data are corrected for vapour pressure deficit (Gibbs 1998), with phase transition indicated by deviation from a straight line relationship (Fig. 4.1).

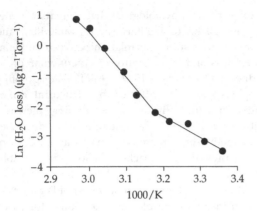

Figure 4.1 Arrhenius plot of the effect of temperature on water loss from a grasshopper (*Melanoplus sanguinipes*).

Note: The plot is biphasic, with a break point at 42°C, and the same break point is obtained when the data are plotted without correction for vapour pressure deficit.

Source: Gibbs (1998).

Study of the biophysical properties of cuticular lipids is difficult, because of the small amounts available and the complexity of the lipid mixtures. However, in recent years, improved technology has been available in the form of Fourier transform infrared spectroscopy (FTIR), used extensively by Gibbs and colleagues to investigate phase behaviour in 5–50 µg samples of surface lipids. For example, lipids extracted from the exuviae of individual grasshoppers showed progressively higher levels of *n*-alkanes and higher melting points during acclimation to higher temperatures (Gibbs and Mousseau 1994). Lipid melting points and transition temperatures are highly correlated in the same individuals (Fig. 4.2). In general, data from model membranes, *in vitro* preparations and intact insects tend to support the lipid melting model for cuticle permeability. However, there are still inconsistencies in the relationships between water loss and the chemical and physical properties of cuticular lipids (Gibbs 2002*b*). For example, apparently adaptive changes in cuticular lipids of *Drosophila mojavensis* during thermal acclimation do not result in reduced water loss (Gibbs *et al*. 1998).

Dramatic regional differences in permeability are seen in the special case of evaporative cooling in desert cicadas (see also Section 6.3.3). *Diceroprocta apache* (Homoptera, Cicadidae) is active at

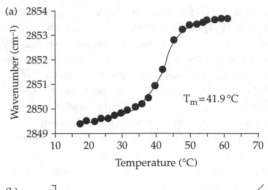

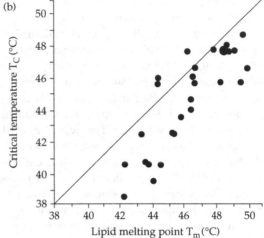

Figure 4.2 Lipid melting temperatures for cuticle of the grasshopper *Melanoplus sanguinipes*, measured by infrared spectroscopy. (a) At the melting temperature (T_m) the lipids are 50% melted, as calculated from the fitted logistic curve. (b) Intraspecific variation in melting points and transition temperatures measured using the same individuals. Solid line is line of equality.

Source: Reprinted from *Journal of Insect Physiology*, **48**, Gibbs, 391–400, © 2002, with permission from Elsevier.

afternoon temperatures approaching 50°C, but its body temperature remains significantly below ambient, owing to extremely high rates of cuticular water loss. This does not involve changes in epicuticular lipids, but is an energy-dependent process occurring via large pores on the dorsal surface (Toolson 1987; Hadley *et al*. 1989). It is also possible for evaporative cooling to occur via the respiratory system in desert grasshoppers (Prange 1990), but this is an emergency mechanism and respiratory water loss is normally minimized, as in insects in general.

4.1.2 Respiration

Conflicting needs of obtaining oxygen and preventing water loss are met in insects by internal tracheal surfaces for respiratory exchange, opening to the exterior via recessed and occludable spiracles (Chapter 3). The role of respiratory water loss in the water economy of insects is currently an active research area, reviewed most recently by Chown (2002). Measurements of respiratory and cuticular water loss have always been closely connected (see above). Earlier studies estimated respiratory transpiration by difference, comparing water losses of untreated insects and those with sealed spiracles (Ahearn 1970; Edney 1977; Loveridge 1980). The solution to the problem of cuticle damage due to spiracle sealing came from allowing unrestrained and resting insects to seal their own spiracles during discontinuous gas exchange (DGC, see also Chapter 3) (Kestler 1985). Evaporative losses can be measured by continuous weighing during such experiments (Kestler 1985; Machin *et al.* 1991), but are preferably measured electronically in conjunction with flow-through respirometry, usually involving CO_2 analysis because of its greater sensitivity (Lighton 1991*b*). A sample trace of simultaneous water loss and CO_2 emission in the lubber grasshopper *Taeniopoda eques* (Orthoptera, Acrididae) is shown in Fig. 4.3. Water loss during the interburst periods is assumed to represent cuticular water loss alone, whereas water loss during the burst periods represents total evaporative losses. Recently, Gibbs and Johnson (2004) described a new method of partitioning cuticular and respiratory water losses: when water loss rates are plotted against CO_2 production, a positive relationship is obtained, and the intercept represents cuticular transpiration.

Respiratory versus cuticular water loss
Published data on the relative contributions of cuticular and respiratory transpiration to total water loss vary widely, both because of technical considerations and because of the strong influence of factors such as temperature and activity on metabolic rate (for discussion see Hadley 1994*a,b*). In general, spiracular control ensures that respiratory losses in resting insects are relatively minor: spiracular transpiration represents less, and

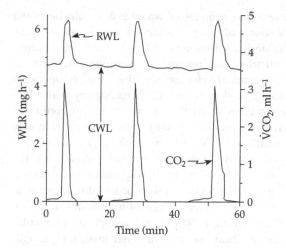

Figure 4.3 Simultaneous recording of water loss (upper trace) and CO_2 emission (lower trace) in the grasshopper *Taeniopoda eques* at 25°C.

Note: Peaks in water loss represent respiratory water loss (RWL); the balance is presumed to be cuticular water loss (CWL).

Source: *Physiological Zoology*, Quinlan and Hadley, **66**, 628–642. © 1993 by The University of Chicago. All rights reserved 0031-935x/93/6604-92108$02.00

usually much less, than 20 per cent of total water loss (Chown 2002). For example, simultaneous isotopic measurement of water and CO_2 losses in *Periplaneta* showed that tracheal water loss averaged only 3.8 per cent of total water loss (Noble-Nesbitt *et al.* 1995): this value is much lower than the 13 per cent measured by Machin *et al.* (1991) using a continuous gravimetric method. Only in species from more arid environments, in which cuticular permeability is reduced, can we expect respiratory transpiration to constitute a greater proportion of total water loss (Zachariassen *et al.* 1987; Zachariassen 1991*a*; Quinlan and Hadley 1993; Addo-Bediako *et al.* 2001). Hadley (1994*b*) pointed out that, in the majority of species examined, respiratory water loss comprised such a small portion of total water loss that changes in its relative contribution would have little effect on the water status of the insects.

There are certain problems with the suggestion that because respiratory water loss constitutes a small proportion of total water loss it is unlikely to be important (Chown 2002). First, the null expectation has never been stated. That is, it is not at all clear what the expected relative contributions of these two avenues of water loss should be in the absence of strong selection for water savings. Second, if selection were to act in concert to reduce cuticular and respiratory transpiration, then no proportional change in the two components might be detected at all. If respiratory water loss constitutes a small, but important, proportion of overall water loss initially, then even a considerable reduction in the latter is unlikely to translate to a particularly large proportional increase in the former. For example, if respiratory transpiration represents 5 units of a total of 100, this amounts to 5 per cent, whereas if it constitutes 5 units of a total of 50 (after a c. 50 per cent reduction in cuticular transpiration and overall water loss), the percentage contribution rises to only 10 per cent. Indeed, expressing the relationship between cuticular and respiratory transpiration as a proportion seems more likely to obscure than to clarify investigations of the importance of respiratory transpiration. As is the case in other fields of physiology (e.g. Packard and Boardman 1988; Raubenheimer 1995), investigations of the components of water loss would benefit substantially from construction and statistical analysis of bivariate plots of the variables involved.

Many of the misgivings regarding the importance of respiratory transpiration have arisen as a consequence of recent work examining the partitioning of water loss in species showing discontinuous ventilation or DGCs. Water conservation was long thought to be the chief adaptive function of DGC in insects. Support for this hypothesis comes from studies such as that of Lighton *et al.* (1993*b*), who measured real-time water loss rates in female harvester ants *Pogonomyrmex rugosus* by both IR absorbance and gravimetric means, and found that water loss during the open phase was 2.8 times higher than cuticular water loss alone. However, the water conservation hypothesis has been questioned for two major reasons. First, as we have seen, respiratory water loss is often too small a component of total water loss to have selective significance. Second, the distribution of DGC among insects is patchy and many species from xeric regions, in particular, apparently do not exhibit cyclic patterns of gas exchange (Lighton 1994; Lighton and Berrigan 1995; Lighton 1996).

An alternative to the water conservation hypothesis was proposed by Lighton and Berrigan (1995), who suggested the DGC may be at least as important in facilitating gas exchange in hypercapnic (high CO_2) and hypoxic (low O_2) environments (Lighton 1996, 1998). Such environments may be encountered by psammophilous and subterranean beetles and ants, taxa which often exhibit DGC (Bosch *et al.* 2000). In such conditions, the DGC leads to increased concentration gradients for both O_2 and CO_2, thus increasing diffusional gas fluxes, and water conservation might be a secondary benefit (Chappell and Rogowitz 2000; Duncan and Byrne 2000). Other possible functions (not necessarily mutually exclusive) have been discussed in Chapter 3 (Section 3.4.4). The topic remains controversial: a recent analysis of water loss in five species of *Scarabaeus* dung beetles exhibiting DGC (Chown and Davis 2003) supports the idea that modulation of DGC characteristics and metabolic rate can be used to alter water loss rate, and the changes are consistent with differences in habitat of the five species.

Flightless insects: does the subelytral cavity conserve water?

Many arid-adapted beetles are flightless and their fused elytra enclose a subelytral cavity above the abdomen. Because the abdominal spiracles open into this space rather than directly to the exterior, the reduced gradient of water vapour pressure is widely assumed to aid in water conservation (for references see Chown *et al.* 1998). However, the vapour pressure of air in the subelytral cavity has never been directly measured. Tracheal interconnections mean that it is possible for the whole respiratory system of an insect to be supplied by tidal air flow via a single open spiracle. In the dung beetle *Circellium bacchus* (Scarabaeidae) the main route for respiratory gas exchange at rest is the right mesothoracic spiracle, which lies outside the subelytral cavity (Duncan and Byrne 2000, 2002) (presumably the left spiracle is involved in other individuals, and both during activity). Similarly, the Namib Desert tenebrionid beetle *Onymacris multistriata* uses mesothoracic spiracles and not the subelytral cavity for gas exchange with the atmosphere (Duncan and Byrne 2000). The subelytral spiracles do, however, play a role in respiration: sampling of air from inside the subelytral cavity of *C. bacchus* shows sequestration of CO_2 and water, which are then expelled through the mesothoracic spiracle, and the DGC patterns of anterior and posterior spiracles are synchronized (Byrne and Duncan 2003).

Respiratory water loss during flight

Respiratory water loss increases dramatically during activity and especially flight (Harrison and Roberts 2000). High metabolic rates during flight require increased ventilation and increased respiratory water loss, but also generate metabolic water. Weis-Fogh's classic study of ventilation in tethered flying locusts (*Schistocerca gregaria*) demonstrated a 10-fold increase in water loss to 8 $mg\,g^{-1}\,h^{-1}$ at 30°C, much of which can be attributed to increased respiratory water loss, although this is offset by metabolic water production. A balance between respiratory losses and metabolic water production can be maintained during many hours of sustained flight, depending on ambient temperature and relative humidity: locusts migrating under desert conditions must fly at high altitude to avoid dehydration (Weis-Fogh 1967*b*).

Studies of respiratory water loss during flight are sparse, but they include work on challengingly small insects. Evaporative water loss has been measured during tethered flight in aphids: Cockbain (1961) mounted *Aphis fabae* (0.8 mg) on fine pins in a wind tunnel, and estimated that only 1 per cent body mass per hour was lost by evaporation during a 6 h flight at 25°C. Forty years later, Lehmann *et al.* (2000) investigated scaling effects on respiration and transpiration during tethered flight in four species of *Drosophila* (0.65–3.10 mg). They used moving visual stimuli to induce the flies to alter their energy expenditure in flight, and measured regular changes in CO_2 production and respiratory water loss. Metabolic water production from the oxidation of glycogen compensated for 23–73 per cent of total water loss, depending on species, but this was much greater than its contribution to water balance in resting flies (Fig. 4.4). An exciting finding is that the fruit flies minimized the risk of desiccation during flight by modulating

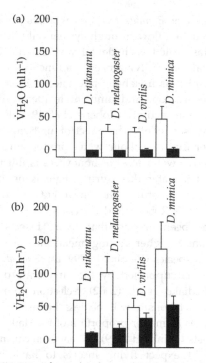

Figure 4.4 Rates of water loss (open columns) and metabolic water production (filled columns) in resting (a) and hovering (b) flies of four *Drosophila species*. Values for water loss during flight are net rates calculated by subtracting resting values (means ± SD).

Source: Lehmann *et al.* (2000).

the opening area of their spiracles according to metabolic needs (Lehmann 2001).

The combination of a nectar diet and high metabolic water production leads to water excess in large flying bees, and respiratory water loss, although extremely high, may not be enough to dissipate the water burden. The first direct measurements of water loss in flight were made on large carpenter bees *Xylocopa capitata* (Hymenoptera, Anthophoridae) during both tethered and free flight. At moderate temperatures bees in free flight evaporated 27 $mg\,g^{-1}\,h^{-1}$ (Nicolson and Louw 1982). Male *X. nigrocincta* evaporate nectar at the nest entrance in order to reduce their water load prior to territorial patrolling (Wittmann and Scholz 1989). Bumblebees in free flight produce water by metabolism faster than they can lose it by evaporation and must excrete urine to prevent water loading (Bertsch 1984; Nichol 2000). Evaporative water loss of *Bombus terrestris* workers could be measured in a flow-through system only after the anus was sealed with dental wax (Nichol 2000). Water balance of flying bees is a function of body mass (and thus metabolic rate), nectar intake and its concentration, and ambient temperature and water vapour activity (Roberts *et al.* 1998; Roberts and Harrison 1999). The effect of air temperature on water flux (metabolic water production minus evaporative water loss) of honeybees is illustrated in Fig. 4.5: water flux during flight is neutral at ambient temperature values near 30°C (Louw and Hadley 1985; Roberts and Harrison 1999).

It has been argued that flying insects have significantly higher resting metabolic rates than flightless species (Reinhold 1999; Davis *et al.* 2000) and this is supported by the large-scale data of Addo-Bediako *et al.* (2002). Selection to reduce resting metabolism will be less intense when it constitutes a smaller proportion of the daily metabolic costs (Reinhold 1999), so that even during rest we might expect flying insects to have higher respiratory transpiration.

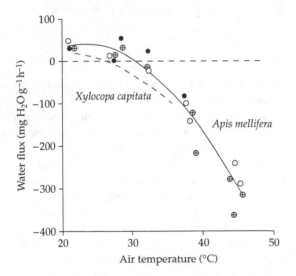

Figure 4.5 Water flux as a function of air temperature for honeybee workers flying in a respirometry chamber.

Note: Water flux is calculated as the difference between metabolic water production and evaporative water loss. Symbols represent honeybees from three different colonies.

Source: Roberts and Harrison (1999); data for *Xylocopa capitata* are from Nicolson and Louw (1982).

4.1.3 Excretion

Homeostasis of the extracellular fluid in insects is achieved by a two-part excretory system. Primary urine, isosmotic to the haemolymph and usually rich in potassium, is secreted by the Malpighian tubules and subsequently modified during passage through the hindgut, although modification of its volume and composition can also occur in downstream tubule segments. Phillips (1981) and Bradley (1985) provided excellent and comprehensive reviews of the functioning of the excretory system as a whole, and a more recent account appears in a wider review by O'Donnell (1997). Hormonal control of the processes involved is covered in detail by Coast *et al.* (2002).

Malpighian tubules: structure and function
Insect Malpighian tubules are the most intensively studied invertebrate excretory organs (O'Donnell 1997). They are slender blind-ending tubes (varying in number from 2 to 200 depending on the species), consisting of a single cell layer with basal and apical membrane foldings which greatly increase the surface area exposed to haemolymph (Bradley 1985). This simple tubular epithelium makes an excellent *in vitro* preparation: Malpighian tubules will survive and secrete for long periods in appropriate Ringer solutions, they respond rapidly to stimulants, and their large cells are suited to impalement with microelectrodes. Ion transport mechanisms involved in fluid secretion are reviewed by Nicolson (1993), Pannabecker (1995) and Beyenbach (1995).

The driving force for Malpighian tubule secretion, once described as a 'common cation pump', is now recognized to be a proton pump or vacuolartype H^+-ATPase (V-ATPase) located on the apical membrane (Fig. 4.6). Its role is evident from the effects of the specific blocker bafilomycin A_1 and from intracellular and luminal pH measurements (Zhang *et al.* 1994; Harvey *et al.* 1998; Beyenbach *et al.* 2000). Electrogenic transport of H^+ into the tubule lumen establishes a proton gradient which energizes secondary transport of cations via apical K^+/H^+ and Na^+/H^+ antiporters. With the exception of bloodsucking insects, K^+ is normally the major cation in the primary urine. The 'common

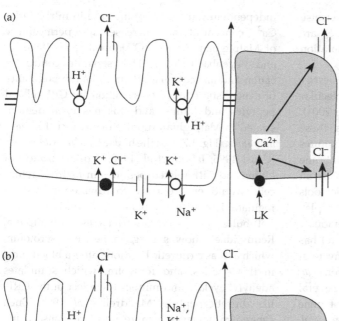

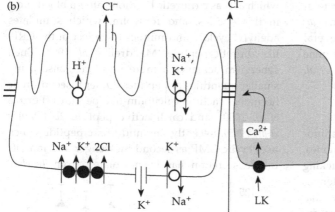

Figure 4.6 Ion transport pathways in Malpighian tubules of (a) *Drosophila* and (b) *Aedes*.

Note: Diagrams illustrate the different ways in which kinins are thought to stimulate the movement of Cl⁻ ions. In *Drosophila*, leucokinin (LK) acts via Ca^{2+} to modulate Cl⁻ channels in the stellate cells (hatched). In *Aedes*, LK enhances Cl⁻ transport through the paracellular pathway.

Source: Reprinted from *Journal of Insect Physiology*, **46**, O'Donnell and Spring, 107–117, © 2000, with permission from Elsevier.

cation pump' is thus equivalent to the parallel operation of proton pump and antiporters. At the basolateral membrane, routes for K^+ and Na^+ entry include channels, $Na^+:K^+:2Cl^-$ or $K^+:Cl^-$ cotransporters, and the Na^+/K^+-ATPase, their relative importance differing according to insect species (Leyssens *et al*. 1994; Linton and O'Donnell 1999). Electrical coupling between the two membranes serves to balance basolateral entry of cations and their apical secretion (Beyenbach *et al*. 2000). The apical H^+-ATPase generates a favourable lumen-positive transepithelial potential for counter-ion, predominantly Cl⁻, transport. Water movement across the tubule is considered to be osmotically coupled to ion transport and ultra-structural observations suggest that it is transcellular (O'Donnell 1997).

Anion transport in Malpighian tubules is less well understood than that of cations, and there is controversy concerning the pathway of passive Cl⁻ transport, which is considered to be transcel-lular in *Drosophila* (O'Donnell *et al*. 1996, 1998) but paracellular in *Aedes* (Pannabecker *et al*. 1993). Tubules of both species contain principal cells interspersed with much smaller stellate cells, and in *Drosophila* peptides of the kinin family are thought to open Cl⁻ channels in stellate cells via an increase in intracellular Ca^{2+} (O'Donnell *et al*. 1998). Even though Cl⁻ channels have recently been identified in stellate cells of *Aedes* tubules (O'Connor and Beyenbach 2001), fast fluid secre-tion may still involve a paracellular shunt pathway for Cl⁻, with stimulation by kinins changing the tubule from a tight to a leaky epithelium

(Beyenbach 1995). In contrast, tubules of the tsetse fly *Glossina morsitans* (Diptera, Glossinidae) are permeable to Cl⁻ ions even before stimulation (Isaacson and Nicolson 1994).

Malpighian tubules of *Drosophila melanogaster* have become a valuable model in molecular physiology (Dow *et al.* 1998; Dow and Davies 2001). Only 2 mm long, with 145 principal cells, these tiny tubules can be studied by standard techniques for measuring fluid secretion rates and membrane potentials, but possess unique advantages as an experimental model because of the genetic tools available for this organism. For example, complex heterogeneity of Malpighian tubules (previously underestimated in this epithelium and others) has been demonstrated by the mapping of physiological domains to the level of single cells (Sözen *et al.* 1997). Cloning and characterization of several V-ATPase subunits have provided the first gene knockouts of V-ATPases in an animal (Dow *et al.* 1997).

Malpighian tubules: hormonal control

Fluid secretion by Malpighian tubules is stimulated, often dramatically, by diuretic hormones. These were initially discovered in blood-sucking insects which experience severe osmotic challenges which are corrected with a fast post-prandial diuresis (Maddrell 1991). Diuretic hormones are apparently present in all insects, including those with far slower rates of water turnover such as desert beetles (Nicolson and Hanrahan 1986): in the latter the function of such hormones may be clearance of the haemolymph with recycling of water (Nicolson 1991). The first corticotropin releasing factor (CRF)-related diuretic peptide was characterized from adult heads of the tobacco hormworm *Manduca sexta* by Kataoka *et al.* (1989). The majority of diuretic hormones characterized to date belong to two neuropeptide families: the CRF-related peptides and the smaller kinins (for review see Coast 1996; Coast *et al.* 2002). The CRF-related peptides share various degrees of homology with vertebrate CRF and act through the second messenger cyclic AMP to increase the rate of cation transport (Beyenbach 1995; O'Donnell *et al.* 1996). The kinins, initially isolated on the basis of myotropic activity on hindgut preparations, act

independently through an increase in intracellular Ca^{2+} concentration to increase anion permeability of Malpighian tubules (O'Donnell *et al.* 1998; Yu and Beyenbach 2001). This separate control of cation and anion transport suggests the possibility of synergistic control of secretion by CRF-related peptides and kinins, and this has been demonstrated in Malpighian tubules of the locust *Locusta migratoria* (Fig. 4.7) and the house fly *Musca domestica* (Coast 1995; Iaboni *et al.* 1998). Co-localization of kinins and CRF-related peptides in neurosecretory cells would ensure their coordinated release to regulate fluid secretion (Chen *et al.* 1994).

Tubules of *Rhodnius prolixus* (Hemiptera, Reduviidae) show synergism between serotonin, which acts as a diuretic hormone after a blood meal in this species, and forskolin, which stimulates adenylyl cyclase and mimics the effect of the CRF-like diuretic peptide (Maddrell *et al.* 1993). Such synergism achieves a more rapid response using smaller quantities of peptides. Other insect diuretic hormones include calcitonin-like peptides (Furuya *et al.* 2000), and cardioactive peptide 2b (Davies *et al.* 1995). Recently, an antidiuretic peptide which uses cyclic GMP as second messenger and inhibits tubule secretion has been isolated from *Tenebrio*

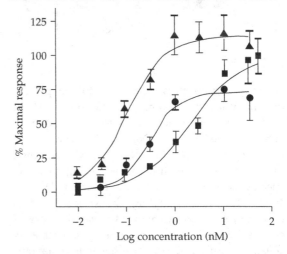

Figure 4.7 Synergistic stimulation of Malpighian tubule secretion by CRF-related peptides and kinins. Locust diuretic peptide (squares) and locustakinin (circles) were assayed separately and together (triangles) on Malpighian tubules of *Locusta migratoria*. Means ± SE ($n = 5$ to 8).

Source: Reprinted from *Regulatory Peptides*, **57**, Coast, 283–296, © 1995, with permission from Elsevier.

molitor (Eigenheer *et al.* 2002). The complexity of control of insect Malpighian tubules was discussed by O'Donnell and Spring (2000), and their ideas are supported by studies of the antagonistic action of synthetic endogenous diuretic and antidiuretic peptides in *Tenebrio* tubules (Wiehart *et al.* 2002), and by the multiple diuretic factors which seem to be present in many insects (for a definitive review see Coast *et al.* 2002). The latter authors stress that physiological relevance *in vivo* is much more difficult to demonstrate than the cellular actions of identified diuretic or antidiuretic factors. As yet, few studies meet the stringent criteria (e.g. haemolymph titres of appropriate timing and concentration) for these factors to be assigned functional roles as neurohormones.

Hindgut

The reabsorptive regions of the insect excretory system consist of an anterior ileum and posterior rectum (structurally more complex), and the transport mechanisms involved and their control have been reviewed by Phillips and colleagues (Phillips *et al.* 1986, 1998), mostly based on the best studied insect hindgut, that of the desert locust *S. gregaria*. For other insects, the mechanisms and the control are less well known.

Malpighian tubules enter the gut at the midgut–hindgut junction, and the hindgut is lined with cuticle which protects the tissue and acts as a molecular sieve (Phillips *et al.* 1986). Primary urine first passes through the ileum, which in locusts appears to be functionally analogous to the mammalian proximal tubule, because it is responsible for bulk isosmotic reabsorption of primary urine. Studies of the ileum, using well characterized *in vitro* preparations initially developed for the rectum (Hanrahan *et al.* 1984), have shown that the driving force for fluid reabsorption is electrogenic Cl^- transport across the apical membrane. The counter-ion is K^+ (generally present at high concentration in the primary urine) and the ileum is also the site of reabsorption of Na^+ (Irvine *et al.* 1988).

The insect rectum is analogous to the more distal segments of the mammalian nephron (Phillips *et al.* 1996*a*). Structural modifications for reabsorption are seen in rectal pads or papillae, derived from a single layer of columnar cells. Current ideas concerning the mechanism of fluid reabsorption are based on ultrastructural studies and micropuncture sampling from various compartments, supported by microprobe analysis of ion distribution (Gupta *et al.* 1980). These techniques show that, as in other epithelia, ion pumping creates osmotic gradients and ultrastructural details lead to net water movement (reviewed by Phillips *et al.* 1986). Ion transport into narrow intercellular spaces (Fig. 4.8) creates locally high osmolalities, so that water

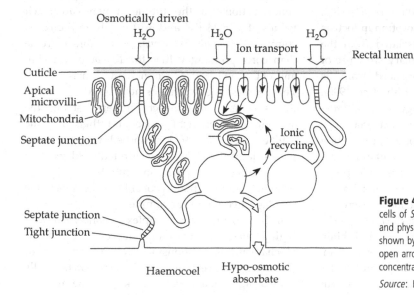

Figure 4.8 Fluid reabsorption by the rectal pad cells of *Schistocerca gregaria*. Anatomy on the left and physiology on the right: ion movements are shown by closed arrows and water movement by open arrows. Note that mitochondria are concentrated at the apical and lateral membranes.
Source: Bradley (1985).

follows. With the resulting build-up of hydrostatic pressure, fluid is forced in the direction of the haemolymph, but reabsorption/recycling of ions converts it into a hypo-osmotic absorbate (Phillips *et al.* 1986). In contrast, the ileum lacks the elaborate lateral membranes which are necessary for ion recycling (Irvine *et al.* 1988), so that in this tissue the absorbed fluid cannot be hypo-osmotic to the luminal contents. In both rectum and ileum, Cl^- ions are actively transported from lumen to cell across the apical membrane. It seems unlikely that a V-ATPase drives Cl^- transport in the locust hindgut, for reasons discussed by Phillips (1996*a*). Proline in the primary urine is also reabsorbed by the rectum and, in fact, is the main respiratory substrate for this tissue.

Antidiuretic factors may be broadly defined as decreasing Malpighian tubule secretion, increasing hindgut reabsorption, or inhibiting whole animal water loss (Gäde *et al.* 1997). Characterization of neuropeptides acting on hindgut reabsorption is far less advanced than of those controlling Malpighian tubules, except in the case of locusts. The rate of faecal water loss decreases 10-fold after locusts are transferred from succulent plant material to dry diets (Loveridge 1974). One candidate for locust antidiuretic hormone is chloride transport stimulating hormone (CTSH), which acts via cyclic AMP to stimulate active Cl^- transport and thus fluid reabsorption in the rectum. However CTSH has been only partially purified (Phillips *et al.* 1986). The major stimulant of ileal reabsorption in locusts is ion transport peptide (ITP), isolated from the *corpora cardiaca*: the complete primary structure of 72 amino acid residues was deduced from its cDNA nucleotide sequence (Meredith *et al.* 1996). ITP has high sequence homology with a large family of crustacean hormones, and acts via cyclic AMP to increase Cl^- transport and apical cation conductances (Phillips *et al.* 1998). Recent molecular studies show that ITP homologues are widespread in insects (Coast *et al.* 2002). The coordinated functioning of the two parts of the locust excretory system was investigated by Coast *et al.* (1999). Diuretic peptides of the CRF-related and kinin families were without effect on ileum or rectum, but their synergistic action on Malpighian tubules was more rapid than ITP stimulation of hindgut

tissues. This separate control and different timing of secretory and reabsorptive processes may permit initial elimination of excess water before fluid recycling predominates.

The hindgut also plays an important role in acid–base regulation (reviewed recently by Harrison 2001). Locusts recover from acute acid challenge (HCl injected into the haemocoel) by ileal and rectal transport of acid–base equivalents, but the contribution of the Malpighian tubules is minor because of proton recycling in fluid secretion (Phillips *et al.* 1994). Feeding state is important in the excretory capacity to remove acid loads, disturbances being less easily dealt with in starved animals (Harrison and Kennedy 1994).

Rectal complex

Another type of structural complexity occurs in the rectal complex (cryptonephric system) of many Coleoptera (both larvae and adults) and larval Lepidoptera (which as adults possess rectal pupillae). It has been best studied in the mealworm *Tenebrio molitor*: both structure and function were described in classic detail by Ramsay (1964). The distal (blind) ends of the Malpighian tubules are not free in the haemolymph, but are held against the rectal wall by a perinephric membrane. The use of ion-sensitive microelectrodes provided direct evidence for K^+ transport into the Malpighian tubules as the driving force for reabsorption. Ion concentrations in the tubule lumen can reach values of 3.35 M K^+ and 3.10 M Cl^- in the rectal complex of *Onymacris plana*, even more extreme than corresponding values for *T. molitor* (Machin and O'Donnell 1991; O'Donnell and Machin 1991). This creates a powerful osmotic gradient which draws water from the rectal lumen first into the surrounding perirectal space and then into the tubule lumen (Fig. 4.9). Reabsorbed water progressively dilutes the highly concentrated fluid secreted by the posterior perirectal tubules, and final osmotic equilibration probably occurs as the tubules leave the rectal complex.

In beetles the rectal complex is concerned with highly efficient dehydration of faeces, and the same mechanism is responsible for water vapour absorption (see Section 4.2.4). Caterpillars normally have a high dietary water intake, but faecal water

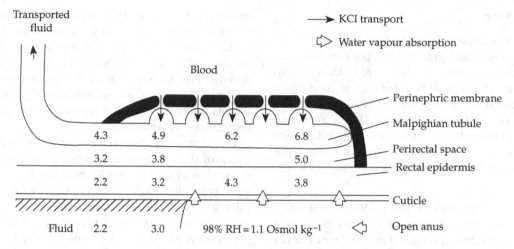

Figure 4.9 Longitudinal section of the rectal complex of *Tenebrio molitor*, showing the various compartments.

Note: Values are osmolalities (Osmol kg^{-1}), and arrows show direction of water movement. Osmolalities decrease from posterior to anterior in all compartments. The rectal lumen is shown partially air-filled during water vapour absorption (see Section 4.2.4).

Source: Machin (1983).

content can be modulated according to water need in *Manduca sexta* caterpillars fed on artificial diet (Reynolds and Bellward 1989). Field observations on the same species showed that faecal water content decreases during the day to offset increased evaporative losses (Woods and Bernays 2000). Faecal water loss in caterpillars is apparently under neuroendocrine control, but the complexity of the rectal complex makes it difficult to distinguish between diuretic and antidiuretic events occurring in different compartments (Gäde *et al.* 1997). Neurohaemal processes of the rectal nerve within the rectal complex showed immunoreactivity to both the CRF-like diuretic hormone of *Manduca* (Mas-DP1) and to leucokinin IV (Chen *et al.* 1994), and Audsley *et al.* (1993) showed that Mas-DP1 stimulates fluid absorption by the rectal complex (i.e. is antidiuretic), probably via a diuretic effect on the cryptonephric tubules. A single peptide can, thus, stimulate both the free segments of the tubules and the rectal complex to accomplish fluid recycling through the excretory system.

Water recycling in the midgut
In unfed locusts some of the primary urine entering the gut moves anteriorly rather than posteriorly and is reabsorbed in the midgut and caecae (Dow 1981): This may reduce the load on the hindgut, which has a relatively small surface area compared to that of the Malpighian tubules (Phillips 1981). Countercurrent flow of fluid in the midgut is important in water recycling in caterpillars (Reynolds *et al.* 1985) and tenebrionid beetles (Nicolson 1991, 1992), and in the latter may be a means of hydrating dry food (Terra *et al.* 1985). The midgut epithelium contains diffuse endocrine cells of largely unknown function (see, for example, Veenstra *et al.* 1995), some of which could be involved in water recycling.

4.2 Water gain

The relative importance of the avenues of water gain discussed below depends largely on the availablity of water in the insect's environment. An extreme example is seen in stored products insects which live in an atmosphere of low a_v, have no access to free water, and consume food of minimal water content (varying with ambient a_v). Periods of fasting may be necessary to survive such conditions (Arlian 1979). Higher a_v enables feeding to continue, but adults of *Sitophilus granarius* (Coleoptera, Curculionidae) consuming tapioca at an a_v of 0.8 gain much more water from metabolism and from passive vapour absorption than from the food itself (Devine 1978).

4.2.1 Food

The water contents of insect diets range from 2 per cent in seeds and grain to 99.9 per cent in xylem sap (Slansky and Scriber 1985). Dow (1986) categorized insect diets according to mechanical type (solid or liquid) and nutrient composition (plant or animal origin) (Fig. 2.1). The solid and liquid plant feeders selected as examples below have to process large volumes of food because plant material is generally suboptimal in nutrient levels.

Solid plant feeders

Caterpillars typically contain 80–90 per cent water, while leaf water content varies from 50 to over 90 per cent across species: it may also vary by 10 per cent on a diurnal basis and may decline by 20 per cent as leaves mature (Slansky 1993). Phytophagous insects tend to adjust their food consumption to maintain constant dry matter intake (see Chapter 2). In evaluating compensatory feeding, care must be taken in comparing foods of different water content, and artificial diets are sometimes diluted with indigestible cellulose to avoid water stress (Slansky and Wheeler 1991; Slansky 1993). It has long been debated whether water intake can be a primary objective of feeding in insects (Edney 1977; Hadley 1994a). Caterpillars at least do not compensate for a decline in leaf moisture levels by increasing their feeding rates (discussed by Slansky 1993).

The growth performance of leaf-chewing larvae is strongly correlated with leaf water content (Slansky and Scriber 1985; Woods and Harrison 2001), but analysis is complicated by covariation of leaf water and nitrogen contents (both being high in young leaves, for example). White (1974) hypothesized that plant water deficits lead to the mobilization of nitrogen, making plants more nutritious for herbivorous insects and leading to episodic outbreaks. Mexican bean beetles, *Epilachna varivestis* (Coccinellidae), feeding on soybean plants (*Glycine max*) were used to test this hypothesis (McQuate and Connor 1990a,b). Both physical and chemical changes in foliage are associated with water deficits in plants, so water-deficient soybean foliage was rehydrated before feeding trials to control for physical differences. Free amino acid concentrations increased markedly in foliage

grown under water deficits, even after re-watering, but third instar larvae avoided this foliage. Trials under growth chamber, glasshouse, and field conditions showed a consistent tendency for larval survival and growth rate to be reduced on the treatment foliage. It is apparent that plant water deficits can have complex effects on the behaviour and ecology of herbivorous insects.

The relationships between dietary water, feeding, and growth have been extensively investigated in caterpillars of *M. sexta* under both laboratory and field conditions. When raised on semi-defined artificial diet in the laboratory, fifth instar caterpillars compensate for food dilution by eating more (Timmins *et al.* 1988) and maintain water homeostasis by modulating faecal water loss (Reynolds and Bellward 1989). Preformed water intake and faecal water loss are the main components of the water budget, metabolic water production and evaporative losses being relatively small, and about half the dietary water intake is incorporated into tissues during rapid growth (Reynolds *et al.* 1985; Martin and Van't Hof 1988). Evaporative losses are much higher in 'wild' larvae (Woods and Bernays 2000). However, evaporative losses and faecal water savings are superimposed on comparatively enormous water fluxes through the alimentary canal. Woods and Harrison (2001) recently compared the effects of hydric stress on performance (growth) and on the mechanisms of water conservation. *Manduca* caterpillars reared on low-water diet grew more slowly and showed both short-term changes in faecal water loss and long-term changes in evaporative loss. The water budgets of caterpillars are now relatively well understood.

Liquid plant feeders

Homopterans are exclusively phytophagous. Xylem and phloem tissues of plants represent substantial food resources but have certain limitations. Xylem sap is the most dilute food consumed by any herbivore. Phloem sap has high and variable concentrations of sugars, but large volumes must still be imbibed in order to meet nitrogen requirements. Xylem feeding insects such as Cercopoidea and Cicadoidea tend to be larger than phloem feeders because they must cope with the negative tension of xylem fluid in the plant and the resistance of the feeding apparatus (Novotný and Wilson 1997). Differences in xylem fluid composition determine the feeding preferences of xylem feeders, and feeding rates of leafhoppers (Cicadellidae) are adjusted in relation to diurnal changes in xylem chemistry, being highest when amino acid conentrations are highest (Brodbeck *et al.* 1993). Xylem sap has an osmolality of only 9–26 mOsmol kg^{-1} (Andersen *et al.* 1992), and the problem of internal flooding is solved in cicadas by osmotic transfer of water from the anterior midgut to the Malpighian tubules in the filter chamber, and production of extremely dilute excreted fluid (Cheung and Marshall 1973). Dehydrated aphids have been reported feeding on xylem rather than phloem (Spiller *et al.* 1990). Homopteran guts are notable for their complexity (Goodchild 1966), and simpler filter chambers are present in some phloem feeders, such as cicadelloid leaf-hoppers (Eurymelidae) (Lindsay and Marshall 1981). Osmoregulation in phloem feeders is discussed below (Section 4.3.2).

Fruit flies must deal with dilute food sources, such as juices oozing from fruit. 'Bubbling behaviour' in *Rhagoletis pomonella* (Diptera, Tephritidae) is a form of excretion of excess dietary water: liquid droplets are repeatedly regurgitated and reingested after evaporation, enabling flies to continue feeding (Hendrichs *et al.* 1992). The consequences of nectar diets for insect water balance depend on body size and flight activity (Nicolson 1998) and excess dietary water is not usually a problem for adult Diptera or Lepidoptera. Ecological aspects of nectar feeding from floral and extra-floral nectaries have been reviewed by Boggs (1987) and Koptur (1992).

4.2.2 Drinking

Drinking is common in insects, and even those with substantial water intake in their food, as shown by Edney's (1977) comprehensive list of arthropods, are known to drink water. Drinking in terrestrial insects is usually measured by simple gravimetric methods, especially when prolonged drinking occurs after dehydration, as in dehydrated desert tenebrionid beetles, *Onymacris plana* (Nicolson 1980). Fog and dew are important water resources for desert insects (Broza 1972; Seely 1979). Loveridge

(1974) demonstrated that *Locusta migratoria* was able to utilize saline water sources, surviving when 0.25 and 0.5 M NaCl were provided with dry food, although initially reluctant to drink. An apparatus for measuring the timing and extent of drinking behaviour, from a plastic straw coated with conducting paint, showed that adult male cockroaches fed dry food consume 50–140 µl of tap water daily (Whitmore and Bignell 1990). Many laboratory studies have investigated the uptake of sugar solutions by nectar or fluid feeding insects, which probably seldom drink free water. Puddling behaviour by male butterflies and moths (see below) is not concerned with the acquisition of water.

Haemolymph appears to play an important role in the control of drinking, although investigations on blowflies and locusts have produced varying results (for discussion see Edney 1977). Responsiveness to water seems to depend on the Cl$^-$ concentration of the haemolymph in the sheep blowfly *Lucilia cuprina* but not in other blowflies (Barton Browne 1968). The initiation and termination of drinking have different controlling mechanisms in locusts: drinking readiness is associated with reduced abdominal volume, but the volume ingested depends on the increase in osmolality (Bernays 1977). Compensatory drinking occurs in locusts which exhibit larger drinking bouts on NaCl solutions than on distilled water (Raubenheimer and Gäde 1993). Since phagostimulatory and volumetric effects can be ruled out, the locusts appear to be compensating for the decreased ability of the imbibed fluids to hydrate the haemolymph. There are strong interactions between hunger and thirst in locusts: the amounts of either water or food consumed are markedly affected by the absence of the other, on both a short and long-term basis, and all deprivation treatments increase the time spent in locomotion (Raubenheimer and Gäde 1994, 1996).

Reproductive harvester termites *Hodotermes mossambicus* (Isoptera, Hodotermitidae) form pairs after swarming flights and drink large volumes of water: 82 and 42 per cent of initial wet mass in males and females, respectively (Hewitt *et al.* 1971). Dilute fluid is stored in large water sacs (salivary reservoirs) which extend into the abdomen, and may be an important water reserve for the founding of the new colony. Conditions are ideal for emergence after the first substantial summer rains, when desiccation is reduced, the soil is moist and soft for excavation, and free water is available (Hewitt *et al.* 1971). Water imbibition and transfer to the water sacs has also been described in workers and reproductives of fungus-growing termites *Macrotermes michaelsoni* (Termitidae, Sieber and Kokwaro 1982). Although salivary gland reservoirs are also used for water storage in cockroaches (Laird *et al.* 1972), in this case the fluid is a hypo-osmotic saliva (House 1980), rather than water stored after drinking.

4.2.3 Metabolism

The importance of metabolic water production in counteracting respiratory water loss in flying insects depends on factors such as body mass, the level of activity, and ambient temperature. At high temperatures, honeybees ferry water to the hive for cooling purposes, carrying up to 65 per cent of their body mass, and the increased metabolic demand produces enough metabolic water to almost balance evaporative losses (Louw and Hadley 1985). Larger bees were discussed in Section 4.1.2. On a more modest scale, aphids maintain water balance during long flights by production of metabolic water, metabolizing glycogen in the early phase and then lipid (Cockbain 1961). Fat-based and carbohydrate-based metabolisms differ considerably in yields of metabolic water (Corbet 1991). This is because these substrates differ not only in the amount of water released per unit weight of energy, but also in the amount of water associated with them in storage (Weis-Fogh 1967b). Because 1 g dry weight of stored glycogen is associated with about 2.5 g water of hydration, metabolism of glycogen yields six times as much water as does that of lipid. This does not happen, however, if sugars in the crop are used directly as fuel for flight.

Metabolic water is also significant in inactive insects which are not feeding or drinking, and in those consuming dry food. Doubly labelled water (DLW) has been used to measure field water balance of tenebrionid beetles in Northern and Southern hemisphere deserts, and either food, metabolism or drinking dominated water influx,

depending on season (Cooper 1982, 1985). Metabolic water suffices for water balance of inactive *Onymacris unguicularis* in the Namib Desert, but active beetles must drink fog water. The DLW technique (see Butler *et al.* 2004) has been considered less accurate for arthropods than for vertebrates, and there have been few recent applications to insects. However, a validation study on bumblebees *Bombus terrestris* has shown excellent agreement between values obtained using respirometry and DLW (Wolf *et al.* 1996).

4.2.4 Water vapour absorption

Certain arthropods lacking access to conventional sources of water from feeding or drinking have developed the ability to extract water from unsaturated air (Machin 1983; O'Donnell and Machin 1988). Different oral or rectal structures are utilized for this purpose, indicating that the mechanisms have evolved independently, often based on pre-existing capabilities. In Tenebrionidae, the rectal complex (Section 4.1.3, Fig. 4.9) has evolved for efficient drying of faeces and water vapour uptake is probably a secondary function. Ticks use oral uptake involving salivary secretions. Most mechanisms for water vapour absorption are based on the colligative lowering of vapour pressure by accumulating solutes, with the exception of oral uptake in the desert cockroach *Arenivaga* (reviewed by O'Donnell and Machin 1988).

Water vapour absorption is a regulated process which only occurs above a critical equilibrium activity (CEA) and only when the animal is dehydrated. The CEA (water activity at which vapour absorption balances passive losses) is discussed by Wharton (1985) and O'Donnell and Machin (1988). Net gain of water does not occur below a particular a_v that is characteristic of the species and stage in question. Steep gradients are involved in vapour uptake, because the a_w of insect haemolymph is usually about 0.995 (equivalent to an osmolality of 300 mOsmol kg^{-1} or 99.5 per cent RH). In *Tenebrio molitor*, the importance of steep solute gradients in absorption is shown by the good agreement between the CEA ($a_v = 0.88$) and the maximum osmolality (6.8 Osmol kg^{-1}) observed in the rectal complex (Fig. 4.9) (Machin 1983).

Uptake thresholds and absorption kinetics are the main factors determining the physiological and ecological significance of water vapour absorption in different groups: together, they determine the a_v range over which absorption is possible and the water deficits which can be recovered in a given time. The finding that water vapour absorption occurs in terrestrial isopods in moist environments has altered perspectives concerning its ecological significance (Wright and Machin 1993). The entire order Psocoptera possesses this ability, regardless of habitat and flight status, so the adaptive value is not readily apparent (Rudolph 1982). Recently, the common soil collembolan *Folsomia candida* has been shown to absorb water vapour from the atmosphere by accumulating myo-inositol and glucose in order to raise its haemolymph osmolality above ambient a_v (Bayley and Holmstrup 1999). These authors point out that water vapour absorption at an a_v of 0.98 is as ecologically relevant for this soil insect as that occurring at much lower humidities in a desert insect.

Eggs of many insects, especially Orthoptera and Coleoptera, take up water in liquid form during development (Tanaka 1986), but diapausing eggs of the tropical stick insect *Extatosoma tiaratum* (Phasmida) can absorb water vapour from vapour activities down to 0.30 (Yoder and Denlinger 1992). Vapour uptake also occurs in the diapausing first instar, although the CEA is then 0.60. To date, this is the only instance of water vapour absorption in insect eggs, although the mechanism remains unknown.

Passive sorption
Wharton (1985) has stressed that the atmosphere is a source of water for all insects, not only those capable of active water vapour absorption. It is a misconception that insect cuticle is asymmetrical to the diffusion of water. Passive absorption (sorption) of water vapour from the atmosphere serves to reduce net losses (Wharton and Richards 1978). For example, when atmospheric a_v is 0.50, roughly half of the water lost by transpiration will be regained without expending energy on absorbing mechanisms. The significance of passive sorption is seldom considered in studies of insect water balance, in part because isotopic techniques are

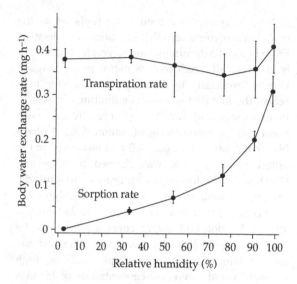

Figure 4.10 Water exchange by transpiration and passive sorption in worker ants, *Formica polyctena*, during the first hour of exposure to different vapour activities at 29°C.

Note: Exchange rates in $mg\,h^{-1}$ for an ant weighing 10.8 mg, with a water content of 77%.

Source: Reprinted from *Comparative Biochemistry and Physiology A*, **83**, Coenen–Stass, 141–147, © 1986, with permission from Elsevier.

required to differentiate between fluxes (Arlian 1979; Coenen-Stass 1986). Its contribution to water gain can be high when atmospheric a_v is high (Devine 1978). Fig. 4.10 demonstrates clearly that net water loss, the difference between transpiration and sorption rates, decreases at high vapour activities. This is because the sorption rate increases while transpiration remains esssentially constant (Coenen-Stass 1986). Fortunately for many laboratory studies of transpiration rates, low ambient a_v results in insignificant levels of passive sorption. In many terrestrial environments, however, passive sorption can be important in the water economy of insects, especially during fasting.

4.3 Osmoregulation

4.3.1 Haemolymph composition

The composition of insect haemolymph was last reviewed by Mullins (1985). The phylogenetic connection was pointed out early in a classic paper by Sutcliffe (1963): in the more advanced insect orders, inorganic ions tend to be replaced as the main haemolymph osmolytes by organic molecules. From being the dominant haemolymph cation, Na^+ is dramatically reduced in many phytophagous insects, especially in the larval stages. This may reflect the fact that sodium is a limiting element in plant tissues and varies geographically and with plant species, tissue and age (Denton 1982). A blood Na^+ concentration of only 0.2 mM measured in the aphid *Myzus persicae* was claimed by Downing (1980) to be the lowest, by an order of magnitude, for any animal. Important organic osmolytes are amino acids, trehalose, and organic acids (Mullins 1985). Haemolymph sugar concentrations can be extremely high, as in the pea aphid *Acyrthosiphon pisum*, which has haemolymph with a high osmolality and a trehalose concentration of 255 mM (Rhodes *et al.* 1997). Analysis of the haemolymph of larval *Onymacris rugatipennis* (Coleoptera, Tenebrionidae) showed approximately equal osmotic contributions from cations, chloride ions, amino acids, and trehalose, all of which participate in osmoregulation (Coutchié and Crowe 1979*b*). The hypothesis of Zachariassen (1996) concerning water balance in xeric insects should be mentioned in this context. He suggested that in herbivorous groups, such as tenebrionids, lower metabolic rates would mean lower haemolymph Na^+ concentrations (due to reduced activity of the Na^+/K^+-ATPase), but elevated amino acid concentrations (due to reduced Na^+-amino acid cotransport), and that this is not true of predators such as carabid beetles. However, it is difficult to disentangle this hypothesis from phylogenetic constraints, carabids belonging to the basal suborder Adephaga. Moreover, dietary input has a marked effect on sodium balance in the carnivorous Dytiscidae (Frisbie and Dunson 1988).

Published data on haemolymph chemistry may differ because information is derived from both whole haemolymph and cell-free plasma, and also because haemolymph is a complex and dynamic fluid, with solute concentrations affected by multiple factors such as diet, feeding, development, hydration state, and parasitism (Mullins 1985). Lettau *et al.* (1977) recorded cycles in haemolymph K^+ activity throughout the day in free-walking cockroaches, using ion-selective microelectrodes for continuous measurements. Wide fluctuations in haemolymph osmolality are evident in field-collected insects, with variation demonstrated on a daily basis (Fig. 4.11) in caterpillars of various species (Willmer 1980) and seasonally in the alpine weta *Hemideina maori* (Orthoptera, Anostostomatidae (formerly Stenopelmatidae), Ramløv 1999). The latter insect is freeze-tolerant and its high haemolymph levels of trehalose and proline during winter have a cryoprotectant function (Neufeld and Leader 1998) (see also Chapter 5). Protein and amino acid patterns in haemolymph are particularly dynamic (proline as a flight fuel has already been mentioned in Section 3.2.1). Holometabolous larvae accumulate abundant storage proteins called hexamerins (consisting of six identical subunits); these are synthesized in the fat body and supply amino acids for synthesis of adult tissues during metamorphosis. Arthropod haemocyanins and insect hexamerins belong to the same protein superfamily and their evolutionary relationships are discussed by Burmester *et al.* (1998). Other important plasma proteins are the vitellogenins synthesized by adult females for egg manufacture. Haemolymph buffering involves proteins and organic acids as well as bicarbonate (Harrison 2001).

Metamorphosis leads to loss of water and changes in the distribution of water and ions between body compartments. Dramatic changes in body mass, water content, and haemolymph volume and osmolality have been described during the pupal–adult transformation of Lepidoptera (Nicolson 1976; Jungreis *et al.* 1982). The haemolymph volume of adult Diptera, Lepidoptera, and Hymenoptera tends to be greatly reduced in preparation for flight. Flight itself may disturb water balance and haemolymph composition. Haemolymph volume increases by 30 per cent as a result of flight in *P. americana* (King *et al.* 1986) but decreases in *Rhodnius prolixus*, probably due to the release of diuretic factors (Gringorten and Friend 1979). However, in these occasional flyers which are better known for fast running, it may be difficult to separate responses to stress and to flight (Corbet 1991). In the honeybee, a short flight increases haemolymph volume, while exhaustive flight reduces it drastically (Skalicki *et al.* 1988).

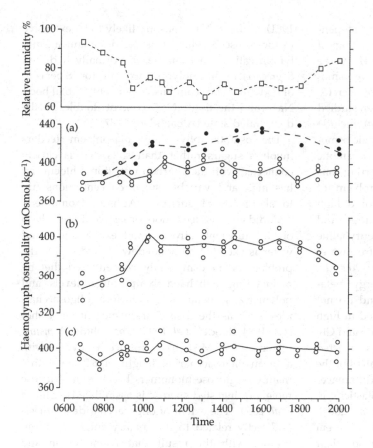

Figure 4.11 Diurnal variation in haemolymph osmolality (mOsmol kg^{-1}) of *Pieris brassicae* (a) third instars, (b) fifth instars, and (c) pupae.

Note: For third instars, values for larvae in natural positions on the host plant (cabbage; open circles) are compared with those for isolated larvae (filled circles). Top panel shows RH throughout the day.

Source: Willmer (1980). *Ecological Entomology* **5**, 271–292, Blackwell Publishing.

4.3.2 Responses to osmotic stress

Larval Diptera inhabiting saline waters exhibit extraordinary osmoregulatory abilities. For example, soldier fly larvae *Odontomyia cincta* (Stratiomyidae) maintain haemolymph osmolalities of 232 and 419 mOsmol kg^{-1} when acclimated to external media of 3 and 5414 mOsmol kg^{-1}, respectively (Gainey 1984). The transition from hypo-osmotic to hyper-osmotic regulation occurs at an isosmotic point of 280 mOsmol kg^{-1}, comparable to that of other aquatic insects. Brine flies (Ephydridae) are superbly tolerant of extreme environments, such as alkaline lakes or hot springs with unusual ion composition (Barnby 1987). Evidence for the hindgut as a site of osmotic regulation in ephydrid larvae includes rectal fluid osmolalities of 8700 mOsmol kg^{-1} in *Ephydrella marshalli* maintained in external osmolalities of 6000 mOsmol kg^{-1} (Marshall *et al.* 1995). In addition, lime glands (modified Malpighian

tubules) store $CaCO_3$ in the alkali fly *E. hians* (Herbst and Bradley 1989). It should be emphasized that physiological tolerance under extreme conditions in the laboratory probably exceeds that in the field: the long-term effects of high salinities are costly, and early instars are certainly less tolerant to prolonged exposure (Herbst *et al.* 1988). Most studies have examined final instars only (owing to difficulties in obtaining haemolymph, especially from dehydrated animals in high salinities).

Mosquitoes (family Culicidae) are an ideal group in which to examine the evolution of saline tolerance: their larvae occur in a huge variety of aquatic habitats, the mechanisms involved in the osmotic physiology of saline-tolerant species have been intensively studied, and their medical importance ensures the availability of good taxonomic information (Bradley 1994). Ancestral species are thought to have had obligate freshwater larvae. Salinity tolerance has evolved at least five times independently in mosquitoes, but only two physiological strategies are involved (Bradley 1994). Some species of *Aedes, Opifex,* and *Anopheles* are osmoregulators, and the rectal salt gland of *Aedes* secretes NaCl-rich hyperosmotic fluid and has been studied in detail. Other genera (*Culex, Culiseta*) include osmoconformers, which accumulate organic osmolytes in their haemolymph above the isosmotic point: this reduces the need for transporting ions and is a cheaper solution to the problem of inhabiting saline environments. Regulation of high haemolymph concentrations of trehalose and proline has been examined in detail in euryhaline *Culex tarsalis,* in comparison with freshwater *C. quinquefasciatus* (Patrick and Bradley 2000*a,b*). Trehalose and proline are used in energy metabolism, have a low molecular mass, and do not disrupt enzyme action when accumulated in high concentrations. Recent comparison of Na^+ and Cl^- uptake mechanisms in *C. quinquefasciatus* collected in California and from acid, ion-poor waters of the Amazon region has revealed population differences in ion uptake, and greater phenotypic plasticity in the Amazonian population (Patrick *et al.* 2002). Adult mosquitoes, of course, face entirely different water and salt challenges as a result of their desiccating environment and infrequent but large blood meals.

During dehydration many terrestrial insects conserve cell water at the expense of the haemolymph, and closely regulate the osmotic and ionic concentration of the dwindling volume of haemolymph. In American cockroaches (*P. americana*), changes in haemolymph osmolality are minimized by sequestration of solutes (especially K^+ and Na^+) during dehydration and their mobilization during rehydration (Tucker 1977; Hyatt and Marshall 1985). The fat body acts as an ion sink by sequestering ions as insoluble urates, and a similar mechanism may account for the uptake and release of haemolymph solutes during pronounced changes in haemolymph volume throughout a dehydration–rehydration cycle in *Onymacris plana* (Nicolson 1980). Tenebrionid beetle larvae can thus survive prolonged dehydration and replenish water deficits by vapour uptake without having to exchange solutes with their environments (Machin

1981). Shifts of Na^+ ions are likely to be extensive in those insects where it is the dominant haemolymph cation: for example, in normally hydrated *S. gregaria,* the haemolymph accounts for 18 per cent of body mass but contains 76 per cent of total body Na^+, and the tissue Na^+ content doubles in the dehydrated state (Albaghdadi 1987).

The osmotic physiology of phloem feeders involves sugars rather than salts and is closely connected with their carbon nutrition. Phloem sap has high and variable sugar concentrations (up to about 0.8 M sucrose). Aphids (Homoptera, Aphididae) must feed more or less continuously to obtain sufficient nitrogen, and excess sugar and water is excreted as honeydew. Aphids solve the problem of an osmotically concentrated diet by maintaining high haemolymph sugar levels, and polymerizing dietary sugars to form oligosaccharides, such as the trisaccharide melezitose (Fisher *et al.* 1984; Rhodes *et al.* 1997). Pea aphids, *A. pisum,* reared on 0.75 M sucrose produce honeydew with a mean oligosaccharide length of 8.2, consisting mainly of glucose monomers because the fructose moiety of ingested sucrose is assimilated (Ashford *et al.* 2000). The extent of oligosaccharide synthesis is directly related to the dietary sucrose concentration, with the result that haemolymph and honeydew osmolalities remain independent of diet concentration (Fisher *et al.* 1984). Aphids are the best-studied phloem feeders, but whiteflies (Homoptera, Aleyrodidae) also use sugar transformations for osmoregulatory purposes. As dietary sucrose concentration increases, the main sugars in excreted honeydew change from glucose and fructose to the disaccharide trehalulose (Fig. 4.12, Salvucci *et al.* 1997). Development of sweet potato whiteflies *Bemisia tabaci* is unaffected by water stress in the host plant, as a result of physiological adjustments such as increased trehalulose in the honeydew (Isaacs *et al.* 1998). Aphids and whitefly, two closely related taxa of phloem feeders, have evolved different enzymatic processes for coping with high sugar concentrations (Ashford *et al.* 2000). Ion regulation in the aphid *Myzus persicae* cultured on a salt marsh plant was found to be very effective, and must involve the gut as aphids lack Malpighian tubules (Downing 1980).

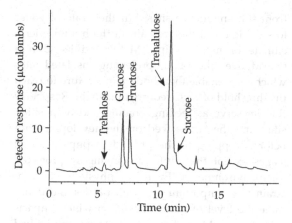

Figure 4.12 HPLC of sugars in honeydew of whiteflies *Bemisia argentifolii*, showing the high proportion of trehalulose on a concentrated diet of 30% (w/v) sucrose.

Source: Reprinted from *Journal of Insect Physiology*, **43**, Salvucci et al., 457–464, © 1997, with permission from Elsevier.

4.3.3 Salt intake

The physiological regulation of salt output (see Section 4.1.3, Excretion) is much better understood than the behavioural regulation of salt intake through feeding (Trumper and Simpson 1994). However, selection of food on the basis of salt content has been demonstrated in nymphs of *Locusta migratoria*: when provided with appropriate artificial diets, they will regulate salt intake, provided it does not compete with stonger regulatory mechanisms for protein and carbohydrate (Trumper and Simpson 1993; Simpson 1994). Salts have been considered either non-stimulatory or feeding deterrents in locusts, depending on concentration, but there is good experimental evidence for phagostimulatory regulation of intake in these insects, as in other animals (Simpson 1994).

A spectacular example of salt acquisition is seen in puddling behaviour of Lepidoptera, which also show sex differences in salt appetite. The habit of feeding at puddles, mud, or dung, thought to compensate for the low sodium content of larval food, is much commoner in males, and males of the skipper butterfly *Thymelicus lineola* (Hesperiidae) transfer large quantities of Na^+ to females during mating (Pivnick and McNeil 1987). Male butterflies are also strongly attracted to nitrogen sources in decaying organic material, and the excretion of

surplus water confirms that dissolved substances, rather than the water itself, are the resource acquired during puddling (Beck *et al.* 1999). The most striking example of Na^+ uptake by puddling is seen in male moths, *Gluphisia septentrionis* (Notodontidae), which ingest huge volumes and void the excess fluid in jets while drinking. Sodium is absorbed across the enlarged ileum of the male moths and transferred at mating to females, and then allocated to the eggs (Smedley and Eisner 1995, 1996).

4.4 Desiccation resistance

At the organismal level, desiccation resistance in insects is generally accomplished in three ways: by increasing body water content, by reducing rates of water loss, or by tolerating the loss of a greater proportion of body water (desiccation or dehydration tolerance) (Gibbs 1999). Each of these organismal traits is, in turn, an integrated measure of multiple physiological processes (e.g. water loss via the cuticle, respiration, or excretion). The costs of physiological regulation can be reduced by behavioural avoidance of desiccating conditions, as in insects with nocturnal or crepuscular activity patterns, or those which select favourable microclimates or form intraspecific aggregations (see below). Clear separation of behavioural and physiological regulation, however, is not realistic, as will become apparent from the examples to follow.

The standard method of measuring desiccation resistance is to record the mass change of insects maintained in dry conditions. Losses represent water and also dry matter metabolized as CO_2, but in most studies respiratory water loss is assumed to be negligible. Useful water balance characteristics are the body water content, maximum tolerable water loss before death, time to maximum water loss, and rate of water loss. Survival time is equivalent to water loss tolerance divided by water loss rate, and is determined by initial water content. Note that insects maintained in dry air experience starvation and desiccation, whereas those maintained at higher a_v experience mainly starvation. Because initial differences in body mass may confound comparisons of absolute rates and tolerances (Packard and Boardman 1988), corrections for initial body mass are often used (e.g. Chown *et al.* 1999; Gibbs and Matzkin 2001).

4.4.1 Microclimates

The importance of microclimate in insect physiological ecology was reviewed by Willmer (1982) and is receiving increasing attention. Soil and underground environments provide refuge from the temperature extremes which occur at the surface (see Chapter 6), while vegetation ensures that terrestrial habitats are varied and complex for small animals like insects. Transpiration accentuates the microclimate around a leaf, and lower leaf surfaces develop a boundary layer of relatively still, cool, and humid air. Both the steepness of the gradients and the rapid changes possible are emphasized by Willmer (1986), who discusses the effects of factors such as leaf size, shape, thickness, reflectance (includes colour and hairiness), and position on the plant on adjacent temperatures. Humidities within the leaf boundary layer have been less easy to measure, and Gaede (1992) used a model developed by Ferro and Southwick (1984), verified by measurements with hygroscopic KCl crystals, to estimate water vapour activities in the microhabitat of a predatory mite, *Phytoseiulus persimilis*: a_v at the leaf surface frequently exceeded 0.90, providing opportunities for active vapour absorption to replace water deficits. The undersurface of leaves is also a favoured oviposition site for herbivorous insects which live permanently on their food plants and fully exploit the heterogeneity of these environments in time and space.

Plant microclimates affect the physiology of insects, hence their behaviour, and translate into effects on ecology and distribution. This is nicely illustrated by the study of Willmer *et al.* (1996) on the distribution patterns of raspberry beetles *Byturus tomentosus* (Byturidae) on their host plant in Scotland. Plant chemistry is much less important than microclimatic effects in determining seasonal and diurnal distribution patterns of raspberry beetles. On 'dry' days with a_v of 0.5–0.7, the a_v was 0.83–0.96 within 5 mm of the leaf undersurface, where stomata are more numerous and there is an unstirred layer of humid air (Willmer 1986). Recently eclosed beetles climb the raspberry canes

from soil emergence sites, but their rate of water loss is high and they remain in the humid microclimate of new leaftips. Mature adults (5 mg) spread over the plant, preferring insolated sites which may enable their body temperature to reach the threshold of 15°C required for flight. Raspberry flowers serve as feeding, mating and oviposition sites, and the larvae feed on the developing fruit before dropping to the ground to pupate, so all stages except the adults are in highly protected microenvironments. Even so, physiological constraints are important for adult beetles and determine the level of floral infestation which impacts on growers. Microclimatic effects on soft-bodied larvae and more vulnerable adults such as aphids are likely to be profound (Willmer *et al.* 1996).

Sometimes insects modify the leaf microenvironment. Nymphs of eugenia psyllids *Trioza eugeniae* develop in pit-shaped galls, and moderate densities of nymphs increase leaf curling, although competition at high densities then cancels out potential microclimatic benefits (Luft *et al.* 2001). Nettle leaf rolls made by *Pleuroptya ruralis* larvae (Lepidoptera, Pyralidae) maintain the a_v above 0.95 (Willmer 1980). Gall-producing and leaf-mining insects create microenvironments within plant tissues (Connor and Taverner 1997), while many insects manufacture microenvironments from other materials. (Benefits also include protection from parasites, predators, and mechanical injury.) Spittle bug (Homoptera, Cercopidae) nymphs live in a mass of foam which protects them from enemies and desiccation. Chinese mantids *Tenodera aridifolia sinensis* (Mantodea, Mantidae) deposit their eggs in exposed locations, but within a specialized case which prevents desiccation of the developing embryos over a six-month period, although a favourable hygric environment is necessary at the time of hatching (Birchard 1991). The barrier function of an ootheca, puparium, or cocoon is useful for immobile eggs or pupae which gain water by metabolism only. In addition, the larval case of the clothes moth *Tinea pellionella* (Lepidoptera, Tineidae) passively absorbs water in high humidities (Chauvin *et al.* 1979), as does the hygroscopic pupal cocoon of the leek moth *Acrolepiopsis assectella* (Plutellidae) (Nowbahari and Thibout 1990). The benefits of manufactured microclimates—and group effects—are seen to an extreme in social insects, many of which have complex nest architectures that permit precise control of both temperature and humidity.

4.4.2 Group effects

Conspecific aggregations serve to create and maintain local microclimates. The increased thermal mass leads to a more stable body temperature because metabolic heat is conserved and convective cooling is reduced. Water loss by evaporation or excretion creates a region of higher vapour pressure. Caterpillars may accentuate these effects by spinning communal webs. Three hypotheses concerning the benefits of aggregation behaviour were considered by Klok and Chown (1999) (see also Chapter 6): avoidance of predation by birds, overwhelming of plant defences by synchronous feeding, or physiological benefits (water and thermal relations). Mopane worms *Imbrasia belina* (Lepidoptera, Saturniidae) are strictly gregarious in the first three instars, with aggregations which may exceed 200 individuals; rates of water loss decrease as aggregation size increases, and body temperatures of aggregations are similar to those of large solitary caterpillars. Aggregation in this species compensates to some extent for the physiological effects of a 4000-fold increase in body mass during the six weeks of larval development (Klok and Chown 1999). Field measurements of haemolymph osmolality also demonstrate more stable water balance in caterpillars living in groups (Fig. 4.11, Willmer 1980).

Coccinellid beetles are known for aggregating in thousands during diapause, behaviour which compensates for modest tolerance of water loss in isolated lady beetles *Hippodamia convergens* (Coccinellidae) (Yoder and Smith 1997). Surprisingly, adult females (8.8 g) of the giant Madagascar hissing cockroach, *Gromphadorhina portentosa* (Blattaria, Blaberidae), also benefit from clustering, along with a mite, *Gromphadorholaelaps schaeferi*, which lives in small groups on its leg bases (Yoder and Grojean 1997). A detailed experimental analysis of the costs and benefits of aggregation was undertaken by Rasa (1997) on a tenebrionid beetle of the southern Kalahari desert, *Parastizopus armaticeps*

(0.26 g). This species aggregates in burrows during the day and emerges at sunset to feed on unpredictably available detritus of *Lebeckia linearifolia* (Fabaceae). During periods of drought, group size increases and evening emergence becomes highly synchronized. Beetles in groups have lower water loss rates than solitary ones, but this benefit is countered by the fact that competition for food is fierce in group burrows. The ideal strategy for these beetles, if they are to balance hygric and nutrititional constraints, is to alternate between grouped and solitary lifestyles, and Rasa's field data showed that marked individuals do behave in this way.

Many female insects choose oviposition sites which have microclimatic benefits (e.g. the undersurfaces of host plant leaves) and some lay their eggs in clusters. Egg clustering in butterflies results in aggregation of early instar larvae, with the potential benefits described above, but egg clustering may also be adaptive by reducing egg mortality due to desiccation. This hypothesis was tested in *Chlosyne lacinia* (Nymphalidae): eggs in large, multilayered batches had greater hatch success than those in smaller, monolayered batches, especially at low a_v (Clark and Faeth 1998). Multiple selective factors probably determine egg clustering, including cannabilism of eggs by siblings (Clark and Faeth 1998).

4.4.3 Dormancy, size, and phylogeny

During the active portion of an insect's life cycle desiccation stress is probably frequent, but short-lived. In addition, the active stages (larvae and adults) have regular access both to free water and to water in their food. In contrast, during dormancy (diapause, quiescence, aestivation), dehydration is likely to be prolonged and access to water and energy resources extremely limited (Lighton and Duncan 1995; Danks 2000). At this time, there is likely to be strong selection both for low metabolic rate, to conserve energy resources, and for any mechanisms that might reduce water loss. Insects overwintering in temperate regions are also subject to desiccating conditions, and some show remarkable resistance to desiccation even when individuals are removed from their plant galls

(Ramløv and Lee 2000; Williams *et al.* 2002). This is not, however, due to the seasonal accumulation of cryoprotectant polyols and sugars, which increases haemolymph osmolality (Williams *et al.* 2002), but in fact has little effect on the water activity gradient between air and insect. The role of reduced water content in overwintering strategies is discussed in Section 5.3.2. Danks (2000) has reviewed the mechanisms of desiccation avoidance and tolerance in dormant insects (his table 1 is a useful summary): one of the reasons for dormancy in the first place is water stress.

Effects of desiccation on eggs and even smaller first instar larvae were recently compared for two species of Lepidoptera, *Grammia geneura* (Arctiiidae) and *M. sexta* (Sphingidae) (Woods and Singer 2001), with an emphasis not on phylogenetic differences but rather on the physiological diversity associated with different oviposition tactics. Female *M. sexta* lay eggs on the underside of leaves of the food plant, while female *G. geneura* deposit theirs in leaf litter or at the base of grasses. Neonates of *G. geneura*, although much smaller, cope with starvation and desiccation very effectively by reducing their metabolic rate and becoming inactive until food is available. Setae covering larval *G. geneura* create a boundary layer and further decrease water loss. Rates of water loss by eggs of both species did not show the expected inverse relation to initial egg mass; nor did large larval size within a species confer resistance to desiccation. Comparing individuals of different body size is not straightforward: third instar *Pieris brassicae* (Lepidoptera, Pieridae) lose water more rapidly than fifth instars, but cuticular permeability might decrease with age (Willmer 1980).

Ants have successfully colonized a wide variety of habitats and vary greatly in body size both within and between species, but several studies show that large size is less of an advantage than might be expected. Desiccation resistance in a large assortment of arboreal and terrestrial ants increased with body size (as dry mass$^{0.55}$), but not as quickly as expected from the surface area to volume relationship (dry mass$^{0.67}$) (Hood and Tschinkel 1990). Habitat accounted for much more of the variation in desiccation resistance, with arboreal ants surviving eight times longer than

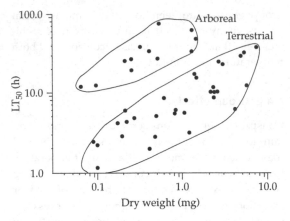

Figure 4.13 Desiccation resistance of arboreal and terrestrial ant species, measured as LT$_{50}$ (h) for groups of 10–22 ants at $a_v = 0$. LT$_{50}$ is time to 50% mortality of the sample. The terrestrial group includes species from the Chihuahuan Desert. Note log scale.

Source: Hood and Tschinkel (1990). *Physiological Entomology* **15**, 23–35, Blackwell Publishing.

terrestrial ants of the same size (Fig. 4.13). This may be due to better waterproofing by epicuticular lipids and more efficient extraction of faecal water, but is not due to differences in water loss tolerance. Arboreal ants lack the considerable microclimatic advantages of underground nests, in which a_v may approach saturation at depths of 50 cm (Hood and Tschinkel 1990). Two species of desert honeypot ants, *Myrmecocystus*, with polymorphic workers showing a great range of size, have water loss rates which increase as dry mass$^{0.31}$, so that again larger ants lose water disproportionately more slowly (Duncan and Lighton 1994). High water contents (about 84 per cent) in *Myrmecocystus* workers may be related to their nectar diet and enable high desiccation tolerance. In harvester ants, *Messor pergandei*, with a body mass range of 1–12 mg, small workers lose water faster and have smaller water reserves than large workers, even after size correction, so their foraging times must be correspondingly reduced (Lighton *et al.* 1994). Finally, when populations of imported fire ants, *Solenopsis invicta*, were sampled across Texas, differences in desiccation resistance of minor workers were not correlated with head width (Phillips *et al.* 1996b). Duncan and Lighton (1994) give a good account of the significance of water relations for ant foraging strategies.

Broad-scale, multi-species assessments of physiological variation in beetles (macrophysiology) have demonstrated good correlations between habitat aridity and desiccation resistance. Comparison of six species of sub-Antarctic weevils (Coleoptera, Curculionidae) belonging to the *Ectemnorhinus* group of genera showed that species from moist habitats had relatively low water contents, high rates of water loss, and reached maximum tolerable water loss faster than those from dry rock face habitats (Chown 1993). Similar relationships are apparent in sub-Antarctic carabid and perimylopid beetles, in which physiological adaptations to restrict water loss are reduced in the smaller carabids, which compensate by inhabiting moister habitats (Todd and Block 1997). Southern African keratin beetles of the genera *Trox* and *Omorgus* (Trogidae) differ from many other beetles in showing little osmoregulatory ability, and the correlation between water loss rates and habitat aridity is partly due to variation in body size (Le Lagadec *et al.* 1998). These authors suggest that there may be strong selection for large body size in desert species. The partitioning of variance in desiccation resistance was also examined in trogids. Most variance in body size, and the physiological traits that are strongly influenced by body size (water and lipid content, maximum tolerable water loss, rate of water loss) is partitioned at the generic level (50–70 per cent), then at the species level (20–50 per cent) (Chown *et al.* 1999). This suggests a certain confidence in past investigations of insect water loss, especially those that provide mass-specific data.

4.5 The evidence for adaptation: *Drosophila* as a model

Variation in physiological traits among populations and species occupying different environments has long been regarded as evidence for adaptation, but this correlational approach has been subject to criticism (Garland and Adolph 1994). The use of phylogenetic information in comparative studies can show whether variation among taxa reflects adaptation to their environment or is, perhaps, an artefact of their evolutionary history. Alternatively, laboratory selection studies are proving a powerful

tool in evolutionary physiology (Bradley *et al.* 1999; Gibbs 1999). Selected populations can be replicated, under strictly defined conditions, and unselected control populations are available for direct comparison with the ancestral condition. Desiccation is a strong agent of selection in small insects, and laboratory populations of *Drosophila melanogaster* which have undergone selection for increased desiccation resistance provide some unexpected information about the physiological traits contributing to this resistance. Analysis of the water budget of five desiccation-resistant populations (D flies) and five control populations (C flies) showed adaptive changes in some mechanisms but not in others (Gibbs *et al.* 1997). D flies are larger than C flies because they contain more water, and they also lose water less rapidly, but there is no difference in water content at death. As for the components of water loss, surface lipids do not differ significantly, and excretion (Fig. 4.14) accounts for less than 6 per cent of total water loss (Gibbs *et al.* 1997, 2003*a*). Reduced water loss rates must, therefore, be achieved through reductions in respiratory losses. Adult D flies contain 34 per cent more water than C flies, and the main storage site

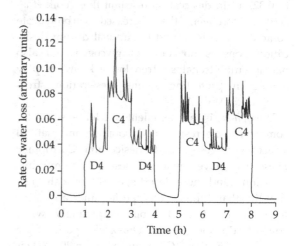

Figure 4.14 Water loss recording for paired control (C) and desiccation-resistant (D) populations of *Drosophila melanogaster* (groups of 20 female flies).

Note: Trace shows both evaporative and excretory water losses; peaks of water vapour recorded by the humidity sensor indicate defaecation by individual flies.

Source: Gibbs *et al.* (1997).

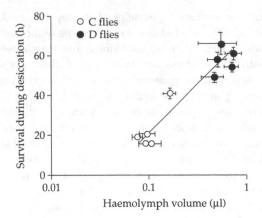

Figure. 4.15 Survival of control (C) and desiccation-resistant (D) *Drosophila melanogaster* during desiccation stress is strongly related to haemolymph volume. Filled circles are means for five D populations and open circles are means for five C populations (± SE). Haemolymph volume (μl) was measured by a blotting technique in 10 flies per population.
Source: Folk *et al.* (2001).

for this 'canteen' of extra water is the haemolymph. Desiccation resistance is strongly correlated with haemolymph volume (Fig. 4.15), which increases significantly from a mean of 0.078 μl in control flies to 0.323 μl in desiccation-resistant flies (Folk *et al.* 2001). Moreover, the increased carbohydrate content which is related to survival during desiccation consists not only of glycogen, but also haemolymph trehalose. Trehalose is known to play a role in protecting against desiccation stress (Ring and Danks 1998).

Drosophila is an excellent model system for comparing the results of laboratory and natural selection for resistance to desiccation. Cactophilic *Drosophila* have invaded deserts on multiple occasions, and these desert species lose water less rapidly than non-desert species and are more tolerant of dehydration but, surprisingly, water contents do not differ in *Drosophila* species from a variety of habitats (Gibbs and Matzkin 2001). This is in marked contrast to the difference in water content between laboratory-selected D and C flies, but the lack of difference between them in dehydration tolerance (Gibbs *et al.* 1997). The mechanistic basis of resistance thus seems to differ in the laboratory and the field, probably because of

differences in selection regimes. Laboratory selection is carried out at constant temperature, in a simple habitat which offers fewer behavioural options. Sex differences are apparent during selection and there are costs involved, such as slower development times in desiccation-resistant *Drosophila* (Chippindale *et al.* 1998). Desert *Drosophila* inhabit necrotic tissues in columnar cacti, with microclimates which offer hygric but not thermal benefits (Gibbs *et al.* 2003*b*). Perhaps desert *Drosophila* do not accumulate water because the additional load may compromise flight performance when it is necessary to fly to the next cactus, whereas D flies in the laboratory remain relatively inactive (Gibbs 1999).

Another advantage of studying water balance in *Drosophila* is that the evolutionary history of the genus is well studied and detailed phylogenetic information is available for use in comparative studies. Gibbs and colleagues (Gibbs and Matzkin 2001; Gibbs *et al.* 2003*a*) have now examined the evolution of water balance in 30 species from mesic and xeric environments. After phylogenetic relationships were incorporated into the analysis, cactophilic *Drosophila* differed from their mesic congeners only in rates of water loss: their greater tolerance of dehydration appears to be an ancestral trait. Thus water conservation is the primary means of increasing survival during desiccation stress. The most recent evidence (Gibbs *et al.* 2003*a*) suggests that reduced water loss is achieved by varying respiratory parameters: compared to mesic species, cactophilic *Drosophila* are less active, have lower metabolic rates, and are more likely to exhibit a pattern of cyclic CO_2 release with possible implications for water savings.

Can the evolution of desiccation resistance be reversed? Relaxation of directional selection in *Drosophila* led to a decline in desiccation resistance after 100 generations, along with flight duration (a correlated character also linked to glycogen reserves) (Graves *et al.* 1992). Reverse selection has recently been examined over a much longer (geological) time-scale in non-drosophilids—a group of sub-Antarctic weevils (Curculionidae) occurring on two Southern Ocean islands (Chown and Klok 2003). Phylogenetic analysis shows a sharp reduction in body size as angiosperms

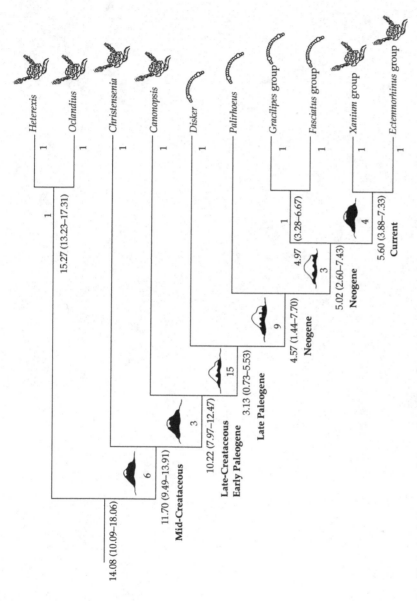

Figure. 4.16 Evolution of body size, habitat use, and diet in the *Ectemnorhinus* group of genera (Curculionidae).

Note: Branch lengths are indicated above the branches and body size values (and 95% confidence intervals) below them. The island schematics indicate extent of past glaciation, diet and habitat use are indicated as either angiosperms (vegetated areas) or algae (rock faces and fellfield), and an approximate geological timescale is also provided.

Source: Physiological and Biochemical Zoology, Chown and Klok, **76**, 634–643. © 2003 by The University of Chicago. All rights reserved. 1522-2152/2003/7605-2139$15.00.

disappeared from the islands due to glaciation and resources declined, even though larger size has water balance advantages. When moist habitats returned, the genus *Ectemnorhinus* returned to angiosperm feeding and increased in size (Fig. 4.16).

The relaxation of selection for water conservation also led to reduced dehydration tolerance and increased water loss rates, presumably because there are costs involved in maintaining these traits (Chown and Klok 2003*b*).

CHAPTER 5

Lethal temperature limits

There is in most animals which have been subjected to experimentation with temperature, a range of several degrees in which the animal is not markedly stimulated ... As the temperature is raised or lowered from such a condition, the animal is stimulated.

Shelford (1911)

Temperature has an effect on most physiological processes. The ways in which insects modify their response to temperature effects, over the range of temperatures at which they are generally active, are known as capacity adaptations. Responses to what otherwise might be lethal effects of temperature extremes have been termed resistance adaptations (for further discussion see Cossins and Bowler 1987). In both cases the term 'adaptation' is something of a misnomer because the responses are often a consequence of phenotypic plasticity or flexibility, or because they might not be adaptive at all. In addition, these responses are often intimately related and cannot be clearly separated. For example, in marine species, and arguably also in several terrestrial groups, lethal temperature limits (typically considered resistance traits) result from insufficient aerobic capacity of mitochondria at low temperatures, and a mismatch between excessive oxygen demand by mitochondria and insufficient oxygen uptake and distribution by ventilation and circulation at high temperatures. These are typically traits associated with capacity (Pörtner 2001). Nonetheless, distinguishing between resistance and capacity responses is often convenient, and both kinds of responses are characteristic of insects. Here, we deal with those responses that occur when potentially lethal temperature extremes are encountered, leaving what are more typically considered 'capacity' responses to the relevant sections in other chapters.

A convenient way to think about the responses of insects to potentially lethal temperatures is the thermobiological scale presented by Vannier (1994) (Fig. 5.1). At both ends of the scale, continuation of the change in temperature results first in knock-down of the insect (stupor in Vannier's terms), then in prolonged coma, and finally in irreversible trauma and death. At low temperatures, insects show a wider variety of responses to sublethal and potentially lethal temperatures than they do at high temperatures, including responses to non-freezing temperatures and those made in preparation for the decline of temperatures below the freezing point of water (Fig. 5.2). It is the ways in which insects alter the relationship between the temperatures they are experiencing and their survival probability, the costs of these changes, and the similarities and differences between the responses to upper and lower lethal temperatures (LLTs) that form the substance of this chapter. How insects alter the temperatures they experience (i.e. thermoregulate) is dealt with in Chapter 6.

5.1 Method and measurement

In previous chapters we have shown that the experimental design and conditions have a major impact on the findings of the experiment or trial in which they are used. This is especially true for the examination of the responses of insects to high and low temperatures. Therefore, before proceeding with a discussion of the ways in which insects cope

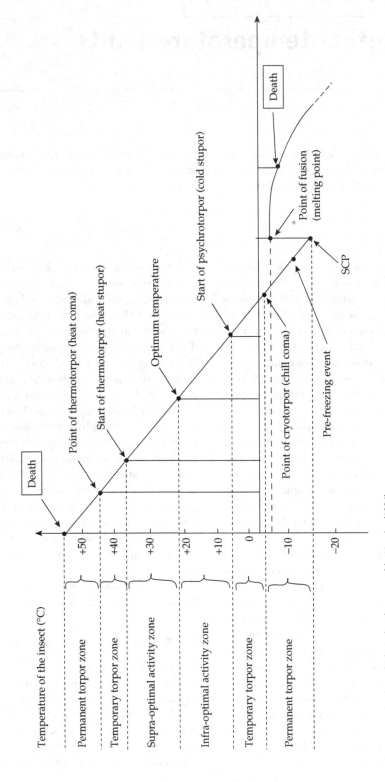

Figure 5.1 The thermobiological scale proposed by Vannier (1994).

Source. Reprinted from *Acta Oecologica*, **15**, Vannier, G, 31-42, © 1994, with permission from Elsevier.

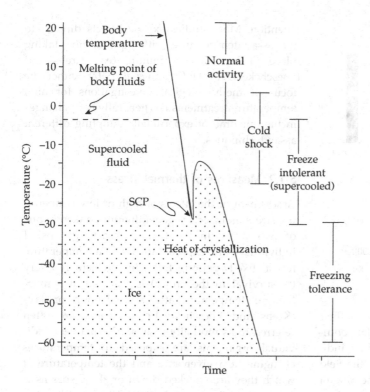

Figure 5.2 Typical responses of insects cooled below their normal activity range.

Source: Lee (1989). *BioScience* **39**, 308–313. © American Institute of Biological Sciences.

with high and low temperature challenges, a treatment of the more significant influences of methods on experimental outcomes is necessary. Here, we deal specifically with experimental protocols, rates of temperature change, and exposure and recovery times, while acknowledging that other factors, such as photoperiod (Lanciani *et al.* 1992), CO_2 anaesthesia, and time of day of the analysis (Lutterschmidt and Hutchison 1997), may also be important. Because acclimation treatments are especially significant, and ecologically relevant, we deal with these separately in the main section on the responses of insects to thermal challenges.

5.1.1 Rates of change

It is widely appreciated that the rate at which insects cool has a profound influence on their ability to survive low temperatures (Miller 1978; Baust and Rojas 1985; Shimada and Riihimaa 1990; Ramløv 2000; Sinclair 2001*a*), although the super-cooling point (SCP) or crystallization temperature (Section 5.3.1) apparently remains largely unaffected by cooling rate (Salt 1966). For example,

Miller (1978) demonstrated that even small deviations ($<0.1°C\,min^{-1}$) from the 'optimal' cooling rate ($0.32°C\,min^{-1}$) have a large influence on survival in *Upis ceramboides* (Coleoptera, Tenebrionidae). Likewise, Shimada and Riihimaa (1990) showed that in the freeze-tolerant *Chymomyza costata* (Diptera, Drosophilidae), altering the cooling rate from 0.1 to $1°C\,min^{-1}$ substantially reduces short-term (1 h) survival of temperatures below $-10°C$ (Fig. 5.3). Cooling rate not only affects low temperature survival in *Drosophila melanogaster*, but also has a marked effect on critical thermal minimum (CT_{min}), a measure of knockdown temperature (Kelty and Lee 1999). Obviously, the use of slower cooling rates in experiments is more ecologically relevant: this is a result of the rapid cold hardening (Section 5.2.3) that takes place as a consequence, and which undoubtedly enhances survival in the field (Kelty and Lee 2001).

This theme of ecological relevance has been raised, *inter alia*, by Baust and Rojas (1985), Bale (1987) and Sinclair (2001*a*), with Baust and Rojas (1985) pointing out that the cooling and subsequent warming (or thawing) rates of an insect's

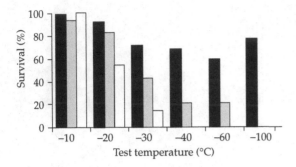

Figure 5.3 Influence of cooling rate on survival of the freeze-tolerant drosophilid fly *Chymomyza costata* at low temperatures. Black bars $0.1°C\,min^{-1}$, grey bars $0.5°C\,min^{-1}$, open bars $1°C\,min^{-1}$.

Source: Data from table 1 in Shimada and Riihimaa (1990).

microhabitat are most relevant to determining their response to cold. Suboptimal experimental conditions provide a measure of response of an individual to these conditions, rather than to the field environment. Thus, it might come as something of a surprise that the large majority of investigations of insect cold hardiness have adopted cooling rates of approximately $1°C\,min^{-1}$ (Block 1990). A largely uniform protocol for experiments was initially adopted to standardize results (Salt 1966), and has clearly been instrumental in facilitating a sound understanding of the physiological and biochemical responses of insects to low temperatures. However, if the responses of insects to their surroundings are to be comprehended within the context of their environmental setting, and compared across regions (Section 5.4), then ecologically relevant rates might be more appropriate in subsequent investigations. Determining these rates and other relevant microclimatic parameters is more important than simply investigating the maximum and minimum temperatures likely to be encountered by a species in its usual environment (Feder *et al.* 1997a; Sinclair 2001b).

Although the rate of heating is also known to have an effect on mortality owing to high temperatures (Lutterschmidt and Hutchison 1997), probably as a consequence of exposure time (Feder *et al.* 1997a), the effects of heating rate on high temperature survival have enjoyed much less

attention. Most studies expose insects directly to the assay temperature with equilibration taking place in less than a minute (e.g. Krebs and Loeschcke 1995a; Hoffmann *et al.* 1997). Rather, the focus of methodological investigations for high temperature treatments has generally been changes in the outcome of experiments adopting different assay techniques.

5.1.2 Measures of thermal stress

Assessment of the effects of high or low temperature stress on insects is generally undertaken in one of two ways. In the first approach, groups of individuals are exposed to a given temperature for a fixed period, following which recovery (% survival in the group) is assessed at a more 'normal' temperature after several hours or days (Krebs and Loeschcke 1995a). These are often referred to as mortality assays. Alternatively, individual organisms are exposed to temperatures changing at a given rate and the temperature at which they are knocked down or show spasms is recorded (Roberts *et al.* 1991; Klok and Chown 1997; Gibert and Huey 2001). The latter method is often modified such that a constant temperature is maintained and the time the insects take to be knocked down is recorded (Hoffmann *et al.* 1997), or time to recovery after a set exposure is examined (David *et al.* 2003). In addition, in a few studies, a combination of temperature change and exposure time has been used (Worland *et al.* 1992). Lutterschmidt and Hutchison (1997) provide a critique of the major methods for the assessment of high temperature limits and refer to the fixed temperature assays involving survival assessments as 'static methods', and the changing temperature assays as 'dynamic methods'. Although Lutterschmidt and Hutchison (1997) come out in favour of the dynamic, critical thermal maximum (CT_{max}) method as a standard measure of thermotolerance, they argue that both methods are useful for assessing tolerance, though they caution that these two methods might be measuring different things.

This idea has been particularly well explored in a variety of *Drosophila* species (Hoffmann *et al.* 2003b). In lines of *D. melanogaster* selected for knockdown resistance to heat in a 'knockdown tube' (Huey *et al.* 1992), knockdown resistance improves substantially compared with control lines. However, there is no effect of selection on knockdown of individuals assessed singly in smaller vials, or on survival in mortality assays (Hoffmann *et al.* 1997). Similarly, selection with and without hardening affects hardening in the 'knockdown tube', but not in the smaller vials. These results have been verified with isofemale lines (lines developed from a single mated female—see Hoffmann and Parsons 1988) and hybrid lines of *D. serrata* and *D. birchii* (Berrigan and Hoffmann 1998). Thus, at the intraspecific level there appear to be no genetic correlations between knockdown and mortality measures of heat resistance. That is, mortality, recovery, and knockdown measures of thermotolerance may be associated with different mechanisms and genes (Gilchrist *et al.* 1997; Hoffmann *et al.* 1997). In support of this idea, McColl *et al.* (1996) have shown that selection for increased resistance to knockdown with hardening results in allelic changes in the stress genes *hsromega* and *hsp68* in *D. melanogaster*. In turn, the absence of correlated changes in other stress resistance traits (high temperature mortality—usually associated with variation in *hsp70* (Feder 1999), desiccation, cold) when knockdown times (with and without hardening) are altered, suggests that these genes are not associated with general stress resistance. These findings support the idea that different mechanisms of thermal tolerance are associated with different genes (Sørensen and Loeschcke 2001).

In a recent comparison of six *Drosophila* species, Berrigan (2000) found significant correlations between thermotolerance assessed using knockdown in the 'knockdown tube', knockdown in small vials, and mortality assays. The presence of correlations in these measures between species, but not within species, suggests that the interspecific correlations are a consequence of correlated selection regimes, and that differences in thermotolerance might be adaptive responses to the environment. A similar, strong correlation between cold stupor (knockdown temperature) and LLT has been found among eight other *Drosophila* species (Hori and Kimura 1998), providing additional support for this idea.

These findings raise the issue of which measure should be used for assessing thermotolerance. To date, most studies have been based on mortality assays (Berrigan and Hoffmann 1998). However, Hoffmann *et al.* (1997) and Sørensen and Loeschcke (2001) have argued that knockdown assays may be more relevant to investigations of the ecology and evolution of organisms because stress levels in these assays are less severe than in those involving mortality. They are, therefore, more likely to be relevant to the survival and fertility of insects in the field. While this is probably true for highly mobile stages (e.g. winged adults, or fast-moving hemipteran nymphs), more sedentary stages (larvae, pupae), or those restricted to a given habitat (galling insects, inhabitants of necrotic fruit) might not have the luxury of a behavioural response to extreme temperatures (Huey 1991). For example, larvae of *D. melanogaster* may be routinely exposed to potentially lethal temperatures for substantial periods (Feder 1996; Feder *et al.* 1997a; Roberts and Feder 1999) with little prospect of escape (Fig. 5.4). In consequence, mortality assays may be more relevant to these stages than to adult flies, which may never experience such high temperatures (Feder *et al.* 2000b). Arguably, there may also be situations where mobile stages cannot respond behaviourally to extreme temperatures. Thus, characterization of the full response of an insect to potentially lethal temperatures could potentially require assessment of several measures of tolerance. Some measures will indicate the thermal limits to activity, while others will provide an estimate of the time for which a given stressful temperature (including subzero temperatures) might be endured. While one of these measures might be more convenient for laboratory purposes than the others (Gibert and Huey 2001), it should be recognized that these measures represent assessments of different aspects of an organism's physiology (Feder 1996; Sørensen and Loeschcke 2001), the ecological relevance of which may be difficult to ascertain without suitable information on the natural history of the species involved (Sørensen *et al.* 2001). A particular problem in this regard, especially in the context of low temperature tolerance, is the exposure time used in any given assessment.

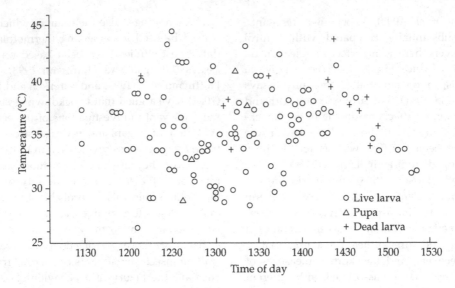

Figure 5.4 Estimated temperatures and condition of *Drosophila* larvae and pupae collected in the field at different times of the day. *Source*: Feder *et al.* (1997). *Functional Ecology* **11**, 90–100, Blackwell Publishing.

5.1.3 Exposure and recovery time

The time for which an insect is exposed to a given potentially lethal temperature has a considerable effect on tolerance, with extended exposure generally resulting in a decline in survival, though often in a complex fashion. The interactions between time and temperature have long been known (see Salt 1961; Asahina 1969; Cossins and Bowler 1987), and are often exploited in high temperature mortality assays to allow probit analyses of the time required for 50 per cent of a sample of insects to die (Berrigan 2000; Hercus *et al.* 2000). However, exposure time is also of considerable ecological relevance given that sublethal temperatures may often become lethal if exposure time is prolonged. For example, at subzero temperatures a liquid cooled beyond its melting point is metastable, and its probability of nucleation (freezing) is a function of temperature, solute concentration, volume, the presence of ice nuclei, and importantly, time (Salt 1961; Sømme 1982; Zachariassen 1991*a,b*; Ramløv 2000). In consequence, the longer an animal remains below subzero temperatures, the greater its risk of freezing. Bale (1987, 1993) has championed the importance of exposure time showing that although the lethal temperatures of some freeze intolerant (Section 5.3.2) species may appear

to be relatively low when assayed over a short period, prolonged exposure results in considerable mortality (see also the contributions of R.W. Salt discussed in Ring and Riegert 1991 and additional discussion in Sømme 1982). Moreover, Bale (1993) has argued that exposure time is an essential component of insect cold hardiness classifications (Section 5.3.1), and in consequence has called for greater knowledge of the microclimates insects are likely to experience. In re-emphasizing the importance of exposure time, both in terms of prefreezing mortality (or chilling injury) and mortality at subzero temperatures, Sømme (1996) pointed out that information on survival of prolonged cold and its relevance to the field situation is comparatively scarce.

However, it is not only exposure time that has an influence on assessments of the survival of potentially lethal temperatures. Warming rates, following a low temperature exposure, and the recovery time after which survival assessments are made, are both crucial components of thermotolerance experimental protocols (Baust and Rojas 1985). The latter is of particular significance. Most assessments of survival in mortality assays are made 24 h after exposure (e.g. Krebs *et al.* 1998), and occasionally after longer periods (Klok and Chown 1997; Jenkins and Hoffmann 1999). Baust and Rojas (1985) point out that in some species, coordinated activity and apparent feeding may persist for longer than a week before the insects die (but see Jenkins and Hoffmann 1999). As a result, several authors have suggested that assessments of fecundity (and possibly fecundity of the F1 generation) are more appropriate measures of the insects' response to the experimental conditions (Baust and Rojas 1985; Bale 1987). The reasoning is that the effects of stress in one stage might not be immediately detectable, but might carry over to another stage, negatively influencing either development, fecundity, survival, or some combination thereof. From an evolutionary (and ecological) perspective, assessments of both survival and fecundity are important (see Endler 1986 for discussion), but in many cases one or the other of these assessments might be difficult or impossible to make. This is likely to be especially true for non-model species that are unwilling inhabitants of the laboratory, but which, nonetheless, constitute the vast majority of the insect fauna. In this instance, survival assessments will be the only straightforward way of assessing the response of the species concerned, and the time over which these assessments should be undertaken would depend substantially on the life history of the species and the stage that is being assessed. For most experiments, however, 24 h should be considered a minimum first assessment time.

5.2 Heat shock, cold shock, and rapid hardening

Cold and heat shock are the stresses inflicted on insects by brief exposures to either low (but non-freezing), or high temperatures, respectively (Lee *et al.* 1987; Lee 1989; Denlinger *et al.* 1991). Injury is generally positively related to the duration and magnitude of the stress, and may eventually culminate in death. Although variation in insect thermal tolerances has long been appreciated, and often ascribed to adaptation to local environments (Shelford 1911; Mellanby 1932; Andrewartha and Birch 1954; Messenger 1959), the complexity of these responses and the nature of their underlying physiological mechanisms have been enjoying

renewed attention. There are undoubtedly many reasons for renewed vigour in this field. Among them is certainly the realization that investigations of model organisms can convincingly confirm (or reject) previously held, but often untested, adaptive hypotheses (Huey and Kingsolver 1993; Feder and Krebs 1998; Feder *et al.* 2000*a*). Furthermore, it is clear that a comprehensive understanding of the effects of global climate change will necessarily mean a thorough comprehension of the responses of insects to changing temperature (Hoffmann and Blows 1993; Cavicchi *et al.* 1995; Coleman *et al.* 1995, see also Chapter 7).

Recent investigations have not only confirmed geographic (and seasonal) variation in basal thermotolerance (Stanley *et al.* 1980; Krebs and Loeschcke 1995*a*; van der Merwe *et al.* 1997; Gibert and Huey 2001) (Fig. 5.5), but have also revealed considerable responses to selection in the laboratory of both basal and inducible thermotolerance (Hoffmann and Watson 1993; Cavicchi *et al.* 1995; Krebs and Loeschcke 1996; Loeschcke and Krebs 1996). Inducible thermotolerance, or hardening, is the increase in tolerance to potentially lethal temperatures that results from a brief exposure of organisms to moderately stressful temperatures

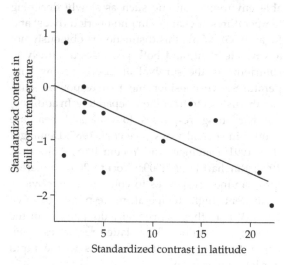

Figure 5.5 Geographic variation in chill coma temperature in 15 species of the *Drosophila obscura*-group of flies, corrected for phylogenetic non-independence.

Source: Physiological and Biochemical Zoology, Gibert and Huey, **74**, 429–434. © 2001 by The University of Chicago. All rights reserved. 1522-2152/2001/7403-00129$03.00

(Lee *et al.* 1987; Chen *et al.* 1990; Dahlgaard *et al.* 1998). In the context of cold, hardening has traditionally been regarded as a slow process that gradually increases an insect's tolerance of low temperatures (Denlinger and Lee 1998). However, in the context of high temperatures, hardening has come to be associated with a more rapid response to short-term, moderately stressful exposures (Denlinger *et al.* 1991; Loeschcke *et al.* 1997; Dahlgaard *et al.* 1998). Over the past several decades a similar, rapid response to cold has been demonstrated in a wide variety of species (Chen *et al.* 1987; Lee *et al.* 1987; Czajka and Lee 1990; Coulson and Bale 1992; Larsen and Lee 1994; McDonald *et al.* 1997). In consequence, here we restrict the term 'hardening' to rapid responses to either moderately high or low temperature exposures, preferring to consider longer-term cold hardiness a 'programmed response to cold' (Section 5.3). This distinction is not just a matter of convenience. Rather, it reflects the fact that hardening is generally a short-term change in thermotolerance (which is often lost within a matter of hours or days), as a consequence of a given, short-term temperature treatment. In contrast, longer-term cold hardening is a programmed set of responses to an often predictable environmental cue such as slowly declining temperatures or changes in photoperiod (Baust and Rojas 1985). While this distinction is obviously not always clear-cut, and both processes are likely to contribute to the survival of insects at low temperatures, we consider the two responses sufficiently distinct to treat them separately. In addition, the hardening responses to cold and heat are similar in several ways (Lee *et al.* 1987; Denlinger *et al.* 1991; Denlinger and Yocum 1998; Wolfe *et al.* 1998; Rinehart *et al.* 2000a; Yocum 2001), whereas 'programmed responses to cold' have no obvious equivalent high temperature response (Chown 2001). Where they occasionally do, such as in the case of aestivation, these latter responses seem closer to each other than either is to the rapid hardening response.

Notwithstanding this distinction, it is evident that both hardening and programmed responses to cold can be considered forms of acclimation (in the laboratory) or acclimatization (in the field). While the mechanisms underlying acclimation have long

fascinated physiologists (Kingsolver and Huey 1998), the evolutionary benefits (or lack) thereof have only recently been carefully examined with the advent of evolutionary physiology (Feder *et al.* 2000a). In consequence, the terminology associated with acclimation must now be interpreted under the broader evolutionary rubric of phenotypic plasticity (Huey and Berrigan 1996). Although there is considerable potential for confusion given the wide variety of uses that characterize the term 'acclimation' (Spicer and Gaston 1999), recent reviews have considerably clarified the terminology.

5.2.1 Acclimation

Terminology
In their discussion of the ways in which evolutionary hypotheses of acclimation can be rigorously tested, Huey and Berrigan (1996) present a useful framework for understanding phenotypic plasticity in the context of evolutionary physiology. In their terminology

(1) *phenotypic plasticity* refers to the malleability of an organism's phenotype in response to, or in anticipation of, environmental conditions experienced (or to be experienced) by the organism. Thus, acclimation responses (or acclimatization in the field) are examples of phenotypic plasticity in physiological traits;

(2) *the norm of reaction* (commonly also termed the reaction norm) is the form that the phenotypic effect takes;

(3) *developmental switches* are a subset of plastic responses and involve an irreversible change in the phenotype in response to environmental conditions experienced during a critical developmental phase;

(4) *cross-generational effects* are phenotypic modifications transmitted either maternally or paternally (e.g. Crill *et al.* 1996);

(5) *developmental pathologies* are environmentally induced pathological modifications of the phenotype that occur during development (e.g. Roberts and Feder 1999);

(6) *labile effects* are acute, rapid modifications of performance as a function of the organism's immediate environment. Performance curves are often used to illustrate these effects, and these

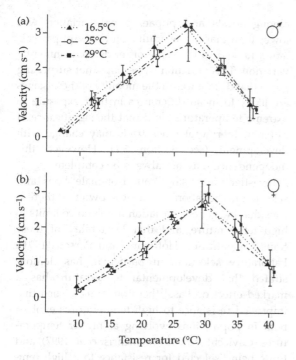

Figure 5.6 The relationship between walking speed and temperature in males and females of three lines of *Drosophila melanogaster* evolving at different temperatures.

Source: *Physiological Zoology*, Gilchrist *et al.* **70**, 403–414. © 1997 by The University of Chicago. All rights reserved. 0031-935X/97/7004-9686$03.00

curves can potentially be modified either by acclimation or by developmental switches (Fig. 5.6).

The evolution of performance curves is dealt with in various sections of this book, and has been reviewed in detail elsewhere (Huey and Kingsolver 1993; Gilchrist 1995). Nonetheless, it is worth noting that any discussion of physiological performance curves immediately shows up the difficulty, and artificial nature, of distinguishing between 'resistance' and 'capacity' adaptations. In some cases, the long- or short-term (plastic) responses of these curves might involve a change in critical limits, but not the optimum, while in others the shape, limits, and optimum values may all change in concert (Huey and Kingsolver 1993).

Beneficial acclimation?

Because physiologists have largely concentrated on the mechanisms underlying acclimation (Kingsolver and Huey 1998), they have often assumed that acclimation is advantageous for the organism concerned when it is subsequently exposed to the conditions under which it was acclimated. This beneficial acclimation hypothesis has recently been questioned on the basis mainly of laboratory selection trials using both *D. melanogaster* and *Escherichia coli* (see Huey and Berrigan 1996; Huey *et al.* 1999 for review). These investigations have not only revealed considerable complexity in the acclimation response, but have also shown that it is not necessarily adaptive. For example, in some cases there appears to be an optimal temperature which results in a phenotype that performs best over all conditions (Tantawy and Mallah 1961; Huey *et al.* 1999; Gibert *et al.* 2001). In addition, acclimation also appears to be associated with considerable physiological costs (Hoffmann 1995) (Section 5.2.2). However, beneficial acclimation has been demonstrated, at least partially, in some insect species not only in the laboratory (Scott *et al.* 1997; Huey *et al.* 1999; Hoffmann and Hewa-Kapuge 2000), but also in elegant field trials. Moreover, the field trials have not only assessed performance, but have also assessed the fitness of individuals characterized by a given phenotype, either by examining survival (Kingsolver 1995*a,b*, 1996), or by directly determining egg-laying ability (Thomson *et al.* 2001).

More recently, the question of the extent to which studies involving development at different temperatures, and under stressful conditions, represent tests of developmental plasticity rather than the beneficial acclimation hypothesis has been raised (Wilson and Franklin 2002). Moreover, Woods and Harrison (2002) argue that exposure to non-optimal temperatures will degrade performance in all environments (which can be called the deleterious acclimation hypothesis), and a focus on fitness as a whole rather than the contribution of the individual trait of interest, both bias tests against findings of beneficial acclimation. They suggest that focusing on the fitness consequences of individual traits is essential for tests of the beneficial acclimation hypothesis and its alternatives. These complexities of acclimation, and the realization that the extent and nature of acclimatization are likely to depend considerably on the degree and predictability of environmental variation, are cogent reminders of the need for a strong inference approach, involving multiple hypotheses, in evolutionary physiology (Huey and Berrigan 1996; Huey et al. 1999; Woods and Harrison 2002).

5.2.2 Heat shock

Basal thermotolerance

The temperature at which heat induces injury and/ or death varies both through space and in time. Because differences in methods of measurement usually assess different traits (e.g. knockdown resistance versus survival), resulting in dissimilar outcomes, it is difficult to reach general conclusions regarding upper thermotolerance limits. However, in insects they generally do not exceed about 53°C (Christian and Morton 1992), and are usually not much lower than 30°C, although these values depend on the trait being measured. There are examples of very low tolerance levels in some species such as alpine grylloblattids, and tolerance may increase dramatically in dormant, virtually anhydrobiotic, stages such as eggs. Ignoring these extreme values, substantial variation in tolerances remains. Geographic variation in thermotolerance levels, in the same direction as that of environmental temperature variation, is typical of both the species and population levels (Andrewartha and Birch 1954; Cloudsley-Thompson 1962; Stanley et al. 1980; Chen et al. 1990; Kimura et al. 1994; Goto et al. 2000), although variation among populations is sometimes less pronounced than that among species (Hercus et al. 2000).

However, the association between climatic conditions and tolerance is not always straightforward, especially because variation in thermotolerance often differs between developmental stages. For example, Krebs and Loeschcke (1995a) found that the rank order of resistance to high temperature stress in seven populations of *Drosophila buzzatii* differed among eggs, larvae, pupae, and adults. High resistance at one stage was not necessarily associated with high resistance at another. Likewise, Coyne et al. (1983) found that the nature of the variation in high temperature tolerance among populations of *D. pseudoobscura* depended on the stage being investigated. While there was significant interpopulation variation

among adults and pupae, pupal variation was lower than that in the adults, and only pupal variation ranked in the same order as environmental variation. This variation is, perhaps, not surprising given that differences in the mobility of these stages are likely to mean differences in their exposure to extreme temperatures, and that the significance of different thermotolerance traits may change with development (see Section 5.1). However, this independence may not always be complete.

Investigations using both isofemale lines and selection in the laboratory have shown that there is considerable genetic variation in tolerance limits to high temperature, and that heritability of these traits is significant (Hoffmann and Parsons 1991). Laboratory selection, in particular, has demonstrated that developmental temperature has a marked effect on basal thermotolerance, such that improved resistance to high temperatures evolves both in populations evolving at higher temperatures (Cavicchi et al. 1995; Gilchrist et al. 1997), and those being selected for resistance to a high temperature treatment (Huey et al. 1992; Hoffmann et al. 1997) (Fig. 5.7). While the majority of studies have concerned *Drosophila*, several other species display similar responses to selection (e.g. Baldwin 1954; White et al. 1970). Because heat shock survival and knockdown temperature are genetically uncoupled (Section 5.1.2), knockdown generally does not respond to laboratory natural selection (Gilchrist et al. 1997), but does show a considerable response to artificial selection for knockdown resistance (Hoffmann et al. 1997). Although the pronounced response to selection in laboratory populations may be a consequence of loss of resistance following adaptation to laboratory conditions (Harshman and Hoffmann 2000; Hoffmann et al. 2001a), the combination of geographic variation of thermal tolerance under natural conditions, and responses to selection in the laboratory suggest that basal thermotolerance is a heritable trait that varies considerably between species and responds strongly to selection.

The physiological basis of this variation in thermotolerance is much less clear. It has been suggested that constitutively expressed heat shock proteins (Hsps, see below) might be responsible for both survival of potentially lethal temperatures and for

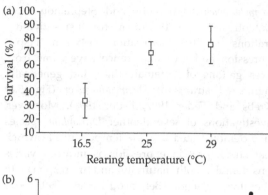

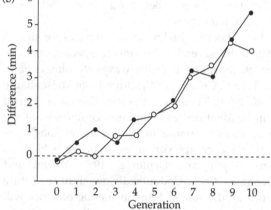

Figure 5.7 (a) *Drosophila melanogaster* adult survival (mean ± SD) of a 38°C treatment for 30 min, following laboratory natural selection at three different temperatures. (b) Responses of knockdown resistance to artificial selection in low rearing density (open circles) and high rearing density (closed circles) lines of *Drosophila melanogaster*.

Source: (a) is from Gilchrist *et al.* 1997 table 1 and (b) is redrawn from Bubli *et al.* 1998.

improved knockdown resistance (McColl *et al.* 1996; Gilchrist *et al.* 1997). Induction of Hsps following thermal stress might also aid in repair of damage (Krebs and Feder 1997). Alternatively, alterations in cell membrane composition (Gracey *et al.* 1996) or changes in allozymes or their concentrations might also be involved (Cossins and Bowler 1987; Somero *et al.* 1996). Because development at higher temperatures often results in an increase in basal thermotolerance but a decline in the response to hardening (Cavicchi *et al.* 1995; Bettencourt *et al.* 1999; Sørensen *et al.* 1999, 2001), one explanation for increased basal thermotolerance is the cost of a low-level, induced stress response (see below and Krebs and Feder 1997, 1998a; Zatsepina *et al.* 2001).

Continuous expression of heat shock proteins reduces survival and fecundity, inhibits growth, and thus affects development time (Krebs and Loeschcke 1994; Feder and Krebs 1998; Feder 1999). It also acts as a substrate sink, and interferes with cellular functioning (Zatsepina *et al.* 2001). In consequence, at high temperatures there would be a considerable premium for reduction of this response, and probably an increase in basal thermotolerance allowing the organisms to cope with what are otherwise potentially injurious temperatures. This basal thermotolerance may be a consequence of constitutively expressed Hsps (Lansing *et al.* 2000), the presence of osmolytes (Wolfe *et al.* 1998) or alterations in membranes and allozymes (Zatsepina *et al.* 2001).

On the other hand, constitutive resistance to high temperatures may not be maintained in individuals that do not routinely experience such temperatures because there are costs associated with supporting a biochemistry associated with these conditions. For example, Watt (1977, 1983) has shown that in *Colias* butterflies (Lepidoptera, Pieridae), adaptive adjustment of enzyme functional characteristics involves a trade-off of kinetic flexibility and efficacy at low temperatures against stability at high temperatures (see Section 6.4.1).

Heat injury

Despite variation in the temperature at which heat induces injury, the response to heat shock is similar in the majority of the insect species that have been examined. High temperature injury generally results from disruption of the function of membranes, especially synaptic membranes (Cossins and Bowler 1987), alterations in the cell micro-environment (e.g. pH), perturbation of protein structure, and DNA lesions (Somero 1995; Feder 1999). In turn, these changes affect development, muscular contraction, and several other processes at higher organizational levels (see Denlinger and Yocum 1998 for comprehensive discussion). The responses of membrane systems to temperature generally involve alterations in the composition of cellular lipids (Gracey *et al.* 1996; Somero *et al.* 1996), while expression of heat shock proteins, which act as molecular chaperones to proteins (Hendrick and Hartl 1993), is now recognized as one of the most widespread and conserved responses to stress, including thermal stress (Lindquist 1986; Feder and Hofmann 1999).

Intense thermal stress can perturb the structure of an organism's proteins. During normal cellular functioning proteins are generally folded, but may be unfolded during transport, synthesis of polypeptides, and assembly of multimeric proteins. Stress may also result in unfolding. In this unfolded state, exposed amino acid side groups, especially hydrophobic residues, can lead to interactions between these 'non-native' proteins and folded proteins, inducing the latter to unfold. The result is irreversible aggregations of unfolded proteins. These unfolded proteins reduce the cellular pool of functional proteins and may also be cytotoxic (Feder 1996, 1999; Feder and Hofmann 1999). Molecular chaperones interact with the unfolded proteins to minimize their harmful effects by binding to the exposed side groups, preventing unfolded proteins from interacting. In an ATP-dependent manner they also release the proteins so that they can fold properly, and may also target proteins for degradation or removal from the cell (Parsell and Lindquist 1993; Feder 1996). Stress proteins, or heat shock proteins, therefore, function as molecular chaperones and have been found in virtually all prokaryotes and eukaryotes (but see Hofmann *et al.* 2000). These heat shock proteins comprise several families that are recognized by their molecular weight, and include Hsp100, Hsp90, Hsp70, Hsp60, and a family of smaller proteins (Denlinger *et al.* 2001).

In insects, the best known of these families is Hsp70, especially because of its dramatic increase in *Drosophila* in response to high temperature stress. Indeed, the history of physiological and biochemical investigations of the role of Hsps in insects is essentially a history of the investigation of the role of Hsps in *Drosophila* and *Sarcophaga crassipalpis* (Diptera, Sarcophagidae) (Denlinger and Lee 1998; Denlinger and Yocum 1998; Feder and Krebs 1998). In the 1960s it was recognized that heat induces puffing of *Drosophila* chromosomes. Subsequently, much of the work demonstrating the importance of Hsps for thermal tolerance, both in the field and in the laboratory, has involved work on *Drosophila* species, and especially

D. melanogaster (Alahiotis and Stephanou 1982; Alahiotis 1983). In the main, conclusive demonstrations of the association between Hsp70 expression and thermotolerance have come from investigations of isofemale lines and genetically engineered strains of *D. melanogaster* (Fig. 5.8) (Krebs and Feder 1997; Feder 1999). Moreover, investigations of several other *Drosophila* species have demonstrated that the temperature at which heat shock protein expression is induced varies considerably, both naturally and in response to laboratory selection (Bettencourt *et al.* 1999; Feder and Hofmann 1999).

Heat shock is also known to induce expression of Hsp70 in several other insect species such as moths, ants, and parasitic wasps (Denlinger *et al.* 1991, 1992; Gehring and Wehner 1995; Maisonhaute *et al.* 1999). Therefore, it seems likely that Hsp70 will be identified as a common component of the heat shock response in most taxa, although the nature and complexity of the response is likely to vary (Joplin and Denlinger 1990; Yocum and Denlinger 1992). Undoubtedly, much of the future work on the role of Hsps in thermotolerance will involve investigation of induced thermotolerance

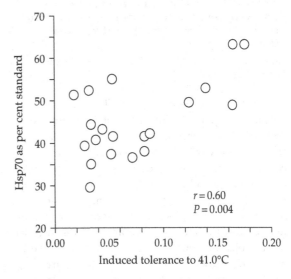

Figure 5.8 Relationship between induced thermotolerance and Hsp70 expression in isofemale lines of *Drosophila melanogaster*.

Source: Krebs and Feder (1997).

(or hardening), as many of the studies have done to date (Krebs and Bettencourt 1999).

Induced tolerance or hardening

It has long been appreciated that injury caused by high temperature can be ameliorated by prior exposure to a sublethal, or moderately high temperature (Hutchison and Maness 1979; Denlinger *et al.* 1991; Hoffmann and Watson 1993). This acclimation response lasts for several hours, but is nonetheless transient (Krebs and Loeschcke 1995*b*). Like basal tolerance, induced thermotolerance responds strongly both to artificial selection (Krebs and Loeschcke 1996), and to laboratory natural selection (Cavicchi *et al.* 1995), and it is clear that this trait shows considerable genetic variation (Loeschcke *et al.* 1994, 1997; Krebs and Loeschcke 1997). However, variation between populations is not always apparent (Hoffmann and Watson 1993; Krebs and Loeschcke 1995*a,b*), although this depends to some extent on the temperature of the heat shock. Likewise, induced thermotolerance differs substantially between life stages in some cases, but not in others (Krebs and Loeschcke 1995*a,b*).

Interspecific differences in induced tolerance also appear counterintuitive in the sense that species from warmer environments often have a reduced response to hardening compared to those from more temperate climates (Chen *et al.* 1990). A reduced response to high temperature acclimation has also been found in populations of *D. buzzatii*, with a population from a low altitude, warm environment, showing little or no response to acclimation compared with one from a cool, high-altitude area (Sørensen *et al.* 2001). Similar results have also been found in laboratory selection experiments. For example, Cavicchi *et al.* (1995) showed that a line of *D. melanogaster* reared at 28°C lost its capacity for induced thermotolerance compared with lines reared at lower temperatures, and this finding was substantiated by Bettencourt *et al.* (1999). This reduced hardening response is probably a consequence of the costs of acclimation, which largely entail a reduction in development rate, fecundity, and survival (Krebs and Loeschcke 1994; Hoffmann 1995; Krebs *et al.* 1998). At least in the case of heat shock, this cost appears to be associated with the expression of heat shock proteins.

Heat shock proteins: benefits and costs

Several studies of genetically engineered *D. melanogaster* have demonstrated both the costs and benefits of Hsp70 expression. The first set of studies involved two strains of *D. melanogaster* produced by Welte *et al.* (1993). One of the strains (excision) possesses the usual 10 copies of the *hsp70* gene, while the second strain (extra-copy) has an additional 12 transgenic copies. The strains differ in no other ways, and the possible effects of mutagenesis, that can make comparative work problematic, are largely controlled for, because of the way in which the strains were engineered (Feder and Krebs 1998).

After a 36°C pretreatment, the extra-copy strain shows both improved survival of 39°C and an increase in Hsp70 relative to the excision strain (Fig. 5.9) (Feder *et al.* 1996). This difference is particularly pronounced at pretreatment (hardening) temperatures that are either relatively low or relatively short in duration (Krebs and Feder 1998*a*). Once a reasonably high hardening temperature is reached (36°C) and the treatment is prolonged, the two strains show similar levels of Hsp70 expression. Thus, after the stage at which the capacity to express Hsp70 reaches full development, *hsp70* copy number no longer enhances inducible thermotolerance, despite differences in Hsp70 levels (Feder 1999; Tatar 1999). Nonetheless, it is clear that the possession of extra copies of *hsp70*, and consequently a higher level of Hsp70, enhances heat shock tolerance.

Further examination of the sites at which Hsp70 is important in providing protection against heat shock has revealed that the gut is especially sensitive to high temperature (Feder and Krebs 1998; Krebs and Feder 1998*a*). By using a second genetically engineered *D. melanogaster* mutant, *mths70a*, which causes the gut to express Hsp70 when reared on medium that contains 2 mM copper, Feder and Krebs (1998) showed that in the absence of hardening the mutant has considerably greater ingestion rates when reared on copper than when the metal is absent. However, with hardening and without heat shock, uptake rates are the same in both treatments (with and without copper). This provides clear evidence that Hsp70 has a considerable role in protecting the gut against heat shock. Using

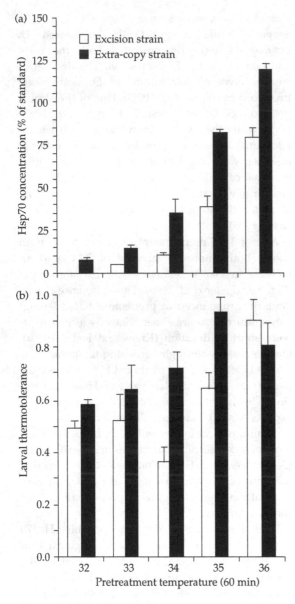

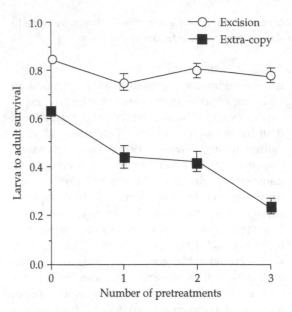

Figure 5.10 Larva to adult survival in excision and extra-copy strains of *Drosophila melanogaster* exposed to repeated hardening treatments, compared to a control strain held at 25°C.

Source: Feder and Krebs (1998).

Figure 5.9 (a) The effect of hardening temperature on Hsp70 accumulation and (b) larval thermotolerance (measured as survival of exposure to a high temperature) in excision and extra-copy strains of *Drosophila melanogaster*.

Source: Reprinted from *Journal of Insect Physiology*, **44**, Krebs and Feder, 1091–1101, © 1998, with permission from Elsevier.

the excision and extra-copy strains Feder and Krebs (1998) also demonstrated that Hsp70 expression promotes the recovery of alcohol dehydrogenase activity following heat shock.

The expression of Hsp70, and presumably its protective role, varies genetically and is conserved across life-cycle stages (Krebs *et al.* 1998). However, the relationship between Hsp70 and thermotolerance is not always readily discernible (Dahlgaard *et al.* 1998; Goto and Kimura 1998), and in some instances tolerance and Hsp70 expression appear to be relatively independent (Goto *et al.* 1998; Lansing *et al.* 2000).

Expression of Hsp70 is not only associated with benefits to the organism. Continuous elevation of Hsp70 levels results in a decline in growth and division of cells, a reduction in larva to adult survival in extra-copy *D. melanogaster* relative to the excision strain when larvae are repeatedly exposed to hardening temperatures (Fig. 5.10) (Feder and Krebs 1998; Feder 1999), and a decline in egg hatch (Silbermann and Tatar 2000). Likewise, in isofemale lines of wild-caught *D. melanogaster* there is a negative relationship between Hsp70 expression and survival at constant 25°C (Krebs and Feder 1997). The constitutive presence of Hsp70 can be harmful because it interferes with cell signalling pathways, impedes normal processing and degradation of unfolded proteins, or directs cellular machinery away from synthesizing other proteins to Hsp synthesis (Zatsepina *et al.* 2001). Although constitutive expression of Hsp70 might also represent a metabolic cost to the organism, so far this idea has not been supported (Krebs and Feder 1998*b*).

The trade-off between the costs and benefits of Hsp70 expression is the most likely explanation for a decline in the expression of Hsp70 and inducible thermotolerance when either laboratory strains (Sørensen *et al.* 1999; Lerman and Feder 2001) or wild populations (Sørensen *et al.* 2001; Zatsepina *et al.* 2001) evolve at high temperatures. These strains experience most of the deleterious consequences, but derive few benefits from Hsp70 expression because they rarely encounter potentially lethal high temperatures (Zatsepina *et al.* 2001). This results in downregulation of Hsp70 expression (Sørensen *et al.* 1999; Lansing *et al.* 2000), and consequently a decline in inducible thermotolerance, although basal thermotolerance generally increases. While the precise mechanism by which Hsp70 expression is reduced remains incompletely known, Zatsepina *et al.* (2001) have suggested that the insertion of two transposable elements, *H.M.S. Beagle* in the 87A7 *hsp70* gene cluster, and *Jockey* in the *hsp70Ba* gene promoter, are responsible for this reduced expression. In addition, there is some evidence that juvenile hormone plays an important role in the development of the stress response. Exposure of *D. melanogaster* to stress results in a decline in JH-hydrolysing activity (Gruntenko *et al.* 2000). Given that JH is known to control reproduction, the continued presence of JH may well explain alterations in fecundity seen in several species in response to stress. Rinehart and Denlinger (2000) also suggested that ecdysteroids may play an important role in the regulation of Hsp90, because ecdysteroids are absent throughout diapause, but are synthesized and released following diapause termination. The time course of these events and those of Hsp90 upregulation are therefore similar. Clearly, the regulation of the stress response by the hormonal system in insects deserves further attention (Denlinger *et al.* 2001).

Heat shock proteins are not only expressed in response to high temperature stress, but are characteristic of the response of insects to virtually all stresses (Burton *et al.* 1988; Goto *et al.* 1998; Feder and Hofmann 1999; Tammariello *et al.* 1999). For example, larval crowding of *Drosophila* also induces Hsp70 expression, and leads to correlated responses such as increases in the ability to survive heat shock and increases in longevity (Tatar 1999; Gruntenko *et al.* 2000; Sørensen and Loeschcke 2001). However, not all stresses that result in expression of heat shock proteins result in a generalized stress response. Tammariello *et al.* (1999) found that although desiccation in *S. crassipalpis* resulted in upregulation of *hsp23* and *hsp70*, this response was less dramatic than the upregulation found in response to heat shock. In consequence, desiccation failed to result in tolerance to high or low temperatures. Likewise, upregulation of *hsp70* has little effect on knockdown temperature (Sørensen *et al.* 2001; Sørensen and Loeschcke 2001), but this may be due to the independence of heat shock survival and knockdown survival (Section 5.2.2). In some species, Hsp70 may also be constitutively expressed and does not increase in response to heat stress, or if it does is sufficiently tissue-specific that overall levels do not change (Salvucci *et al.* 2000). Irrespective of this variation, the most significant questions that remain in Hsp research are whether the responses that have largely been induced in the laboratory are relevant to the field, and whether variation in Hsp expression is related to the variation in thermotolerance found between populations and between species.

Field temperatures, heat shock, and Hsps
The importance of determining the relevance of laboratory studies to the field situation was recognized by Feder and his colleagues early on in their investigation of heat shock in *Drosophila* species. In a series of studies they demonstrated that necrotic fruit in the sun could rapidly reach temperatures that were not only high enough to induce the heat shock response (Fig. 5.11) (Feder 1997), but were also either lethal to larvae and pupae (Feder *et al.* 1997*a*) (Fig. 5.4) or induced developmental defects in the emerging adults (Roberts and Feder 1999). These effects resulted largely from an inability of ovipositing females to respond to cues that might indicate past (and thus future) high fruit

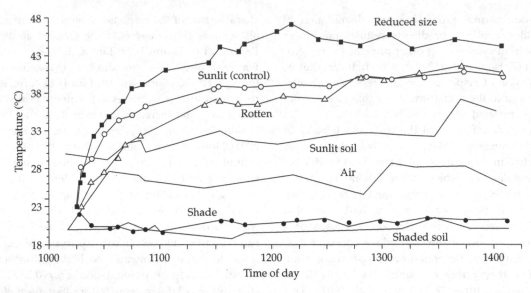

Figure 5.11 Temperatures of necrotic peaches exposed to full sunlight and to shade, and air temperature measured at the same time. *Source*: Feder *et al.* (1997). *Functional Ecology* **11**, 90–100, Blackwell Publishing.

temperatures (Feder *et al.* 1997*b*). Feder and colleagues also showed that under field conditions (or simulated field conditions) enhanced expression of Hsp70 associated with the transgenic extra-copy strains could increase larval survival (Roberts and Feder 2000) and reduce developmental abnormalities in emerging adults (Roberts and Feder 1999). Although it seems likely that behavioural avoidance reduces the exposure of adults to high temperature, some adults also show enhanced levels of Hsp70 expression in the field, indicating exposure to potentially lethal temperatures (Feder *et al.* 2000*b*).

In a remarkable investigation of *Chrysomela aeneicollis* (Coleoptera, Chrysomelidae), Dahlhoff and Rank (2000) showed that at low temperatures (20°C) the beetles do not express Hsp70, but that expression increases rapidly with temperatures above 24°C. Temperatures close to 35°C are routinely experienced by beetles in the field, and consequently Hsp70 expression in field-collected beetles is also high. However, this expression varies with both latitude and altitude such that field collected beetles from colder environments express lower levels of Hsp70. Moreover, the northern, low temperature population, which expresses considerably greater levels of Hsp70 under heat stress in the laboratory, has a phosphoglucose isomerase

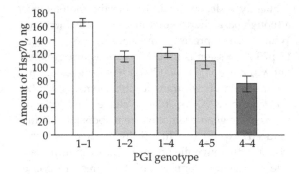

Figure 5.12 Correlation between phosphoglucose isomerase genotype and Hsp70 expression in *Chrysomela aeneicollis* (Coleoptera, Chrysomelidae).

Note: Genotypes with high Hsp70 expression are from cold environments.

Source: Dahlhoff and Rank. (2000). *Proceedings of the National Academy of Sciences of the USA* **97**, 10056–10061. © 2000 National Academy of Sciences, U.S.A.

(PGI) allele that is more temperature sensitive (i.e. subject to unfolding at lower temperatures) than the southern, high temperature population. Within a given population, PGI genotypes that are more thermally labile also express greater levels of Hsp70 than those that are less so (Fig. 5.12). This study, therefore, provides one of the first demonstrations of the ways in which changes in the molecular chaperone system promote survival of populations that require different genotypes to function efficiently under local environmental conditions. Recent work on *Drosophila* is expanding knowledge of the relationship between relevant genetic markers and clinal variation in high and low temperature stress resistance, and body size (Weeks *et al.* 2002).

Alternatives to Hsps
Thermotolerance may involve homeoviscous adaptation, metabolic rate depression, enhanced stability of proteins as a consequence of their structure (i.e. allelic changes), constitutively expressed and induced heat shock proteins, and the synthesis of protective osmolytes (Zatsepina *et al.* 2001). In aphids and whiteflies, the latter is particularly important, and involves the accumulation of polyhydric alcohols in response to high temperatures (Wolfe *et al.* 1998; Salvucci *et al.* 2000). Sorbitol accumulates to levels as high as 0.44 M within 3 h of exposure to high temperatures in *Bemisia argentifolii* (Hemiptera, Aleyrodidae), and appears to serve the same protective role as heat shock proteins in these insects (Wolfe *et al.* 1998; Salvucci 2000; Salvucci *et al.* 2000). That is, at physiological concentrations sorbitol increases the thermal stability of proteins by stabilizing their structure and preventing heat-induced aggregation, thus maintaining catalytic activity at high temperatures (Salvucci 2000). If the insects are deprived of nutrients, sorbitol production declines and heat shock proteins might assume greater importance in protecting proteins against thermal stress. Nonetheless, it appears that sorbitol is routinely produced as a rapid response to high temperature, via an unusual synthetic pathway involving fructose and an NADPH-dependent ketose reductase. Usually, sorbitol biosynthesis involves a substantially different pathway, is often slower than that found in whiteflies, and generally takes place as a longer-term response to cold (Storey and Storey 1991). However, in some insects, polyhydric alcohols are also produced during rapid cold hardening (Lee *et al.* 1987).

5.2.3 Cold shock

Cold shock, or direct chilling injury (Chen *et al.* 1987), is a form of injury that results from rapid cooling in the absence of extracellular ice formation. It increases in severity both with an increase in cooling rate and the absolute temperature of exposure, and is also thought to be a significant cause of injury during freezing. Direct chilling injury is usually distinguished from the consequences of a long-term exposure to low temperatures, which is known as indirect chilling injury. In both cases, the absence of extracellular ice formation distinguishes these kinds of injury from those associated with nucleation or freezing of an insect's body fluids. However, these two forms of injury are quite distinct, or at least in the way that insects respond to them.

Direct and indirect chilling injury
In *Frankliniella occidentalis* (Thysanoptera, Thripidae), rearing of individuals at 15°C (with or without cold hardening) prolongs the duration of their survival at −5°C compared to those maintained at 20°C prior to the −5°C exposure (Fig. 5.13). In contrast, hardening, or induced tolerance to cold, provides no improvement in survival of a prolonged −5°C stress irrespective of the temperatures at which the thrips were kept, but substantially improves survival of a rapid transition to −11.5°C (Fig. 5.14). Thus, McDonald *et al.* (1997) argued that these two physiological responses to cold are very

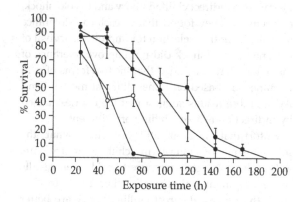

Figure 5.13 Increases in duration of survival (mean ± SE) of a −5°C treatment in *Frankliniella occidentalis* following rearing at 15°C compared to 20°C.

Note: The upper lines are data for hardened and non-hardened thrips at 15°C, whereas the lower lines represent thrips reared at 20°C.

Source: Reprinted from *Journal of Insect Physiology*, **43**, McDonald *et al.*, 759–766, © 1997, with permission from Elsevier.

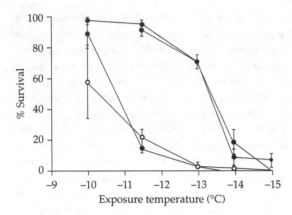

Figure 5.14 The improvement of survival (mean ± SE) of an acute low temperature treatment after hardening in *Frankliniella occidentalis* thrips reared at 15 and 20°C.

Note: The upper lines indicate thrips that were hardened and the lower lines thrips that were not pre-treated.

Source: Reprinted from *Journal of Insect Physiology*, **43**, McDonald *et al.*, 759–766, © 1997, with permission from Elsevier.

different, suggesting that the injuries themselves may well be distinct. A reversal in the relative abilities of *F. occidentalis* and another thrip species (*Thrips palmi*) to tolerate acute vs. chronic cold provides further support for this idea (McDonald *et al.* 2000). Chen and Walker (1994) reached a similar conclusion based on their investigation of the responses of *D. melanogaster* selected for greater tolerance to indirect chilling injury and to cold shock, respectively. They found that in each case tolerance increased, but that selection for improved survival of the one form of shock did not improve tolerance of the other. They concluded that the two forms of cold tolerance are based on rather different mechanisms. Again, these results suggest that the injuries caused by indirect and direct chilling are different.

Unfortunately, it is still not clear what the mechanisms of long-term chilling injury are, although prolonged exposure to cold often results in developmental abnormality (Lee 1991). In contrast, the causes of direct chilling injury are better known (Denlinger and Lee 1998; Ramløv 2000). Chilling induces fluid-to-gel phase transitions in membranes, which result in separation of membrane proteins and lipids, change membrane permeability, and cause a decline in the activity of membrane bound enzymes. This damage to the plasma membrane is thought to have a considerable effect

on neurons and on neuromuscular transmission (Hosler *et al.* 2000). Support for this idea comes from several sources, but in particular from demonstrations that muscle contraction patterns during eclosion are altered following cold shock in *S. crassipalpis* (Yocum *et al.* 1994), and that adult flies fail to behave in a normal way if they have been exposed to cold shock as pharate adults (Kelty *et al.* 1996). Direct chilling injury also results in a decrease in enzyme activity, in protein structural changes and denaturation (Ramløv 2000), and possibly also in an increase in oxidative stress. In house flies, cold resistance is associated with an increase in superoxide dismutase, an enzyme responsible for converting oxygen free radicals into hydrogen peroxide and hydroxyl radicals, which in turn are rendered less toxic by glutathione (Rojas and Leopold 1996). Glutathione levels decline during prolonged cold exposure, further supporting the idea that oxidative stress contributes to cold shock. At the whole organism level, chilling injury not only leads to an increase in mortality, but is also associated with a decline in reproductive output (Coulson and Bale 1992).

Basal and induced cold tolerance

The relationship between basal and induced cold tolerance has yet to be fully explored. Misener *et al.* (2001) found that in the presence of cycloheximide, a protein synthesis inhibitor, basal tolerance, that is, survival time of a −7°C cold shock following 25°C, is substantially reduced in *D. melanogaster* adults compared to untreated controls. By contrast, cycloheximide treatment has no effect on adult flies that were hardened at −4°C for 2 h prior to cold shock. These results suggest that basal and induced responses to cold are distinct.

In contrast, in both *D. melanogaster* and *D. simulans*, the response to artificial selection for cold tolerance is markedly reduced in flies hardened prior to cold shock compared to those that were not acclimated. On this basis Watson and Hoffmann (1996) argued that the selection and hardening responses are based on similar mechanisms. Similarly, a field-based study of *D. melanogaster* revealed that the magnitude of the acclimation response did not differ between populations from cold and warm areas, although flies from the colder area had both a greater basal tolerance to a −2°C cold exposure and a greater tolerance following hardening at 4°C (Hoffmann and Watson 1993). However, this response was not consistent, and differed when a cold stress of −5°C was applied. Thus, it is not entirely clear what the relationship is between basal and induced cold shock tolerance and how it varies under different conditions. What is certain is that short-term acclimation at a sublethal temperature can considerably enhance tolerance of cold shock.

Rapid cold hardening

Traditionally, the response of insects to cold has been considered in the context of winter cold hardening or the 'programmed response' to an extended period of cold. However, beginning with the demonstration of rapid cold hardening in *S. crassipalpis* and several other insects by Lee *et al.* (1987), it has now become clear that in a variety of insect species a short-term exposure to a sublethal cold temperature can substantially reduce cold shock mortality (Table 5.1).

In *S. crassipalpis*, chilling at 0°C for as little as 30 min can increase survival of a −10°C cold shock, although maximal protection usually follows a 2 h pretreatment (Lee *et al.* 1987) (Fig. 5.15). This pretreatment protection lasts for several hours although it declines within 20 days if flies are held at 0°C (Chen and Denlinger 1992). Curiously, exposure to an intermittent pulse of 15°C renews low temperature tolerance after 10 days at 0°C, suggesting that naturally occurring temperature cycles might be important for the maintenance of cold tolerance (see also Kelty and Lee 1999, 2001). Tolerance is, nonetheless, lost rapidly if the flies are returned to 25°C (Chen *et al.* 1991). Similar kinds of responses have been found in most of the other insects in which rapid cold hardening has been examined (Table 5.1). However, the rapid cold hardening response tends to decline with adult age and, like induced tolerance to high temperatures, can vary substantially between developmental stages (Czajka and Lee 1990). Generally, an increase in the chilling temperature reduces the hardening effect, although in several species it has become

Table 5.1 Insect species in which rapid cold hardening has been demonstrated

Species	Order	Family	Investigators
Dacus tyroni	Diptera	Tephritidae	Meats 1973
Sarcophaga crassipalpis	Diptera	Sarcophagidae	Lee *et al.* 1987
Xanthogaleruca luteola	Coleoptera	Chrysomelidae	Lee *et al.* 1987
Oncopeltus fasciatus	Hemiptera	Pyrrhocoridae	Lee *et al.* 1987
Drosophila melanogaster	Diptera	Drosophilidae	Czajka and Lee 1990
Musca domestica	Diptera	Muscidae	Coulson and Bale 1990
Sarcophaga bullata	Diptera	Sarcophagidae	Chen *et al.* 1990
Blaesoxipha plinthopyga	Diptera	Sarcophagidae	Chen *et al.* 1990
Drosophila simulans	Diptera	Drosophilidae	Hoffmann and Watson 1993
Culicoides variipennis	Diptera	Ceratopogonidae	Nunamaker 1993
Danaus plexippus	Lepidoptera	Danaeidae	Larsen and Lee 1994
Musca autumnalis	Diptera	Muscidae	Rosales *et al.* 1994
Spodoptera exigua	Lepidoptera	Noctuidae	Kim and Kim 1997
Frankliniella occidentalis	Thysanoptera	Thripidae	McDonald *et al.* 1997
Rhyzopertha dominica	Coleoptera	Bostrichidae	Burks and Hagstrum 1999
Cryptolestes ferrugineus	Coleoptera	Cucujidae	Burks and Hagstrum 1999
Oryzaephilus surinamensis	Coleoptera	Cucujidae	Burks and Hagstrum 1999
Sitophilus oryzae	Coleoptera	Curculionidae	Burks and Hagstrum 1999
Tribolium castaneum	Coleoptera	Tenebrionidae	Burks and Hagstrum 1999
Phytomyza ilicis	Diptera	Agromyzidae	Klok *et al.* 2003

Note: None of these species is tolerant of freezing in their extracellular fluids.

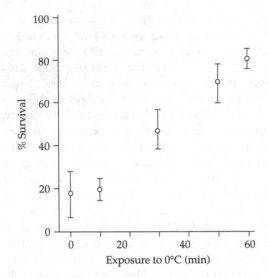

Figure 5.15 The induction of tolerance to a −10°C cold shock after short chilling pre-treatments at 0°C in *Sarcophaga crassipalpis*.

Source: Reprinted with permission from Lee *et al. Science* **238**, 1415–1417. © 1987, with permission from AAAS.

clear that a short pretreatment at a high temperature can also confer resistance to cold shock (Chen *et al.* 1991; Sinclair and Chown 2003). In consequence, it appears that the mechanisms providing protection against heat and cold are similar, although there is only partial overlap of these mechanisms.

At least in pharate adults of *S. crassipalpis*, rapid cold hardening is associated with a threefold increase in glycerol levels to 81.4 mM. Although this change is insufficient to have a colligative effect on cold hardiness (see Section 5.3.2), glycerol probably plays an important role in protecting membranes against low temperature damage associated with phase transitions (Lee *et al.* 1987). However, glycerol is not produced in response to brief pretreatment at a high temperature in this species (Chen *et al.* 1991), nor is it produced in response to cold shock in *D. melanogaster* and *Lymatria dispar* (Lepidoptera, Lymantriidae) (Yocum *et al.* 1991; Denlinger *et al.* 1992; Kelty and Lee 1999). Rather, in response to cold shock, these species, *S. crassipalpis*, several *Drosophila* species, and *Leptinotarsa decemlineata* (Coleoptera, Coccinellidae) upregulate heat shock protein synthesis, including the 92, 78, 75, 72, 70, 45, and 23 kDa proteins (Burton *et al.* 1988; Denlinger *et al.* 1992;

Denlinger and Lee 1998; Goto and Kimura 1998; Yocum *et al.* 1998; Yocum 2001). It is likely that at low temperatures these molecular chaperones fulfil a role similar to the one they assume at high temperatures, by providing chaperoning functions, and removing proteins denatured by low temperature stress. Their protective role must also extend further because both thermotolerance and rapid cold hardening provide protection from the negative effects of high and low temperature shock on fecundity in *S. crassipalpis* (Rinehart *et al.* 2000b).

The relevance of rapid cold hardening to the field situation is only now being explored. Kelty and Lee (2001) demonstrated that during thermoperiodic cycles identical to those likely to be experienced in the field, *D. melanogaster* demonstrates rapid cold hardening during the cooling phase of the cycles that is only partly lost during the subsequent warming phase. Moreover, during subsequent cooling phases tolerance of low temperatures is further improved. However, it is not only survival of a cold shock that improves following rapid cold hardening, but also the ability of flies to tolerate low temperatures that normally cause knockdown, although this improvement ceases after the first thermal cycle. Kelty and Lee (2001) argued that rapid cold hardening provides subtle benefits in the field, allowing adult flies to remain active for longer than otherwise would be possible. It has also been suggested that rapid cold hardening in Antarctic springtails and mites contributes to survival of what would otherwise be lethal temperatures (Worland and Convey 2001; Sinclair *et al.* 2003a). However, in this case it is the supercooling point (SCP) that is altered, rather than a lethal temperature above the SCP, as is found in all of the freezing intolerant insects examined to date.

5.2.4 Relationships between heat and cold shock responses

Given that both polyols and heat shock proteins are expressed in response to cold and heat shock, and that pretreatment at a high temperature also increases tolerance of cold shock, it is tempting to imagine that the responses to heat and cold are identical, or at least very similar. However, this cross-tolerance is only partial and is asymmetric in the sense that cold hardening does not generally improve survival of heat shock. Moreover, cold tolerance can be induced with as little as a 10 min exposure to 0°C in *S. crassipalpis*, whereas a 30 min exposure to 40°C is required to provide protection against injury at 45°C (Chen *et al.* 1991).

Furthermore, although heat shock proteins are synthesized rapidly in response to both cold shock and heat shock, there are considerable differences between these two responses. First, the time course of the response differs. That is, during heat shock, Hsps are synthesized during the stress, while in response to cold shock Hsps are only produced once the animals have been returned to a higher temperature (Goto and Kimura 1998; Rinehart *et al.* 2000*a*). This suggests that heat shock proteins do not contribute directly to rapid cold hardening, but provide protection against low temperature injury (Denlinger and Lee 1998, but see also Minois 2001). Second, the duration of the response differs dramatically between the two forms of shock. Usually, synthesis of Hsps in response to high temperature is brief and ceases almost immediately on cessation of the stress (Yocum and Denlinger 1992), while in response to low temperature, Hsp synthesis may continue for days (Yocum *et al.* 1991). Third, during heat shock, normal protein synthesis is almost entirely replaced by stress protein synthesis, whereas following a cold shock normal protein synthesis and the production of stress proteins occur concurrently.

The fact that the responses to cold and heat shock are different is also illustrated by a comparison of related species from geographically disjunct areas that have dissimilar climates. Although several such studies have been undertaken (e.g. Goto and Kimura 1998), the comparison of flesh flies from tropical and temperate areas made by Chen *et al.* (1990) is one of the most comprehensive. While all the species show an inducible tolerance to heat shock, only the species from temperate and alpine areas show rapid cold hardening (Fig. 5.16). As might be expected, basal tolerance of cold is greater in the temperate and alpine species than in the tropical ones, but this is true also of basal heat tolerance. Although this appears somewhat unusual, it should be kept in mind that mid-latitude areas are often characterized by very high temperatures

(Sømme 1995), and that global variation of absolute maximum temperatures is much less than that of absolute minima (Section 5.4.2).

Clearly, the relationship between tolerance of temperature extremes and the synthesis of heat shock proteins is intricate (Goto *et al.* 1998). One of the reasons for this complexity undoubtedly emerges from the fact that stress proteins are involved in many different processes in the cell and are not synthesized only as a response to stress. In insects this has been most clearly illustrated in investigations of Hsp regulation during diapause (see Denlinger 2002 for a recent review). Expression of *hsp70* and *hsp23* certainly do contribute to thermotolerance, as has been shown by their almost immediate expression on entry into diapause in *S. crassipalpis*, where cold tolerance and diapause are linked (Joplin *et al.* 1990), but delayed expression in *L. dispar*, where cold tolerance and diapause are developmentally separated (Denlinger *et al.* 1992). However, *hsp70* is also upregulated on entry into diapause in *S. crassipalpis* in the absence of any stress (Rinehart *et al.* 2000*a*). Likewise, *hsp23* is upregulated in response to heat and cold shock in non-diapausing *S. crassipalpis*, whereas in diapausing individuals stress does not cause greater expression of *hsp23* which is highly upregulated at the onset of diapause (Yocum *et al.* 1998). In contrast, *hsp90* is downregulated during diapause, but continues to respond to heat and cold shock (Rinehart and Denlinger 2000).

Given that the continued expression of heat shock proteins is known to be deleterious (Section 5.2.2), their continued upregulation during diapause initially appears remarkable. However, cell cycle arrest plays an important role in diapause in *S. crassipalpis*. Therefore, if the majority of negative effects of Hsp expression have to do with reduced cellular growth and differentiation, Hsps may have little adverse effect during diapause and may even assist in the maintenance of diapause (Yocum *et al.* 1998; Rinehart *et al.* 2000*a*), as well as serving to protect diapausing individuals from thermal and other stresses. The downregulation of *hsp90* at the onset of diapause, and its upregulation following diapause termination, or in response to heat or cold shock, is also readily comprehensible within this framework. Hsp90 keeps unstable proteins ready

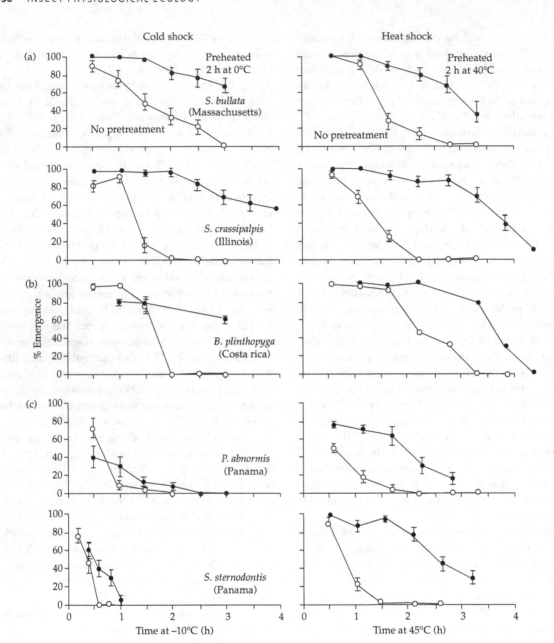

Figure 5.16 The effects of exposure to −10°C or 45°C either following direct transfer from 25°C or following a 2-h treatment at 0°C and 40°C, respectively, on flesh flies from (a) temperate, (b) alpine, and (c) tropical environments.

Source: Chen *et al.* (1990). *Journal of Comparative Physiology B* **160**, 543–547, Fig. 1. © Springer.

for activation until they are stabilized during signal transduction (Rutherford and Lindquist 1998). Thus, given relative cell inactivity during diapause, Hsp90 is unlikely to be required, but because of its ability to stabilize proteins, it remains responsive to

thermal stress. During the non-diapausing state this responsiveness to denatured proteins may also be the cause of the expression of phenocopies, or developmental abnormalities that resemble specific mutations (Denlinger and Yocum 1998), in response to heat shock. Rutherford and Lindquist (1998) have pointed out that thermal stress can divert Hsp90 from its normal functions to partially denatured proteins, thus compromising the ability of Hsp90 to buffer developmental variation. In consequence, the widespread variation in morpho-genetic pathways that is usually suppressed during development can now be expressed. Selection can subsequently lead to expression of these traits even when Hsp90 function is regained, thus leading to considerable evolutionary change in what are otherwise highly conserved developmental processes.

5.3 Programmed responses to cold

The preparation of insects for a cold, winter period has traditionally been the primary focus of research on the response of insects to low temperatures. As we have mentioned previously, this can be considered a long-term (weeks to months) programmed response to declining temperatures, prolonged periods of cold, or, less commonly, to changing photoperiod or dietary cues. Although this programmed response to cold is not always associated with diapause, the two 'programmes' are often intimately related, and this relationship (or the lack thereof) has been explored in detail by Denlinger (1991, see also Denlinger and Lee 1998; Denlinger 2002).

It is widely appreciated that the response of insects to cold is complex, and differs between diapausing and non-diapausing individuals (Lee and Denlinger 1985), individuals at different times of the year (van der Merwe et al. 1997), and in different ontogenetic stages (Vernon and Vannier 1996; McDonald et al. 2000; Klok and Chown 2001), and between populations in different years (Kukal and Duman 1989). Nonetheless, in their response to temperatures below the melting point of their body fluids, insects have regularly been classified either as freeze intolerant (freeze avoiding, freeze susceptible), or freezing tolerant (freeze tolerant) (Lee 1991; Sømme 1999). The former group of species cannot survive the formation of ice within their bodies, and therefore have evolved a suite of measures to prevent ice formation. In contrast, freezing tolerant species can withstand ice formation,

usually only in the extracellular fluids, and in turn have a suite of characteristics that enables them to survive such ice formation. Although variation about these 'strategies' has long been recognized, and has become a recurrent theme in recent reviews, this classification system is still widely adopted, though in a considerably modified form. Perhaps the most significant change to these categories has been the addition of cryoprotective dehydration as a third strategy by which insects can survive subzero temperatures (Holmstrup et al. 2002; Sinclair et al. 2003b). In this instance, the few species that adopt this strategy lose water to the surrounding environment, so resulting in an increase in the concentration of their body fluids and a decline in their melting point (to equilibration with the ambient temperature). In effect, they cannot freeze.

In this section, we examine the modified cold hardiness classification, as well as the characteristics of each of the major classes of cold hardiness. In doing so, we recognize that the field of insect cold hardiness has grown substantially since Asahina and Salt's work in the 1960s (Asahina 1969; Ring and Riegert 1991), and that even a review of the recent reviews amounts to a formidable task. Thus, we provide only a brief overview of the major characteristics of freeze intolerant and freezing tolerant species. Additional information can be found in several large reviews and books (Sømme 1982, 1999; Zachariassen 1985; Cannon and Block 1988; Block 1990; Lee and Denlinger 1991; Storey and Storey 1996; Denlinger and Lee 1998; Lee and Costanzo 1998; Ramløv 2000; Duman 2001).

5.3.1 Cold hardiness classifications

The classification of insect cold hardiness strategies has essentially revolved around whether the temperature at which ice nucleation, or freezing of the body fluids, starts to take place in the extracellular spaces (the crystallization temperature, T_c, or SCP) represents the LLT for the insect. If it does not, then clearly the animal has some measure of freezing tolerance, whereas if the SCP and LLT are equivalent, the animal is freeze intolerant. In Section 5.2.3, we showed that in many species the LLT is much higher than the SCP, and thus there is considerable pre-freeze mortality. Likewise, in Section 5.1.2 we have pointed out that the period over which an insect is exposed to subzero temperatures also has an influence on its ability to survive these conditions. It has been shown that some species can only tolerate partial extracellular ice formation (i.e. they die if ice formation goes to equilibrium) (Sinclair *et al.* 1999), others can tolerate freezing even though they have very low SCPs (Ring 1982), and yet others can switch between a strategy of freezing tolerance and freeze intolerance (Kukal and Duman 1989). Bale (1993, 1996) first proposed that the dichotomous, freezing tolerant–freeze intolerant classification should be modified to encompass the complexity of the response of insects to subzero temperatures. Recognizing the importance of prefreeze mortality and exposure time, Bale (1993, 1996) proposed that the following classes of cold hardiness should be recognized within the freeze intolerant category (Fig. 5.17):

1. *Freeze avoiding* species show little to no mortality at temperatures above their SCPs, even after exposure over a full winter season, and include

species such as the goldenrod gall moth *Epiblema scudderiana* (Lepidoptera, Olethreutidae).

2. *Highly chill tolerant* species can survive prolonged exposure to low, subzero temperatures but there is some mortality above the SCP. In the Antarctic springtail, *Cryptopygus antarcticus*, the SCP is approximately −25°C in winter, but more than 20 per cent of the population does not survive winter cold.

3. *Moderately chill tolerant* species also have a relatively low SCP, but here survival at low temperatures above the SCP is low. As an example of this class Bale (1993) used *Rhynchaenus fagi* (Coleoptera, Curculionidae), which has an SCP of −25°C in winter, but which shows less than 30 per cent survival after 50 days at −15°C.

4. *Chill susceptible* species can survive temperatures below 0–5°C, but show considerable mortality after very brief exposures to relatively high subzero temperatures. For example, *Myzus persicae* (Hemiptera, Aphididae) has an SCP of c.−25°C, but mortality increases from 0 to 100 per cent after just a few minutes below −5°C.

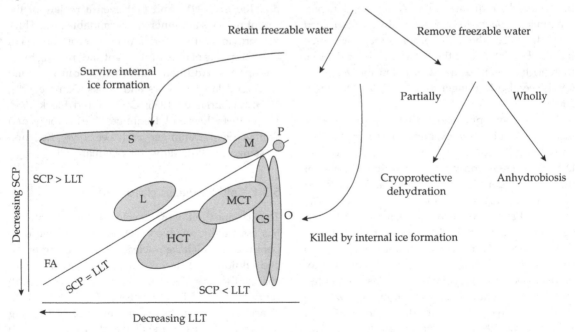

Figure 5.17 Schematic diagram illustrating the classes of cold hardiness characteristic of insects.

Note: CS chill susceptible, FA freeze avoiding, HCT highly chill tolerant, L freezing tolerant with low SCP, M moderate freezing tolerance, O opportunistic, P partial freezing tolerance, S strong freezing tolerance.

Source: Modified from Sinclair (1999).

5. *Opportunistic* species generally show considerable mortality at or just above 0°C when exposed to such temperatures for more than a few days. In environments with cold winters, these species usually avoid low temperatures by seeking thermal refugia.

Clearly, the species discussed in Section 5.2.3 that show rapid cold hardening, and an upregulation of polyhydric alcohols and heat shock proteins in response to cold, fall somewhere between the chill susceptible and moderately chill tolerant species. In addition, this classification illustrates the necessity of understanding the microenvironment within which an insect overwinters (Bale 1987; Sinclair 2001*a*).

Although there has been some debate regarding these new classes of cold hardiness, their utility has generally been widely recognized, especially with regard to better understanding the responses of insects to their natural environments. Indeed, in their investigation of a freezing tolerant caterpillar that dies at relatively high temperatures, Klok and Chown (1997) suggested that this freezing tolerant category could also usefully be broadened to recognize the variety of responses characterizing freezing tolerant species. Sinclair (1999) did just that. He recognized that some species survive only partial ice formation, others die only a few degrees below their SCP, while others can tolerate many degrees of freezing. Based on a quantitative investigation of the relationship between the SCP and LLT Sinclair (1999) recognized four classes of freezing tolerance:

1. *Partial freezing tolerance.* Here, individuals can survive some formation of ice in their bodies, but do not survive if ice formation goes to equilibrium at or above the SCP. This class is difficult to distinguish from freeze avoidance, because mortality may have a similar cause. A New Zealand lowland weta, *Hemideina thoracica* (Orthoptera, Anostostomatidae), is one example of such a species.
2. *Moderate freezing tolerance.* These species freeze at a relatively high temperature and die less than 10°C below their SCP, in some instances can survive relatively long periods in a frozen condition, and generally occur in relatively mild climates. *Pringleophaga marioni* (Lepidoptera, Tineidae)

caterpillars from sub-Antarctic Marion Island have a mean SCP of −5°C and show 100 per cent mortality at −12.5°C (Klok and Chown 1997).
3. *Strong freezing tolerance.* In these species the LLT is substantially lower than the SCP. These include the 'classic' freezing tolerant species such as the goldenrod gall fly, *Eurosta solidaginis* (Diptera, Tephritidae), that freezes at −10°C, but dies below −50°C, and Arctic carabids that freeze at −10°C, but can survive temperatures lower than −80°C (Miller 1982).
4. *Freezing tolerant with low SCP.* These insects have extremely low SCPs, yet, can survive freezing a few degrees below their SCP. For example, *Pytho deplanatus* (Coleoptera, Pythidae) has a SCP of −54°C, but can survive freezing down to −55°C (Ring 1982).

The combination of Bale's (1993) and Sinclair's (1999) categories of programmed responses to cold (Fig. 5.17) provides a useful classification against which to assess the strategy of any given insect species. Although it might be argued that investigation of many more insect species might simply result in a cloud of points on Fig. 5.17, we are of the opinion that there are unlikely to be five million ways in which insects respond to cold (see Chapter 1).

Having recognized the continuum of classes of insect cold hardiness we now explore the characteristics of the two traditional categories, recognizing that there are likely to be characteristics that are unique to each of the nine classes identified above, and which are also shared between them. We limit our discussion of the third strategy, cryoprotective dehydration, to a small section owing to the fact that it has only been recorded, among insects, in the collembolan *Onychiurus arcticus* (Worland *et al.* 1998; Holmstrup *et al.* 2002).

5.3.2 Freeze intolerance

Freeze-intolerant species have been known for a considerable period (see Salt 1961). In part, these species rely on the fact that small volumes of water can be cooled well below their melting point before spontaneous nucleation, or freezing, takes place (Lee 1989). Although this implies a relationship between volume and SCP (Lee and Costanzo 1998), in practice several other factors mean that this relationship is often not found. In preparation for winter, the SCPs decline in freeze-intolerant species, often over a period of several weeks (Rickards *et al.* 1987) (Fig. 5.18). This increase in supercooling capacity is partially the result of the removal of ice nucleating agents (INAs) from the gut and haemolymph and various tissues, and partly the result of the accumulation of either one or several low

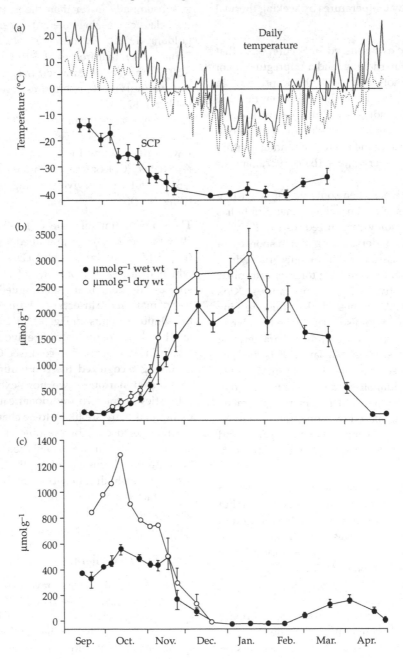

Figure 5.18 Alterations in (a) SCP, (b) glycerol content, and (c) glycogen content in larvae of the freeze-intolerant moth *Epiblema scudderiana* over winter.

Source: *Insects at Low Temperatures*, 1991, pp. 64–93, Storey and Storey, with kind permission of Kluwer Academic Publishers.

molecular weight cryoprotectants, although dehydration (i.e. reduction in water content) may also be important (Zachariassen 1985; Block 1990; Ramløv 2000).

Cryoprotectants

The low molecular weight cryoprotectants include polyhydric alcohols (polyols), such as glycerol, sorbitol, mannitol, threitol (Sømme 1982), sugars, such as trehalose and occasionally glucose and fructose, and amino acids, such as proline (Duman *et al.* 1991; Ramløv 2000). The polyols often increase to molar levels over winter, and in the beetle *Ips acuminatus* (Coleoptera, Scolytidae), molar levels of ethylene glycol, a compound toxic to humans, can be found (Gehrken 1984). These low molecular weight cryoprotectants either stabilize membranes and proteins, or act in a colligative manner to prevent ice formation, that is, they depress the freezing point in a way that is related to the number of molecules in solution (Ramløv 2000). In general, they decrease the melting point by $1.86°C \, molal^{-1}$ and the SCP by approximately twice that amount (Zachariassen 1985). To effectively control ice formation, these compounds are soluble in aqueous solution, do not perturb protein structure, counteract the denaturing effects on proteins of high ionic concentrations, cold and dehydration, and are non-toxic and non-reactive even at high concentrations (Ramløv 2000). The synthesis and breakdown of these low molecular weight substances has been reviewed in detail by Storey (Storey and Storey 1991, 1996; Storey 1997).

Antifreeze proteins

Several freeze-intolerant species also produce antifreeze proteins (AFPs, or thermal hysteresis proteins) (Duman 2001; Walker *et al.* 2001). These proteins lower the non-equilibrium freezing point of water, but do not affect the melting point. In consequence, the melting and freezing points differ, and this is known as thermal hysteresis. Although AFPs have been described from about 40 species of insects, the sequences of AFPs from only three insect species have been published, and their structures are discussed in detail by Duman (2001). Antifreeze proteins increase the extent of supercooling in a non-colligative manner (see Duman 2001 for a description of the mechanism) and may also inactivate various ice nucleators (Ramløv 2000; Duman 2001). In addition, they are also important in preventing inoculative freezing (i.e. contact with external ice resulting in spontaneous nucleation). Although the wax-coated, hydrophobic cuticle is thought to provide an effective barrier to inoculative freezing, many instances are known where this barrier has been penetrated (Ramløv 2000; Duman 2001), and thus where external contact with moisture can increase the SCP (Fig. 5.19). Antifreeze proteins are accumulated in the haemolymph and in epidermal cells, and together with cuticular changes play a large role in preventing inoculative freezing (Duman 2001).

Inoculative freezing

While these changes may be sufficient to prevent inoculative freezing in some species, in others the

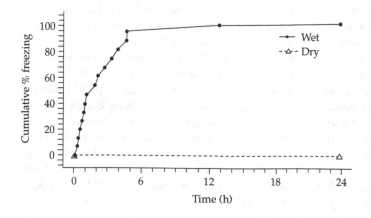

Figure 5.19 Duration of supercooling in wet and dry (lower line) monarch butterflies.

Note: The wet butterflies freeze spontaneously as a consequence of external inoculation.

Source: Reprinted from *Journal of Insect Physiology*, **40**, Larsen and Lee, 859–864, © (1994), with permission from Elsevier.

presence of surface bacteria (see below) may greatly enhance inoculative freezing, which can readily be seen via an increase in whole-body SCPs, and it appears that in this case inoculation takes place via the spiracles (Lee *et al.* 1998). Inoculative freezing as a result of host plant contact has also been observed in several phytophagous species (review in Klok and Chown 1998*a,b*). Generally, the insect's SCP is increased as a result of host plant contact, leading to elevated mortality, and at least in *Embryonopsis halticella* (Lepidoptera, Yponomeutidae), this effect is more marked at lower temperatures. However, host plant contact sometimes leads to a reduction in insect mortality (Butts *et al.* 1997), and, paradoxically, by providing protection from free water, and consequently ice crystals, mining or ensheathed feeding may promote survival of insects at temperatures above the SCP of their hosts (Connor and Taverner 1997; Klok and Chown 1998*b*).

External inoculation is not the only route by which heterogeneous ice nucleation can take place. Ice nucleators in the tissues, haemolymph, and gut lumen can also initiate freezing, and consequently these nucleators are removed or masked prior to the onset of winter (Zachariassen 1985). Several studies have now shown that haemolymph protein ice nucleators and lipoprotein ice nucleators are either reduced in quantity or removed during winter. In those cases where they are reduced, AFPs mask the ice nucleator activity, thereby permitting supercooling (Duman 2001).

The role of nucleators in the gut and the effect of their removal, via gut clearance, prior to winter have not been clearly defined. Although it has long been known that the presence of food (or its contaminants) in the gut can initiate freezing (Salt 1961), and that ingested gram-negative bacteria within the Pseudomonadaceae and Entereobacteriaceae can cause considerable elevation of the SCP (Denlinger and Lee 1998; Lee *et al.* 1998), the effect of gut clearance on SCPs is controversial. In several species, the presence of food in the gut has a marked effect on SCPs, which is often removed if the animals are starved or are in a stage where feeding does not occur (Cannon and Block 1988; Duman *et al.* 1991). Moreover, in species where starvation does not alter SCPs it has often been shown that individuals do not completely evacuate their guts

(Parish and Bale 1990; Klok and Chown 1998*b*). By contrast, Baust and Rojas (1985) point out that many species do not evacuate their guts prior to overwintering, that in some cases diet manipulations only have short-term effects on SCPs, that starvation or the absence of food in the gut sometimes has no effect on supercooling capacity, and that peritrophic membranes might prevent the nucleator potential of the gut contents from being realized. They also note that nucleation may take place elsewhere in the body, or via external contact with ice, and that water consumption can cause a decrease in supercooling capacity. Together these apparently contradictory results suggest that nucleation can be initiated in a variety of sites and that the process is likely to be complicated by the life histories of the species involved. For instance, in species that overwinter in the soil, such as *L. decemlineata*, soil water content has a significant influence on insect water content, which in turn affects both the SCP and mortality (Costanzo *et al.* 1997). Moreover, the physico-chemical attributes of the soil also have a pronounced influence on mortality. Nonetheless, it is clear that the gut and its contents have a significant role to play in nucleation, at least in some species.

This role has been best demonstrated in springtails, and particularly the Antarctic springtail *C. antarcticus*. In this species, in summer, fieldcollected individuals, and individuals fed moss turf homogenate, SCP frequency distributions are essentially high and unimodal, if somewhat left-skewed (Fig. 5.20) (Sømme and Block 1982). If individuals are starved, the SCP distributions either become bimodal or have a unimodal distribution at low temperatures. Such changes also occur over a seasonal basis in other springtail species (Sømme and Block 1991). Although bimodal SCP distributions have generally been ascribed to the two stage process of gut clearance followed by the development of a cryoprotectant system in preparation for winter cold, they have also been documented in response to changing acclimation temperatures in species where gut clearance does not take place (Klok and Chown 1998*b*). While it has been suggested that in summer-acclimated or acclimatized individuals this bimodality would disappear, its underlying causes have not been carefully

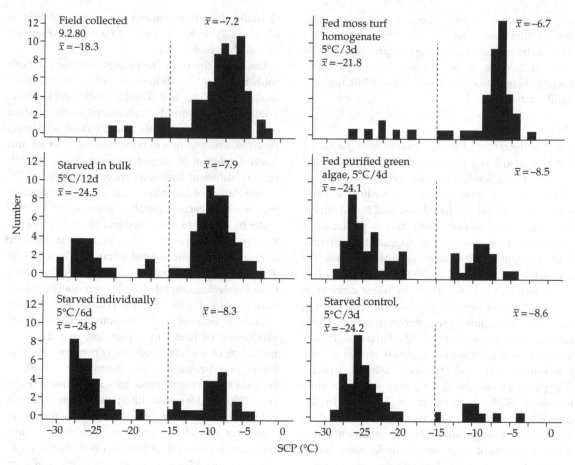

Figure 5.20 The effect of feeding on SCP distributions in the Antarctic springtail *Cryptopygus antarcticus.*
Source: Sømme and Block (1982). *Oikos* **38**, 168–176, Blackwell Publishing.

explored. Indeed, in at least one species, *Hydromedion sparsutum* (Coleoptera, Perimylopidae) it appears that bimodality in SCP distributions represents either a transition from a summer strategy of moderate freezing tolerance to a winter one of moderate chill tolerance, or a bet-hedging strategy, resulting in freeze intolerant and freezing tolerant individuals, in response to continual freeze–thaw cycles (Bale *et al.* 2001).

Supercooling point frequency distributions
Irrespective of the cause of the bimodality of insect SCP distributions, it is often considered a stumbling block for conventional statistical analyses (Worland and Convey 2001). Block (1982) suggested that the

bimodal distributions can be arbitrarily divided into high and low groups, and that the ratio of the numbers of individuals in each of these groups provides a useful measure of bimodality. Similarly, Rothery and Block (1992) provided a method for quantifying the variation in SCPs once a bimodal distribution has been divided into two groups. Unfortunately, neither method allows the significance of the bimodality to be determined. In contrast, and for a somewhat different purpose, Tokeshi (1992) developed a method for identifying the significance of bimodality in frequency distributions, and a means for determining the modal values for each of the groups in the distribution. In a similar vein, the transformations that are used for normalizing highly skewed macroecological data (Williamson and Gaston 1999), might also prove to be useful for allowing comparisons of SCPs between two or more samples where SCP frequency distributions are not bimodal but retain a significant skew.

The role of water

Like the ingestion of some kinds of food particles, drinking also has a significant effect on SCP, often causing a wholesale elevation of the SCP (Block 1996). Given that there is a relationship between water volume and the likelihood that it will supercool (Lee and Costanzo 1998), this is perhaps not surprising. However, it does suggest that altering water content might contribute substantially to the avoidance of freezing. Indeed, it is now widely appreciated that a reduction in water content is characteristic of several freeze-intolerant species, and that this decline in water content causes a substantial lowering of the SCP (Zachariassen 1991b). Such a lowering of water content is typical of both freeze-intolerant and freezing tolerant species (Ring and Danks 1994), and may serve not only to lower SCPs, because of an increase in the concentration of cryoprotectants, but also to reduce mechanical damage caused by ice formation and to limit the extent of water lost by desiccation during extended cold periods. Extreme desiccation resistance in *Embryonopsis halticella* larvae may be responsible for their very low SCP, which is substantially lower than the environmental temperatures they are likely to experience (Klok and Chown 1998b).

Lundheim and Zachariassen (1993) have shown that the cold hardiness strategy adopted by an insect is likely to have a considerable influence on its ability to resist desiccation. Their rationale is straightforward. In insects that overwinter where they are exposed to ice, the haemolymph of frozen individuals will be concentrated by the formation of ice until the vapour pressure of the liquid fraction is equivalent to that of ice at the same temperature. In consequence, no water loss will take place. Insects that do not freeze, but that supercool, will not be in vapour pressure equilibrium with the ice, and will continue to lose water. Thus, freezing tolerant insects should be more desiccation resistant than freeze-intolerant ones, and this is often found to be the case (Ring 1982; Lundheim and Zachariassen 1993).

The similarity of the responses of insects to both cold and dry conditions is being increasingly recognized (Ring and Danks 1994; Block 1996). The basic biochemical mechanisms (accumulation of sugars and/or polyols and thermal hysteresis proteins) are common to both freezing tolerant and freeze-intolerant strategies (Section 5.3.4), though serving different functions in each. Similarly, the accumulation of low molecular weight compounds has been implicated in the absorption of atmospheric water by Collembola (Bayley and Holmstrup 1999), and in the protection of cells against osmotic damage during extreme dehydration (Danks 2000). Thermal hysteresis proteins are also well known from the desiccation tolerant, but generally not cold hardy beetle *Tenebrio molitor* (Walker *et al.* 2001). Although similarities in responses, such as the production of heat shock proteins, may be more indicative of a generalized stress response because levels of expression differ (Goto *et al.* 1998), or because there is little cross-tolerance (Tammariello *et al.* 1999), it does seem likely that the response to cold hardiness may represent upregulation of pathways already extant for protection against desiccation (Storey and Storey 1996). Similarity in injuries caused by low temperature, especially ice formation, and by desiccation provide additional support for this idea.

5.3.3 Cryoprotective dehydration

In the springtail, *Onychiurus arcticus*, dehydration forms a critical component of its overwintering strategy (Holmstrup and Sømme 1998; Worland *et al.* 1998). Fully hydrated animals have a limited supercooling capacity of about −6.5°C. This suggests that the species would be unable to survive temperatures of as low as −20°C that are encountered in its microhabitat. However, when exposed to subzero temperatures its body water content declines significantly, with the final value depending on the temperature, but capable of being as low as <10 per cent of the initial value. This is in keeping with the increase in the vapour pressure difference between supercooled water and ice that is

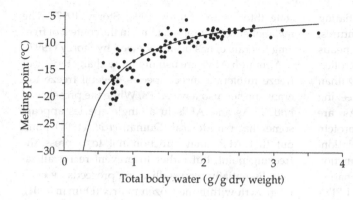

Figure 5.21 The relationship between total body water content and melting point in individuals of *Onychiurus arcticus*.

Source: Reprinted from *Journal of Insect Physiology*, **44**, Worland *et al.*, 211–219, © 1998, with permission from Elsevier.

characteristic of declining temperatures (Holmstrup and Sømme 1998). The decline in body water content results in a substantial lowering of the melting point (Fig. 5.21), such that it equals the ambient temperature (therefore the animal is not supercooled). There is therefore no risk of tissue ice formation or inoculation from the environment. Trehalose, a known anhydroprotectant that is thought to stabilize membranes and proteins, also increases in this springtail in response to cold, and together with dehydration results in a decline in the melting point of the body fluids. This cryoprotective dehydration (which is not the equivalent of prolonged supercooling) is clearly the most significant strategy for the survival of the low temperatures characteristic of this species' Arctic habitats (Holmstrup and Sømme 1998; Worland *et al.* 1998).

5.3.4 Freezing tolerance

Freezing tolerant species are capable of surviving extracellular ice formation, and apparently, in a few instances, intracellular ice formation in the fat body (Denlinger and Lee 1998). In consequence, they must be able to resist the injuries that are associated with this process. This is done, in part, by controlling ice formation at relatively high temperatures. This means that the site of ice formation, ice growth rate (and hence osmotic equilibration time), and the quantity of ice formed (i.e. the concentration of the non-frozen fluids and the extent of dehydration) (Ramløv 2000) can be controlled, so limiting injury.

Injury due to freezing
As soon as the temperature of an organism has declined below its melting point there is a risk of ice formation. Crystallization may take place either by aggregation of water molecules into an ice nucleus (homogeneous nucleation) or via their aggregation around some substance or irregularity (heterogeneous nucleation) (Ramløv 2000). When freezing takes place, additional water is added to the nucleus or nuclei, and effectively the animal begins to desiccate. The removal of water from the solution causes an increase in solute concentration. The progressive concentration of the body fluids may lead to protein denaturation, changes in pH, and alterations of membrane potential and transport properties. In addition, cellular shrinkage may occur owing to removal of water from the cells and this may damage the cell membrane to such an extent that it cannot recover following thawing (the critical minimum cell volume hypothesis) (Denlinger and Lee 1998). The extent to which this damage is incurred depends on the rate of cooling (Ramløv 2000), which may account for higher order (whole-individual) effects of cooling rate (Section 5.1.1). Of course, there may also be injury due to mechanical damage of cells by sharp ice crystals and their growth during the freezing process.

Inoculation at high subzero temperatures
Although many freezing tolerant species can survive very low temperatures, virtually all of them freeze at relatively high subzero temperatures (Section 5.3.1). These high SCPs are generally the result of inoculation of freezing by extracellular

INAs (Ramløv 2000; Duman 2001). By initiating freezing at relatively high subzero temperatures, INAs, which function as sites of heterogeneous nucleation, ensure that ice formation is controlled and restricted to the extracellular spaces (Duman *et al.* 1991), thus protecting cells from freezing damage. The best known of these INAs are the haemolymph protein (PIN) and lipoprotein (LPIN) ice nucleators. Knowledge of the relationship between PIN and LPIN structure and function is limited, although there is some information on the size and structure of both PINs and LPINs (Duman 2001).

Not all freeze-tolerant species make use of haemolymph PINs and LPINs to initiate inoculative freezing. In *Eurosta solidaginis* it appears that crystals of calcium carbonate, uric acid and potassium phosphate may ensure crystallization at relatively high subzero temperatures (Lee and Costanzo 1998). In some species, such as the highly freeze-tolerant *Chymomyza costsata*, external inoculation by ice crystals at relatively high subzero temperatures (−2°C) is essential for the development of freeze tolerance, and this is true also of several other species (Duman *et al.* 1991). In some instances, cooling in the absence of ice results in a depression of the SCP, but complete mortality on freezing. In *E. solidaginis* it appears that external inoculation via frozen gall tissue is essential for induction of freezing early on in the winter season, whereas endogenous nucleators become more important as the galls they inhabit dry out (Lee and Costanzo 1998).

Cryoprotection and recrystallization inhibition
Freezing tolerant species are also characterized by high levels of low molecular weight cryoprotectants. The colligative cryoprotectants, such as glycerol and sorbitol, reduce the percentage of ice that can be formed, thus preventing the intracellular volume from declining below a critical minimum, and lower the rate of ice formation (Storey 1997). In addition, they may also protect proteins against denaturation (Duman *et al.* 1991). The non-colligative cryoprotectants, such as trehalose and proline, interact directly with the polar head groups of lipids to stabilize the bilayer structure, thus preventing an irreversible transition to the gel state (Storey and Storey 1996; Storey 1997). The metabolism of cryoprotectants, in the context of freezing tolerance, has been reviewed by Storey (1997).

Although AFPs are usually more characteristic of freeze intolerant insect species, several freeze tolerant species also have AFPs. While the presence of both INAs and AFPs in a single species appears somewhat paradoxical, Duman *et al.* (1991) point out that AFPs may function not to depress the freezing point, but rather to prevent recrystallization. Essentially recrystallization proceeds by crystal growth within the frozen matrix (Duman 2001), and can cause damage to tissues either during thawing or temperature changes (Ramløv 2000). The danger of this is particularly severe at relatively high subzero temperatures. AFPs inhibit recrystallization by preventing ice crystal growth and this function may be particularly important in species that spend long periods frozen, or in freezing tolerant species from more temperate areas, where the risk of recrystallization is high.

5.4 Large-scale patterns

Geographic variation in the tolerance of both high and low temperatures has been documented in several insect groups at both the species and population levels. For example, latitudinal variation in cold hardiness has been documented between populations of *Drosophila* (Parsons 1977) and ants (Heinze *et al.* 1998), and between species of *Drosophila* (Goto *et al.* 2000) and swallowtail butterflies (Kukal *et al.* 1991). Likewise, geographic variation in upper lethal temperatures has been found among several *Drosophila* species and populations (Tantawy and Mallah 1961; Levins 1969; Goto *et al.* 2000). However, most of the work on geographic variation in thermal tolerance has been undertaken over relatively small spatial scales, extending mostly to the limits of a country (though see Tantawy and Mallah (1961) and Levins (1969) for early exceptions, and Hoffmann *et al.* 2002 for later, more detailed work). More recently, with the development of macroecology (see Chapter 1) and the realization that climate change may have profound effects on the distribution of diversity (Chapter 7), there has been a resurgence of interest in large-scale patterns in the temperature limits of animals, including insects (e.g. Gibert and Huey 2001).

5.4.1 Cold tolerance strategies: phylogeny, geography, benefits

The apparently clear distinction between freeze intolerance and freezing tolerance has prompted questions regarding the relative benefits of these strategies under different circumstances, the coexistence of species with different strategies in relatively similar microhabitats, and the reasons why some species adopt one strategy over another. Traditionally, it has been thought that the higher insect orders contain more freezing tolerant species than the other orders, but that there is otherwise no clear evidence of phylogenetic constraint (Block 1982; Duman *et al.* 1991). In addition, it has also been thought that freezing tolerance is a characteristic of Arctic species from particularly cold environments (Block 1995; Bale 1996), and Duman *et al.* (1991) have suggested that freezing tolerance is unlikely if insects are exposed to high subzero temperatures and periodic thaws occur during winter. Nonetheless, Duman *et al.* (1991) point out that if insects are routinely exposed to nucleators, such as external ice crystals in damp environments, or those associated with feeding, then freeze tolerance may be a strategy that ensures survival of subzero temperatures, particularly aseasonal freezing events. That the likelihood of inoculation and variability in temperature might be important in influencing the strategy adopted by a species has been suggested for alpine insects and for insects inhabiting more maritime environments in the Northern Hemisphere (Sømme and Zachariassen 1981; van der Laak 1982).

Recently, these arguments have been explored in considerable detail by Sinclair *et al.* (2003c), based on the compilation of a large quantitative database on SCPs and LLTs by Addo-Bediako *et al.* (2000). It appears that freeze intolerance is the most common strategy among insects and is widespread within the Arthropoda, characterizing entire lineages within this phylum, including the springtails, mites, ticks, and spiders (Fig. 5.22). Most millipedes are freeze intolerant as are the few scorpions and terrestrial isopods that have been investigated.

Within the insects, in terms of the absolute numbers of species, freeze intolerance is dominant, although freezing tolerance has now been documented from many orders (Fig. 5.22). The majority of orders examined to date rely on either one or the other of the strategies, suggesting mutual exclusivity. However, six orders include both strategies, while two orders (Mecoptera, Siphonaptera) include only freeze-intolerant species, and two (Blattaria, Neuroptera) include only freezing tolerant species (Table 5.2). Within the insects there is a strong phylogenetic component to the SCP, and LLTs, with most variation being explained at the family and generic levels (Addo-Bediako *et al.* 2000). However, insufficient data are available to examine the extent to which this applies throughout the insects, or to attempt a similar sort of comparison at the level of each of these strategies. In sum, it appears that freeze intolerance is basal within the arthropods, and that freezing tolerance has evolved many times within different taxa.

The geographic distribution of the strategies was also examined by Sinclair *et al.* (2003c). They found quantitative support for the idea espoused by Klok and Chown (1997) that freezing tolerance is more common in Southern than in Northern Hemisphere species. However, this tolerance is of quite a different form to that found in many Northern Hemisphere insects. The large majority of Southern Hemisphere freezing tolerant species are moderately freezing tolerant, with mortality generally occurring less than 10°C below the freezing point. In contrast, many of the Northern Hemisphere species are strongly freezing tolerant, tolerating very low subzero temperatures (Section 5.3.1). Sinclair *et al.* (2003c) concluded from this that the specific nature of cold insect habitats in the Southern Hemisphere, which are characterized by oceanic influence and climate variability around 0°C must lead to strong selection in favour of freezing tolerance in this hemisphere. Thus, there are two main scenarios where it would prove advantageous for insects to be freezing tolerant. In the first, characteristic of cold continental habitats of the Northern Hemisphere, freezing tolerance allows insects to survive very low temperatures for long periods of time, and to avoid desiccation. These responses tend to be strongly seasonal, and insects in these habitats are only freezing tolerant for the overwintering period. In contrast, in mild and unpredictable environments, characteristic of habitats influenced by the Southern Ocean, and by strong El Niño Southern Oscillation events (see Sinclair 2001a), freezing tolerance allows insects which habitually have ice nucleators in their guts to survive summer cold snaps, and to take advantage

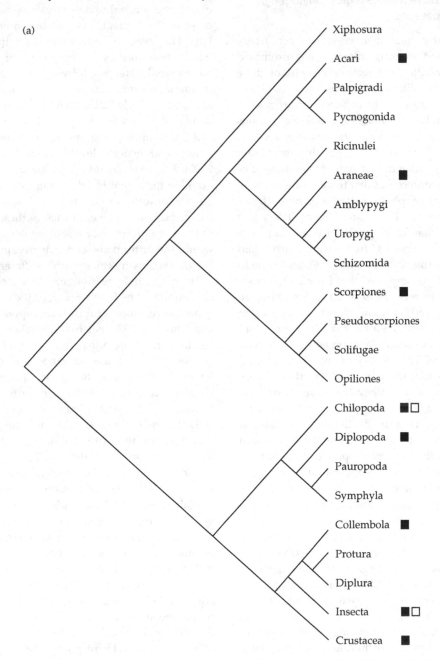

Figure 5.22 Phylogenetic distribution of known cold tolerance strategies of larvae/nymphs and adults of (a) terrestrial arthropod taxa and (b) insect orders. Filled squares are freeze intolerant, open squares are freezing tolerant.

Source: Redrawn from Sinclair *et al.* (2003c).

(b)

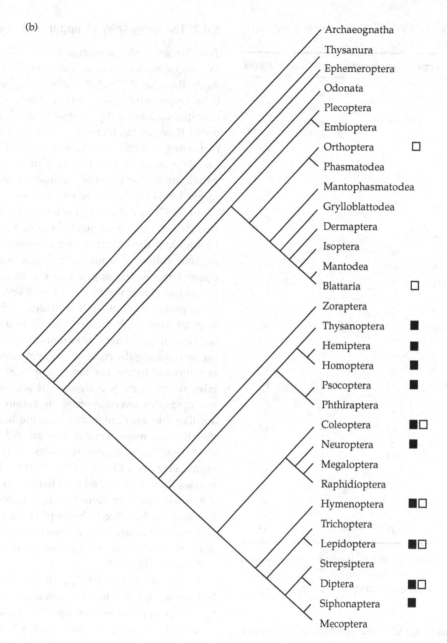

Figure 5.22 (*Continued*).

of mild winter periods without the need for extensive seasonal cold hardening.

Based on these conclusions, Sinclair *et al.* (2003*c*) suggested that in unpredictable environments, freeze-intolerant species should hedge their bets, thus resulting in low SCPs in some individuals irrespective of the season (see Klok and Chown 1998*b*; Bale *et al.* 2001 for evidence of this), and that freeze-intolerant species should also show rapid cold hardening. Rapid cold hardening has indeed been demonstrated in many freeze-intolerant species (Table 5.1). In the only freeze tolerant species that has been examined to date, the caterpillar *P. marioni* (Lepidoptera, Tineidae) from

Table 5.2 Cold hardiness strategies (where explicitly determined) for 30 orders of Hexapoda

Order	Eggs	Larvae/nymphs	Pupae	Adults
Thysanura	×	×	−	×
Collembola	×	s	−	s
Protura	×	×	−	×
Diplura	×	×	−	×
Archaeognatha	×	×	−	×
Zygentoma	×	×	−	×
Ephemeroptera	×	s	−	×
Odonata	×	n	−	×
Mantodea	s	×	−	×
Blattaria	×	f	−	f
Isoptera	×	×	−	×
Zoraptera	×	×	−	×
Plecoptera	s	n	−	×
Embioptera	×	×	−	×
Orthoptera	s	f	−	f
Phasmida	×	×	−	×
Grylloblattaria	×	×	−	×
Dermaptera	×	×	−	×
Psocoptera	s	×	−	×
Phthiraptera	×	×	−	×
Thysanoptera	×	s	s	s
Hemiptera	s	s	−	s
Neuroptera	×	s	×	×
Coleoptera	×	f,s	×	s,f
Strepsiptera	×	×	×	×
Diptera		s,f	s	s,f
Mecoptera	×	×	×	s
Siphonaptera	×	×	×	s
Trichoptera	×	×	×	×
Lepidoptera	s	s,f	s	s,f
Hymenoptera	s	s,f	s	s,f

Notes: s = freeze intolerant; f = freezing tolerant; n = no cold hardiness; − = life stage not present; × = no published data.

Source: Taken from Sinclair *et al.* (2003c).

Marion Island, rapid cold hardening is absent, as might be expected from a species that can freeze to a moderate degree and thaw repeatedly with little risk (Sinclair and Chown 2003). Sinclair *et al.* (2003c) also suggested that the differences between climates in the two hemispheres and species responses to them may have profound implications for the large-scale distribution of species richness and the responses of species to climate change.

5.4.2 The geography of upper and lower limits

The largest-scale investigation of temperature tolerances undertaken to date has been that of Addo-Bediako *et al.* (2000), who compiled data on both upper and lower lethal limits to test the climatic variability hypothesis underlying Rapoport's Rule (or the increase in species ranges with increasing latitude—see Gaston *et al.* 1998). They found pronounced differences in the extent of variability in these limits. Latitudinal variation in upper lethal limits, though significant (a range of about 30°C), is much less pronounced than spatial variation in LLTs (a range of about 60°C) (Fig. 5.23). In addition, the extent of the variation at a given latitude differs considerably. While variation in upper lethal limits appears to be roughly similar across the globe, variation in lower lethal limits is more pronounced at high latitudes. This variation with latitude might be due mostly to a latitudinal increase in the variety of situations either promoting or reducing the risks of low-temperature injury encountered by insects. For example, at those latitudes where there is a significant accumulation of snow, species overwintering in subnival habitats are likely to encounter less extreme temperatures than those in more exposed habitats. Addo-Bediako *et al.* (2000) also found pronounced hemispheric asymmetry in LLTs with Southern Hemisphere species showing reduced cold hardiness, compared with their Northern Hemisphere counterparts, and this may well be due to hemispheric differences in absolute minimum temperatures recorded in continental areas (with the exception of continental Antarctica) (Fig. 5.24).

An investigation of CT_{max} and CT_{min} in dung beetles across a 2400-m altitudinal gradient in South Africa provided evidence for similar patterns at a regional scale (Gaston and Chown 1999a). Here, CT_{max} shows considerably less variation with altitude than does CT_{min}, a pattern similar to that found in the global study. Thus, thermal tolerance range increases with an increase in latitude or altitude, and in both cases this change is associated with an increase in climatic variability with latitude/altitude. Thus, at both spatial scales, interspecific variation in upper limits is much less marked than that in lower limits.

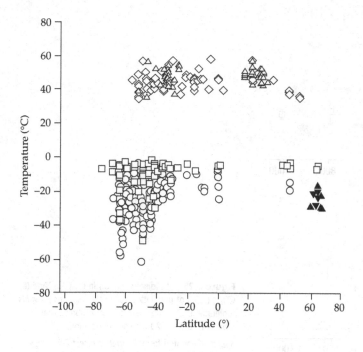

Figure 5.23 Latitudinal variation (northern latitudes are negative) in SCPs (freezing intolerant (○) and freeze tolerant insect species (□) and upper thermal limits (critical thermal maxima (△) and upper lethal temperatures (◇).

Note: Data for some freezing intolerant Antarctic Collembola (▼) and Acari (▲) are shown for comparative purposes.

Source: Addo-Bediako *et al.* (2000).

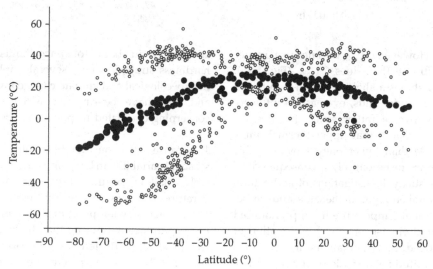

Figure 5.24 Latitudinal variation in absolute maximum and minimum temperatures (open circles) and annual mean daily temperature (closed circles) across the New World.

Source: Gaston and Chown (1999). *Oikos* **84**, 309–312, Blackwell Publishing.

Although spatial variation in upper and lower lethal limits, and the extent to which acclimation has an effect on them, is less variable at the intraspecific level (e.g. Hoffmann and Watson 1993), some studies have suggested that variation is greater in lower than in upper lethal limits. For example, intraspecific variation in critical thermal limits (CT_{max} and CT_{min}) in weevil species from sub-Antarctic Marion and Heard Islands varies substantially across altitudinal gradients (Klok and

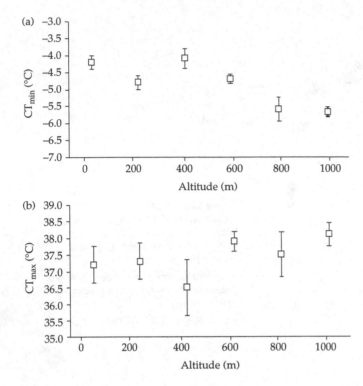

Figure 5.25 Altitudinal variation in (a) CT_{min} and (b) CT_{max} ($\pm$SE) of populations of *Bothrometopus parvulus* (Coleoptera, Curculionidae) on sub-Antarctic Marion Island (47°S) ($n = 20$ for each population).

Source: Reprinted from *Journal of Insect Physiology*, **47**, Chown, 649–660, © 2001, with permission from Elsevier.

Chown 2003). However, CT_{min} declines to a greater extent with altitude than does CT_{max} (Fig. 5.25). Indeed, CT_{max} shows a slight, but non-significant, increase with altitude. Thus, even at the intraspecific level, and over a rather small spatial scale, there is some evidence that upper thermal limits are less variable than lower ones. However, this variation appears to be entirely a consequence of phenotypic plasticity. Investigations of acclimation that were undertaken as part of the same study indicated that treatment temperature has a pronounced effect on short term (seven days) acclimation of the CT_{min} (indeed, the acclimation effect accounts for all of the altitudinal variation), although there is no effect on CT_{max}.

Such decoupling between upper and lower lethal limits has been documented in several other studies, mostly investigating interspecific differences (e.g. Chen *et al.* 1990; Kingsolver and Huey 1998; Hercus *et al.* 2000, but see Hoffmann *et al.* 2002), and has also been documented in the responses of *Drosophila* species to selection (Gilchrist *et al.* 1997; Hoffmann *et al.* 1997). The apparent decoupling of upper and lower lethal limits, and the hemispheric

asymmetry in both tolerance limits and cold hardiness strategies have several implications for macroecological and climate change investigations, and also have bearing on widely held ideas concerning correlated responses to stress.

1. Because of considerable inter-annual and seasonal variability in low temperatures, and a relatively shallow latitudinal gradient in low temperature extremes in the Southern Hemisphere, and because low temperatures seem more likely to be geographically limiting than high temperatures (Kukal *et al.* 1991; Jenkins and Hoffmann 1999; Chown 2001), physiological range limitation might be less stringent in the Southern than in the Northern Hemisphere. In consequence, geographic range size gradients of southern ectotherms should be much less steep than those of northern species. There should also be a less pronounced shift in range sizes near the Tropic of Capricorn than at the Tropic of Cancer because the interplay between the geographic and climatic determinants of range size should be less pronounced (Gaston and Chown 1999*c*). Moreover, if 'zonal bleeding' is important

as a factor leading to the establishment of species richness gradients (Chown and Gaston 2000), then species richness gradients should be less steep in the Southern than in the Northern Hemisphere. Of course, these predictions might also apply to vertebrates, and indeed in these groups the evidence is suggestive (Gaston *et al.* 1998).

2. The information on tolerances and climatic variation supports the idea that towards higher latitudes annual climatic variation increases and that, at least in some species, this also true of the thermal tolerance range (Addo-Bediako *et al.* 2000). Thus, a major assumption of Rapoport's Rule appears to hold, although whether this translates into larger geographic ranges at high latitudes for these insect species is not well known because coupled data on tolerance limits and latitudinal ranges are rare (see Gaston and Chown 1999*a* for an altitudinal example).

3. If absolute temperatures are less important in constraining the ranges of Southern than Northern Hemisphere species, and opportunistic utilization of warm spells is the norm in the former hemisphere, then climate change might be expected to have much less of a direct effect on high latitude insects in the Southern than in the Northern Hemisphere. While there is evidence for pronounced recent changes in the geographic ranges of Northern Hemisphere species (Parmesan *et al.* 1999), information for the south is scanty.

4. Although a generalized stress response, involving cross-tolerance to several stresses may be present at the biochemical level in insects (Hoffmann and Parsons 1991), this general response does not appear to hold for higher levels of organization, as decoupling between upper and lower lethal temperatures (Huey and Kingsolver 1993) and differences in measures of heat resistance (Section 5.1.2) clearly demonstrate.

Clearly, there is much to be gained from understanding the relationship between upper and lower lethal limits at several scales. At the cellular level, these limits might either arise from problems associated with oxygen delivery and demand (Pörtner 2001), or with the effects of temperature on neuronal functioning and protein folding (Feder and Hofmann 1999; Hosler *et al.* 2000; Wu *et al.* 2002), and the effects might differ at either end of the temperature spectrum. In turn, these effects will not only influence whole-organismal tolerances, but potentially also the distribution of species. These interactions are explored further in Chapter 7.

CHAPTER 6

Thermoregulation

These tropical beetles are unable to sustain the body in flight at the temperature of an English laboratory.

Machin *et al.* (1962)

Our treatment of the thermal physiology of insects can broadly be divided into 'resistance adaptations', which concern the lethal effects of extreme temperatures and are the subject of the previous chapter, and 'capacity adaptations', which concern the effects of ambient temperature on physiological performance (Prosser 1986; Cossins and Bowler 1987). Figure 6.1 illustrates how the relationship between body temperature (T_b) and performance in ectothermic organisms is bounded by their critical thermal limits. Maximal performance occurs at an optimal body temperature, and the thermal performance breadth is the range of T_b permitting a certain level of performance (Huey and Stevenson 1979). Performance curves may be shifted by acclimation, changing in position or shape, but the extent of thermal acclimation is often rather modest, one reason being that thermoregulation alters the relationship between environmental and body temperatures (Kingsolver and Huey 1998). The thermal sensitivity of performance in ectotherms, especially locomotor performance, has become an important research focus in evolutionary physiology, and flight performance and thermoregulation of insects are inextricably linked. Adequate power output for flight is usually available for only part of the temperature range experienced by insects (Josephson 1981), but thermoregulation prolongs flight activity in suboptimal thermal environments and enhances flight performance in more favourable conditions.

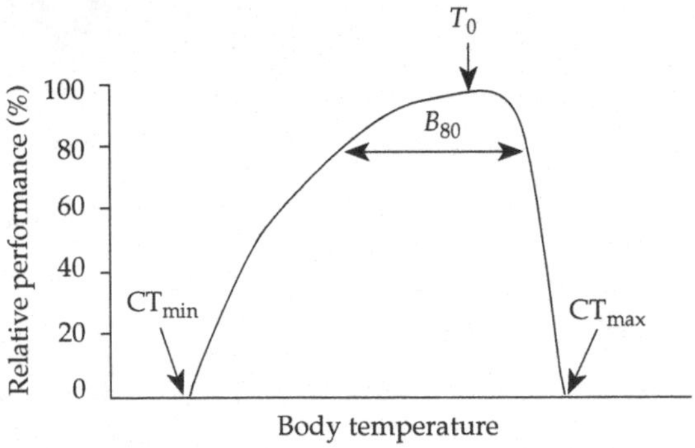

As with studies of insect flight (Dudley 2000), anthropomorphic bias and technical limitations have focused attention on large but unrepresentative insects, most of them heterotherms. There are numerous examples of impressive endothermic abilities being recorded for large tropical insects in the absence of information on their natural history, and thus the ecological significance of the endothermy (e.g. Morgan and Bartholomew 1982). The mechanisms involved in insect thermoregulation were comprehensively reviewed by Heinrich (1993), using a taxon-based approach. His conclusion that research in the field had already reached the stage of redundancy may well have been true for the mechanistic details, but in recent years the research approach has broadened considerably into the thermal ecology of insects. In consequence, we emphasize the ecological and evolutionary implications of thermoregulation. Small insects are the

Figure 6.1 Relationship between body temperature and performance in ectotherms, showing the optimum temperature for performance (T_0), the 80% performance breadth (B_{80}), and CT_{min} and CT_{max}.

Source: Huey and Stevenson (1979).

majority: the average adult insect body length is around 4–5 mm (Dudley 2000). It follows that most insect species (and, of course, all larval stages) are ectotherms. However, ectothermy does not preclude sophisticated thermoregulation.

6.1 Method and measurement

Miniaturization of electronic data loggers is facilitating the collection of detailed information on microclimate parameters. Rather than measuring air and substrate temperatures, these devices are often used to record operative (environmental) temperature (T_e), which gives an integrative thermal index of microclimate and provides an estimate of the T_b achieved without metabolic heating or evaporation (Bakken 1992). For studies of insect thermal biology, field measurements of T_e are usually obtained from thermocouples implanted in dead insects (known as T_e thermometers). These have the same size, shape, and radiative properties as live animals, so react similarly to air temperature, air movement, and solar radiation. Bakken (1992) discussed measurement of T_e and its applications in detail. Physical models such as copper pipes are widely used to measure T_e in field studies of reptile thermoregulation, and the attributes of such models (size, reflectance, orientation, and substrate contact) have been shown to have surprisingly little effect on estimates of T_e (Shine and Kearney 2001). Corbet et al. (1993) measured the minimum black globe temperature for flight activity in various social bees, arguing that for interspecific comparisons this method was more appropriate than the use of taxidermic models of different sizes.

The extent of thermoregulation is commonly assessed by linear regression of T_b on ambient temperature (T_a), with a zero slope indicating perfect thermoregulation (T_b independent of T_a) and a slope of one indicating thermoconformity. The 'thermoregulatory performance index' used by Bishop and Armbruster (1999) in their study of Alaskan bees was defined as the slope of the relationship between thoracic temperature (T_{th}) and T_e. They argued that T_e is more realistic as a reference temperature and permits comparison of insects captured under different microclimatic conditions.

A simple regression of T_b on T_a has sometimes been criticized as simplistic: by examining T_b for various categories of T_a, it is possible to assess an insect's thermoregulatory ability in more detail—its ability to raise T_b above the minimum threshold (at low T_a), to maintain T_b near the performance optimum (at intermediate levels), or to avoid overheating (at high T_a), as demonstrated for grasshoppers by Willott (1997). Dreisig (1995) modelled these three successive levels of behavioural regulation (basking, graded, and heat-avoidance phases) and showed that Hipparchia semele (Lepidoptera, Satyridae) passed through all three phases as T_e increased. For ectotherms showing a graded series of temperature-modulating behaviours, a polynomial regression may be the best fit describing the relationship of T_b to T_a or T_e (see Stone 1993; Schultz 1998).

The methods commonly used to evaluate thermoregulation in ectotherms, particularly lizards, have been criticized by Hertz et al. (1993). Low variance in T_b measurements may reflect a thermally homogeneous environment, not careful thermoregulation, and T_b versus T_a regressions suffer from the limitations of measuring only T_a. These authors advocate the collection of three kinds of data—T_b measurements for a representative sample of animals, the distribution of operative temperatures T_e at the study site (which may require many models), and laboratory measurements of preferred T_b to give an independent indication of the target T_b that the animal attempts to achieve. Effective thermoregulation is indicated by field-active T_b being closer to preferred T_b than T_e is, and field T_b near average T_e merely represents thermoconformity. Unfortunately this approach is not applicable to flying or endothermic insects, in which there are dramatic changes from rest to flight and T_b differs among body segments. However, close correspondence between field-active T_{th} and laboratory-measured optimal temperature has been recorded in dragonflies (Marden 1995b; Marden et al. 1996).

Thermocouples and thermography
Thermocouples are easy to construct and use, and have generated a wealth of information on insect thermoregulatory abilities (Heinrich 1993).

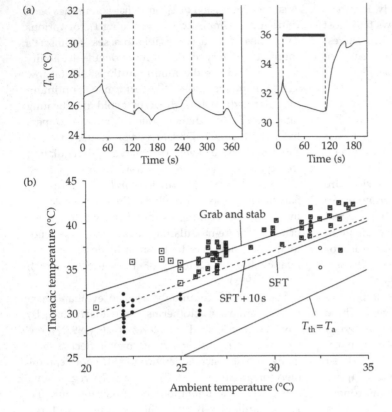

Figure 6.2 (a) Changes in T_{th} in an anthophorid bee, *Creightonella frontalis*, at the end of tethered flight, showing both brief and prolonged rises in T_b. Periods of flight shown by solid bars. (b) Thoracic temperatures (T_{th}) of female *C. frontalis* measured by the 'grab and stab' method, compared with stable flight temperatures (SFT).

Note: SFT + 10 s (dashed line) indicates body temperatures 10 s after cessation of tethered flight.

Source: Stone and Willmer (1989*a*).

However, Stone and Willmer (1989*a*) have criticized the 'grab and stab' method of measuring insect body temperatures on the basis that there may be rapid and unpredictable changes in T_b when a flying insect is captured. Although it has commonly been assumed that passive cooling takes place between capture and insertion of the thermocouple, T_b can rise at the end of a period of tethered flight, either briefly or on a more prolonged basis (Fig. 6.2a) (Stone and Willmer 1989*a*). Honeybees flying on a roundabout show increases in surface temperature of the thorax immediately after flight (Jungmann *et al.* 1989). 'Post-flight warm-up' during the interval between grabbing and stabbing is the result of heat production continuing after convective cooling has stopped, and can lead to significant errors in the slope of the T_{th}/T_a regression used to determine the extent of thermoregulation (Fig. 6.2b). Beetles close their elytra on capture and this may also lead to potential overestimates of T_{th} (Chown and Scholtz 1993). When more than one body segment is sampled, the common practice of measuring T_{th} first, because it has the largest gradient (e.g. Coelho and Ross 1996) may also be a source of error. Watt (1997) has further criticized the 'grab and stab' method for the following reasons: (1) its apparent simplicity encourages anecdotal observation; (2) only active members of an insect population are sampled; and (3) each data point comes from a killed or injured insect. The use of chronically implanted thermocouples provides more balanced and reliable information on the thermal ecology of insects (e.g. Ward and Seely 1996*a*). Shallow insertion of implanted thermocouples is recommended by Stone and Willmer (1989*b*) to minimize tissue damage (and is justified because temperature gradients within the thorax are unlikely). The appendix to the latter paper also provides an indication of the likely extent of heat loss along the thermocouple wire, which can be significant for small insects.

Infrared thermography (Stabentheiner 1991) avoids the problems associated with invasive measurements of body temperature. Continuous recording of the IR radiation emitted by an unrestrained insect is converted to surface temperatures using careful calibration, and the thermal behaviour is recorded on videotape. The surface temperature of a bee heated to 20°C above T_a is approximately 1°C lower than that just beneath the cuticle, and relative measurements can be made with an accuracy of ±0.25°C. Provided all of the insect is in focus, thermal imaging can give simultaneous remote measurements of different body regions (Schmaranzer and Stabentheiner 1988; Farina and Wainselboim 2001). However, the IR camera remains in a fixed position and this technique can not be used for flying insects. It is ideal for studies of bees visiting an artificial feeder or leaving the hive entrance (see Section 6.6.2), and extension cables have increased the field possibilities (Kovac and Schmaranzer 1996). Infrared thermography also has advantages when the 'grab and stab' method might be risky for researchers: it has been used to show how the Japanese honeybee *Apis cerana japonica* kills its hornet predators by engulfing them in a cluster heated to about 47°C, which is lethal to the hornets but not the bees (Ono *et al.* 1995).

6.2 Power output and temperature

Temperature effects on digestion and growth rates were considered in Section 2.6. Here, we are concerned with the thermal sensitivity of muscle performance, especially in flight. Short reaction times are vital for prey capture and predator evasion. The speed of terrestrial and aerial locomotion is strongly dependent on temperature, even in caterpillars (Joos 1992). Thermal biology also has important consequences for mate acquisition tactics in ectotherms, often because of improved aerial performance (Willmer 1991*a*). For example, male *Hybomitra arpadi* (Diptera, Tabanidae; about 85 mg) hovering at mating sites in Canada maintain a mean T_{th} of 40°C, fuelled by highly concentrated crop contents (Smith *et al.* 1994). Oviposition is also governed by temperature, often via its effect on flight (Watt 1968; Toolson 1998). Stridulation of katydids employs the flight muscles (Stevens and Josephson 1977) and singing of cicadas uses tymbal muscles (Josephson and Young 1979): both are

high-performance muscle systems and strongly temperature-dependent.

Thermoregulation physiologists are naturally most interested in the thorax of insects, which is packed with muscle and the warmest region of the body (Schmaranzer 2000). Flight requires a substantial investment in flight muscle—up to 60 per cent of body mass in dragonflies (Marden 1989), and a minimum of 12–16 per cent—which is expensive to build and maintain, and can vary considerably in size within individuals and within species, with resulting effects on flight performance (reviewed by Marden 2000). Flight polymorphisms, in which female insects invest in either flight muscle or fecundity, have already been discussed in Chapter 2, and trade-offs between mobility and egg mass are important constraints in butterfly design (Marden and Chai 1991).

Coleoptera, Diptera, and Hymenoptera (three of the four largest insect orders, therefore, the majority of insect fliers) have asynchronous flight muscle in which multiple contractions originate from a single nervous impulse. This muscle type, also known as myogenic or fibrillar, is stretch-activated and operates at high contraction frequency by mechanical deformation of the metathorax (Josephson *et al.* 2000*a*). The frequency at which asynchronous flight muscle operates most efficiently is determined by temperature and the resonant frequency of the wing–thorax system. Asynchronous muscles of both beetles and bumblebees have been studied using work-loop techniques (see Josephson *et al.* 2000*a*), but the basalar muscles of the cetoniine beetle *Cotinus mutabilis* (Scarabaeidae), which depress the hindwing and comprise one-third of its flight muscle mass, make a better preparation (Josephson *et al.* 2000*b*). Mechanical power output in flight muscle is the product of the work per cycle and the cycle frequency. The optimal cycle frequency of the beetle basalar muscle and the power output at that frequency increase with increasing muscle temperature (Fig. 6.3). Average measurements for mechanical power output at the wingbeat frequency (94 Hz) and temperature (35°C) of free flight were 127 W kg^{-1} muscle (Josephson *et al.* 2000*b*). This value is about twice that measured for synchronous dorsoventral muscle of the sphingid moth *Manduca sexta* under similar conditions

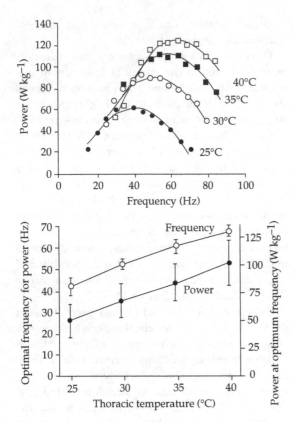

Figure 6.3 Effect of temperature on performance of the basalar muscle of *Cotinus mutabilis* (Scarabaeidae). (a) The effects of changing cycle frequency on mechanical power output at different temperatures. (b) The effects of temperature on the optimal frequency for power output and the power output at the optimal frequency. Both increase linearly with muscle temperature, with Q_{10} values of 1.4 and 1.6, respectively.

Source: Josephson *et al.* (2000*b*).

(Stevenson and Josephson 1990), suggesting an evolutionary advantage to asynchronous operation. High-frequency operation is achieved in asynchronous muscle without high rates of Ca^{2+} cycling and without large investment in sarcoplasmic reticulum, so more volume can be allocated to myofibrils. This leads to greater efficiency and greater power output. Mechanical properties of synchronous muscle of *Schistocerca americana* and asynchronous muscle of *C. mutabilis* were compared by Josephson *et al.* (2000*a*). The efficiency of muscle contraction can be estimated by comparing mechanical work with metabolic rates measured

during flight, and values are low for both synchronous and asynchronous muscle (Dudley 2000). These low efficiencies (5–10 per cent) lead to high rates of heat production, especially in large insects.

For the synchronous muscle preparation of *M. sexta*, maximal mechanical power output is produced by temperatures of 35–40°C and cycle frequencies of 28–32 Hz (Stevenson and Josephson 1990). This is in excellent agreement with the T_{th} (35–42°C) and wing beat frequency (24–32 Hz) observed during hovering flight. It can also be calculated that muscle temperatures must be at least 30–35°C to reach the minimum power threshold for flight in this moth (Stevenson and Josephson 1990).

There are few studies of the relationship between temperature and power output in small or weak fliers, with muscles that operate at low temperatures. Marden (1995*a*) compared muscle performance in *M. sexta* with that of 12-mg males of a winter-flying ectothermic moth (*Operophtera bruceata*) (Geometridae), which fly over a very broad range of temperatures and can have flight muscle temperatures near zero. Thermal sensitivity of flight muscle of the two moth species (shortening velocity, tetanic tension, and the resulting instantaneous power output) is illustrated in Fig. 6.4. Muscle of *O. bruceata* contracts more slowly but generates more force than that of *M. sexta*. As a result, the maximum instantaneous power output of *O. bruceata* at 15–20°C is as high as that of *M. sexta* up to about 35°C. Note that maximum instantaneous power output obtained from such force–velocity data is about twice that achieved during normal contraction cycles (Stevenson and Josephson 1990). Muscle tension is generally less sensitive to temperature than twitch duration (Josephson 1981). In addition to possessing cold-adapted muscle, males of *O. bruceata* (females are wingless) have the morphological advantages of very low wing loading and a high ratio of flight muscle mass to total body mass (Marden 1995*a*). Geometrid moths, in general, are capable of immediate flight, regardless of T_a, and do not require preflight warm-up (Casey and Joos 1983).

The higher wingbeat frequencies associated with asynchronous flight muscle result in greater force production, and this permits reduction in wing

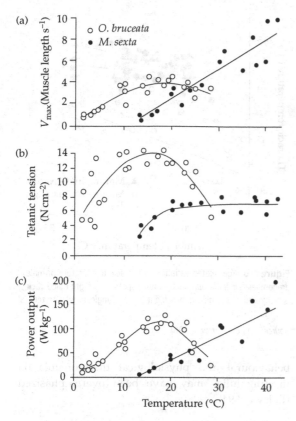

Figure 6.4 Flight muscle physiology of moths: a comparison of ectothermic *Operophtera bruceata* (Geometridae) and endothermic *Manduca sexta* (Sphingidae). (a) Maximum shortening velocity V_{max}, (b) tetanic tension, and (c) instantaneous power output are plotted as a function of flight muscle temperature.

Source: Marden (1995a).

area relative to body mass, as in the conversion of one pair of wings either to elytra or to halteres (Dudley 2000). Multiple independent origins of asynchronous flight muscle are most evident in the major endopterygote orders of Coleoptera, Diptera, and Hymenoptera, which all show strong evolutionary trends towards miniaturization (Dudley 2000). Josephson (1981) stressed that most insect behaviour involves muscles operating more or less at ambient temperature (due to small size). Wingbeat frequency is typically inversely related to body size, varying from 5 Hz in large butterflies to 1000 Hz in minute ceratopogonid flies, with enormous variation between species of the same body mass but different morphology (Dudley

2000). Temperature sensitivity was evident in only some of the beetle species studied by Oertli (1989) and was attributed to limited changes in the resonant properties of the flight system. Power output is varied by changing both the frequency and amplitude of muscle contraction, but it was long assumed that insects do not modify heat production by varying wingbeat frequency (e.g. Joos *et al.* 1991). Recently, however, several insects have been shown to decrease wingbeat frequency and metabolic heat production (MHP) at high T_a (Section 6.5.2).

Muscle physiology of dragonflies

Dragonflies are superb flying machines, but adults emerge with immature flight muscles which increase 2.5-fold in mass during maturation, resulting in equally dramatic changes in aerial competitive ability. Marden (1989) attached small weight belts to territorial male dragonflies, decreasing the ratio of flight muscle mass to total mass, and measured declines in their mating success. Vertical force production during brief flight attempts provides a measure of changes in flight performance with ontogeny (Marden 1995b). The optimal thoracic temperature, corresponding to peak performance, increases from 35 to 44°C during maturation in *Libellula pulchella* (Libellulidae), while the thermal sensitivity curve narrows (i.e. the thermal breadth decreases) (Fig. 6.5). Field data show an increase in time spent flying from 2–32 per cent, and the T_{th} of 29–40°C measured during the flight of tenerals increases to 38–44°C during routine territorial patrolling flight and to 40–45°C during high-speed chases (Marden *et al.* 1996). Age, T_a, and level of exertion all have significant effects on T_{th} (Fig. 6.6). The weak flight and relatively sedentary existence of tenerals contribute to their lower T_{th}, but also conserve energy. Horizontal bars in Fig. 6.5 show the close correspondence between field-active T_{th} and laboratory-measured optimal temperature in each age class; for mature dragonflies the match improves if field T_{th} for high-speed flights is used. Maturational changes are also evident at the ultrastructural level as muscle cells increase in diameter and the fractional cross-sectional area of mitochondria increases threefold (Marden 1989). This work on the ontogeny

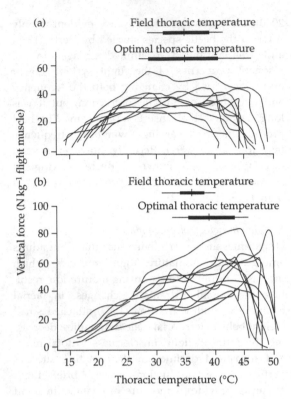

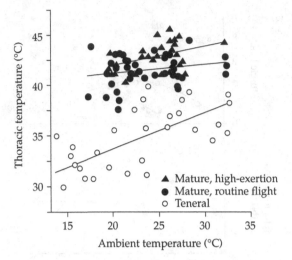

Figure 6.6 Age-related variation in thoracic temperature of male dragonflies (*Libellula pulchella*). Routine patrolling flight (filled circles) is compared with high-exertion flight (filled triangles) and with teneral individuals (open circles).

Source: Marden *et al.* (1996).

Figure 6.5 Vertical force production of (a) teneral and (b) mature dragonflies (*Libellula pulchella*) as a function of thoracic temperature.

Note: Grouping the dragonflies into age classes shows that the thermal sensitivity curve narrows and performance increases with maturation. Field thoracic temperatures are compared with optimal thoracic temperatures: horizontal bars show the mean (vertical line near centre) ±SD (thick bar) and range (thin bar).

Source: Marden *et al.* (1996).

of muscle physiology in dragonflies is a superb integrative analysis extending across all levels of biological organization from ecology and flight performance through to cellular and now molecular mechanisms (for review see Marden *et al.* 1998). Marden *et al.* (1996) assumes that endothermic warming is the most derived condition in dragonflies, and that evolution of endothermy was probably preceded by selection on the thermal sensitivity of performance. A phylogenetic distinction between ectothermic perchers (Libellulidae) and endothermic fliers (Aeschnidae) was made by Heinrich and Casey (1978), but does not hold for all dragonflies, and the dichotomy between

behavioural and physiological thermoregulation in dragonflies may have been overemphasized (Polcyn 1994).

6.3 Behavioural regulation

The biophysical ecology and heat exchange of insects have been reviewed by Casey (1988, 1992). Heat transfer routes of convection and radiation are the most important, because conduction will be negligible when only tarsi are in contact with the substrate, metabolic heat production (MHP) is insignificant in small insects, and evaporation is costly and generally assumed to play a minor role. Hence, the importance of behaviour that takes advantage of air temperature, air movement, and solar radiation (which are integrated in measurements of T_e). All behavioural mechanisms depend on the availability of solar radiation to provide thermal diversity in the environment in both time and space (Stevenson 1985). In this landmark paper, Stevenson constructed a series of heat-transfer models to provide quantitative estimates of the relative importance of various behavioural and physiological mechanisms in changing body temperature in ectotherms. He concluded that behavioural mechanisms were easily dominant and, of these, the times of seasonal and daily activity have the greatest impact on T_b, followed by microhabitat selection and then postural adjustments. Usually a complex interplay between these factors is involved, as the examples below will demonstrate.

6.3.1 Microhabitats and activity

The significance of microclimate (= climate of the microhabitat, see Section 4.4.1) in the thermal and water balance of insects was reviewed by Willmer (1982). The critical parameters are temperature and the moisture content of the air, continuously modified by solar radiation and air movement. Steep gradients are common near the ground (both above and below the surface), and climate can be greatly modified in boundary layers, especially when vegetation is present to transpire, provide shade and slow air movement. It is flight which enables insects to use the three-dimensional structure of vegetation and its associated microhabitats. The thermoregulation literature is replete with studies recording activity times, body temperatures, and microclimate parameters for insect ectotherms.

Desert beetles
Flightless tenebrionid beetles are abundant in arid and semi-arid habitats around the world. Many are diurnally active and black in colour, and their thermal biology has attracted much attention from ecophysiologists. Their preferred body temperatures and thermal tolerances are broadly correlated with habitat, those of some North American arid-land tenebrionids (mainly *Eleodes* species) (e.g. Kenagy and Stevenson 1982; Parmenter *et al.* 1989) being considerably lower than those of Namib desert tenebrionids of the genus *Onymacris*, which exhibit high preferred T_b (among the highest of any ectotherms) in both the laboratory and field, and high upper lethal temperatures (Roberts *et al.* 1991). Mechanisms of thermoregulation in *Onymacris* species occupying different sandy habitats were investigated using continuous measurement of field T_b with implanted thermocouples (Ward and Seely 1996b). High preferred T_b was maintained primarily by shuttling behaviour,

climbing on bushes, and burying in the sand (a depth of 20 cm is enough to maintain constant T_b day and night). There was little evidence of basking or of stilting behaviour; although *Onymacris* species have long legs, stilting may be effective only in a narrow range of wind velocities. Microhabitat shifts were thus more important than postural adjustments in controlling T_b (Ward and Seely 1996a). Although the beetles are exposed to high surface temperatures, sand is a readily available thermal refuge providing access to a broad range of T_a. Comparative phylogenetic analysis shows perfect coadaptation of preferred and field T_b in *Onymacris* (Ward and Seely 1996b). Many desert tenebrionids show biphasic activity to avoid high midday temperatures. However, this is not the case in all biphasic desert beetles. For example, in keratin-feeding *Omorgus* species (Trogidae) of the Kalahari desert, the relationship between temperature and surface activity is more complex; a dawn peak of activity, at moderate temperatures and high humidity, when feeding predominates, and a sunset peak for social interactions when temperatures are still high (Scholtz and Caveney 1992).

Thermal respites for heat-tolerant ants
Small body size and the inability to fly restrict many ant species (Formicidae) to severe surface temperatures (Willmer 1982), but a mosaic of microclimates can be utilized even in sandy desert habitats, and ants are important desert fauna throughout the world. Extreme heat tolerance is seen in thermophilic desert ants which forage at nearly lethal temperatures for insect prey that has already succumbed to heat stress. The Namib Desert ant *Ocymyrmex robustior* begins foraging when sand surface temperatures reach 30°C, but when they exceed 51°C it forages intermittently, pausing in thermal refuges such as the shade of a grass stalk or running up stems to cool off (Marsh 1985). The frequency and duration of thermal respite behaviour increases with sand temperatures over 51°C, and refuge temperatures are 7–15°C lower than those an ant would experience on the sand surface. Live mass of *O. robustior* is about 4 mg and its low thermal inertia ensures rapid heat exchange. Even more extreme heat tolerance is seen in the synchronized midday foraging of the

Saharan ant *Cataglyphis bombycina*, which occurs during a brief thermal window with its upper limit set by the critical thermal maximum of 53.6°C and the lower limit by the retreat of a lizard predator into burrows (Wehner *et al*. 1992). Again, thermal respite behaviour is vital during foraging. Heat shock proteins are accumulated by *Cataglyphis* species prior to heat exposure (Gehring and Wehner 1995) (Section 5.2.2). Most ants are too small for direct measurement of T_b using thermocouples, and T_b is estimated from T_e thermometers placed at ant height and in other appropriate locations. However, T_b has been measured directly in a larger thermophilic species, *Melophorus bagoti*, which is the Australian equivalent of *Ocymyrmex* and *Cataglyphis* (Christian and Morton 1992). Both large and small workers of *M. bagoti* maintain T_b of 45–46°C through much of the day. Size effects may be insignificant in the steep temperature and convection gradients next to the surface.

Interspecific interactions lead to temporal resource partitioning in ants occupying different thermal niches. Subordinate species in a Mediterranean grassland are forced to be active in more stressful conditions during the day and to forage at temperatures closer to their critical thermal limits; however, this risk-prone strategy has foraging benefits (Cerdá and Retana 2000). Similarly, in an Argentinian ant community (Bestelmeyer 2000), subordinate species are active at temperature extremes (both high and low) when they have almost exclusive access to resources. The balance between stress and competition in ant assemblages is thought to be responsible for the unimodal relationship between species richness and the abundance of dominant ants (Andersen 1992). During stressful conditions, richness is low and most species are thermal specialists. As stress declines, more species become active, and under the most favourable conditions numerically abundant and aggressive dominant ants exclude other species. This unimodal relationship between species richness and the abundance of dominant ants is convergent across three continents (Africa, Australia and North America). However, modelling work by Parr (2003) has demonstrated that interactions between stress and competition are not necessarily required to explain this unimodal relationship.

Rather, it may also be the simple consequence of the constraints imposed by abundance frequency distributions, themselves the outcome of a variety of processes (Tokeshi 1999).

Sun patches in forests

Short-term selection of sunlit or shaded substrates is probably the most common mechanism for control of body temperature in insects (May 1979). Tiger beetles in the genus *Cicindela* (Cicindelidae) are active diurnal predators of open sandy habitats which maintain high T_b around 35°C through a combination of basking, stilting, and shuttling between sun and shade (Dreisig 1980). However, a few *Cicindela* species inhabit forest floor environments: *C. sexguttata* adults aggregate in light gaps to maintain their preferred T_b of 33°C, behaviour which increases prey capture rates and intraspecific encounters, but also the risk of predation (Schultz 1998). Robber flies (Asilidae) are ambush predators which spend 98 per cent of their time perched (O'Neill *et al*. 1990), and those in a tropical rainforest can be grouped into light- and shade-seeking species: The former bask but the latter, which belong to a more recent lineage, do not thermoregulate at all and their foraging and activity patterns differ accordingly (Morgan *et al*. 1985). This occupation of either sunlit or shaded habitats occurs on a long-term basis compared to shuttling behaviour, and is also evident in damselflies in a tropical forest (Shelly 1982) and in butterflies (Srygley and Chai 1990). Herrera (1997) examined the varied insect pollinators of summer-flowering *Lavandula latifolia* in a mosaic of shade and sun on a Spanish forest floor. Compared to hymenopterans, the dipteran pollinators (mainly hoverflies; Syrphidae) tended to be restricted to shade and had significantly lower T_{th}, lower temperature excesses (the difference between T_b and T_a) and higher T_{th}/T_a regression slopes. Insolation of plants thus determines the pollinator assemblages to which they are exposed. Individual plants of *L. latifolia* show characteristic temporal patterns of exposure to direct sunlight, and patterns of insect visitation vary accordingly (Herrera 1995a). Microclimate effects on pollinator activity at flowers were reviewed by Corbet (1990). Physical factors affect plant reproduction through both the hygrothermal balance of pollinators and the floral rewards on offer, and these effects depend on the local microclimate within the flower. Temperature excesses as high as 8°C inside flowers are of critical importance for small ectothermic pollinators crawling deep into the flowers (Herrera 1995*b*). The thermal benefits of flower basking have also been demonstrated for insect pollinators visiting heliotropic and other Arctic flowers (Kevan 1973, 1975), and for scarab pollinators resting inside endothermic flowers in neotropical forests (Seymour *et al.* 2003).

Males of the speckled wood butterfly *Pararge aegeria* (Nymphalidae) perch in sun spots in woodland habitats and defend these territories against patrolling intruders. The contest takes the form of a short spiral flight and is usually won by the owner of the sun spot. Patrollers have a lower T_{th} than perchers, due to flying through shade, but their darker wings have basking benefits (Van Dyck and Matthysen 1998). This butterfly has been widely used in behavioural research, and Stutt and Willmer (1998) have elegantly separated temperature and ownership asymmetries by releasing pairs of butterflies in sun spots after manipulating their body temperatures: contests were won by the warmer butterflies. Sun spots can be considered as limited solar energy resources.

6.3.2 Colour and body size

The rate and extent of radiant heat gain depend on thermal mass and surface reflectance of insects (in combination with behavioural adjustments of posture and orientation). In a significant paper, Willmer and Unwin (1981) measured the reflectance, heating rates, and temperature excess of 41 insect species (half of them Diptera) exposed to insolation. Darker insects heat faster and achieve a greater temperature excess. Larger insects heat more slowly but achieve a greater temperature excess. Thus, although larger insects can become active earlier in the day than small insects, they may risk overheating in full sun. Indeed, small insects (the majority) show remarkably rapid rates of heat gain and loss (Fig. 6.7). The temperature excess depends mainly on size, but reflectance plays a larger role as size increases.

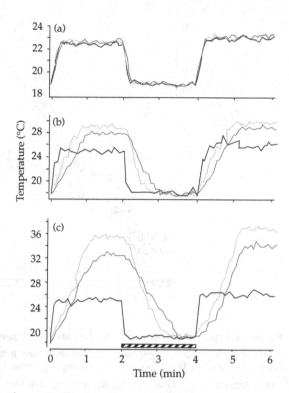

Figure 6.7 Thoracic temperature as a function of time for insects of various sizes in sunlight (radiant flux 840–900 W m^{-2}) and in shade (hatched bar). Two-min intervals of sun and shade. Darkest lines show air temperatures, and palest lines show the less reflective insect of each pair. The insects were (a) *Xyphosia* sp. and *Hilara* sp. (Diptera 3 mg); (b) *Rhagonycha fulva* and *Cantharis livida* (Cantharidae 26–27 mg); (c) *Crabro cribrarius* and *Cerceris arenaria* (Sphecidae 90–100 mg).

Source: Willmer and Unwin (1981). *Oecologia* **50**, 250–255, Fig. 3. © Springer.

The importance of size and colour in behavioural thermoregulation is clearly illustrated by Whitman's (1987) study of a large unpalatable desert grasshopper, *Taeniopoda eques* (Acrididae), which uses postural changes and vertical movements between vegetation and soil to maintain a preferred temperature of 36°C for most of the day (Fig. 6.8). These massive grasshoppers (females up to 10 g) are able to warm rapidly soon after sunrise, but later restrict T_b to a level far below the midday ground temperature of 65°C. In this species the possession of toxins is associated with large size and black warning colouration, both of which contribute to temperature homeostasis.

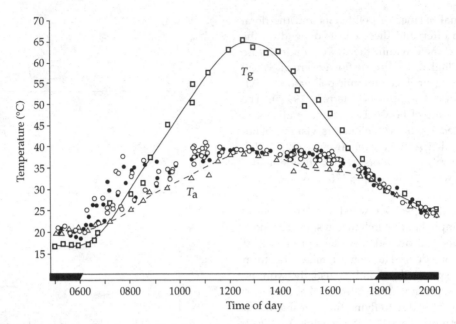

Figure 6.8 Behavioural thermoregulation in a large black desert grasshopper, *Taeniopoda eques.*

Note: The figure compares air temperatures T_a (triangles), ground temperatures T_g (squares) and thoracic temperatures of *T. eques* (open circles for males, filled circles for females) during a typical hot Chihuahuan Desert day.

Source: Reprinted from *Animal Behaviour*, **35**, Whitman, 1814–1826, © 1987, with permission from Elsevier.

Modification by wind

Convection can greatly diminish the thermal significance of colour, as shown by examples from tenebrionid and coccinellid beetles. Most species of *Onymacris* are black, but white elytra evolved on a single occasion (Ward and Seely 1996*b*). The thermal implications of elytral colour were examined by Turner and Lombard (1990), who measured higher subelytral temperatures in black *O. unguicularis* than in white *O. bicolor*, but the differences between the two species decreased at wind speeds of 2 m s^{-1} and above. Colour affects only one avenue of heat exchange, the absorption of direct short-wave and visible radiation, but it is convective conditions which are most important in the thermal physiology of these desert beetles. Colour polymorphism in the two-spot ladybird beetle, *Adalia bipunctata* (Coccinellidae), which occurs in red and black (melanic) morphs, has been attributed to thermal melanism. Non-melanic forms have a higher reflectance and lower temperature excess when illuminated, and the temperature excess of both morphs is proportional to body mass, so that the effect of colour decreases with size

(Brakefield and Willmer 1985). The heat exchange of ladybirds has been modelled and empirically tested (De Jong *et al.* 1996). Warm-up curves under irradiation are shown in Fig. 6.9: the maximum temperature excess was significantly higher for melanic beetles, but the rate of warming was not. Wind had a substantial effect, with a greater predicted drop in T_b of melanic beetles because convective heat loss is proportional to the temperature excess.

Aggregation increases effective body size of caterpillars

Aggregation, with or without tent construction, influences the thermal ecology of caterpillars by increasing their effective body size. Colonies of eastern tent caterpillars (*Malacosoma americanum*, Lasiocampidae) thermoregulate as an aggregate on or within the tent, which provides protection and a choice of temperatures. Although most foraging activity occurs at low T_a when behavioural thermoregulation is impossible, high T_b during inactive periods in the tent increases rates of food processing in the caterpillars (Casey *et al.* 1988;

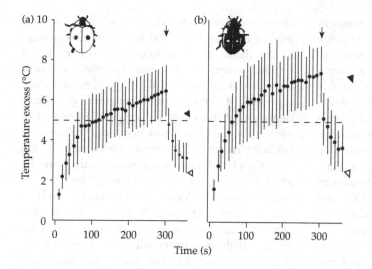

Figure 6.9 Thermal melanism in the two-spot ladybird *Adalia bipunctata* (Coccinellidae).

Note: Average warm-up curve at T_a of 3°C for (a) red and (b) black (melanic) morphs. At time zero, ladybirds were put under the lights (675 W m^{-2}), and a fan was switched on at the point indicated by the vertical arrow. Temperature excess is given as mean ± SD, with predicted values shown by arrowheads on the right (upper values without wind, lower values with wind).

Source: De Jong *et al.* (1996).

Knapp and Casey 1986). All the factors influencing the thermal relations of caterpillars—behaviour, setae, colour, aggregation, and tent making (Casey 1993)—converge in *M. americanum*. Caterpillar setae increase the effective diameter of the animals and reduce convective but not radiative heat transfer (Casey and Hegel 1981; Kukal *et al.* 1988). Dark colouration enhances heat gain during communal basking. Silk tents retard heat exchange and facilitate the clumping of caterpillars to form a composite body with a reduced surface area to volume ratio. A dramatic example of the latter benefit is seen in the anomalous emperor moth, *Imbrasia belina* (Saturniidae), which increases in body mass from 4 mg to 16 g during larval development (Klok and Chown 1999). Aggregations of instars II–III maintain similar T_b to solitary individuals of instars IV and V, and their T_b is much higher than operative temperatures of single models of the early instars. The caterpillars do not bask, but the increase in their effective body mass enhances heat gain (see also Stevenson 1985), as well as decreasing water loss (Section 4.4.2). Bryant *et al.* (2000) compared the thermal ecology of gregarious and solitary nettle-feeding nymphalid butterfly larvae: exposed gregarious larvae of two species (*Aglais urticae* and *Inachis io*) maintained high and regulated T_b, while the concealed larvae of two other species were essentially thermoconformers. The incorporation of larval thermoregulation into models of development time allowed

these authors to estimate that thermoregulation extends the range of *A. urticae* and *I. io* in Britain by 200 km northwards (Bryant *et al.* 2002). Whether thermoregulatory patterns, and the other physiological benefits of aggregation such as reduced water loss (see Chapter 4), are a cause or a consequence of the evolution of gregariousness is not entirely clear. Gregariousness in caterpillars probably evolved as a consequence of the advantages of egg clustering (Courtney 1984; Clark and Faeth 1998), with physiological benefits in the caterpillars perhaps maintaining this behaviour (Klok and Chown 1999). However, several other hypotheses have been proposed to explain caterpillar aggregations, including predator avoidance (Sillén-Tullberg 1988) and the advantages associated with group feeding (Fitzgerald 1993; Denno and Benrey 1997).

6.3.3 Evaporative cooling in ectothermic cicadas

Behavioural thermoregulation in cicadas is interesting in that it may be supplemented with physiological mechanisms at the different extremes of T_a: evaporative cooling and endothermy. Evaporative cooling, long assumed to play a minor role in the heat balance of insects because of their small water reserves, is the only means by which insects can reduce T_b below T_a. Exceptions do occur in hot environments, such as the evaporative cooling which occurs by cuticular and respiratory routes in cicadas and grasshoppers, respectively (Prange 1990; Toolson 1987) and by an oral route in honeybees (Section 6.5.3). With an abundant source of water from plant xylem, desert cicadas are able to sing during midday heat that their enemies can not tolerate. Hadley *et al.* (1991) used a flow-through system to measure body temperature, water loss, and gas exchange simultaneously in a Sonoran Desert cicada, *Diceroprocta apache*, while it was feeding on perfused twigs. Evaporative losses increased dramatically above T_a of 38°C, and higher body water contents in the cicadas were associated with greater depression of T_b below T_a. The 13-year periodical cicada, *Magicicada tredecem*, shows limited cooling ability, which may represent the ancestral condition in cicadas (which are apparently tropical in origin), while the more efficient cooling mechanism of *D. apache* may be a derived evolutionary response which works well in low humidities (Toolson and Toolson 1991) (see also Section 4.1.1). Endothermy in cicadas is associated less with flight than with reproductive behaviour, and enables male cicadas to sing in the rain or after sunset, when bird predators are inactive (Sanborn 2000; Sanborn *et al.* 1995). Singing involves intense activity of the tymbal muscles in the first abdominal segment (Josephson and Young 1979), unlike stridulation in katydids which depends on flight muscles (Stevens and Josephson 1977).

6.4 Butterflies: interactions between levels

Butterflies—a relatively small component of the Lepidoptera—are almost exclusively diurnal so their thermal ecology differs greatly from that of moths. Butterflies fly with elevated T_{th}, which depends mainly on the balance between radiative heat gain during basking and convective heat loss during flight. Basking postures are dorsal, lateral, (less common), and reflectance basking, in which the dorsal wing surfaces are assumed to reflect heat on to the body, but the latter is somewhat controversial (cf. Heinrich 1990; Kingsolver 1987). Convective heat exchange also occurs during basking and the wings have both absorptive and

shelter functions. Most radiant heating occurs at the pigmented wing bases, and heat is then transferred to the thorax by conduction and convection of trapped warm air, as demonstrated in classic experiments by Wasserthal (1975). Butterfly wings represent a compromise between competing demands of selection for crypsis, reproductive success, and thermoregulation, and colour and surface structure of the wings (Schmitz 1994) are important for all these functions. In many butterfly species females are conspicuously larger than males (owing to selection for increased fecundity), with marked thermoregulatory and behavioural consequences. The reduced wing loading of males permits flight at a lower wingbeat frequency and thus at a lower T_{th}, but males are also more vulnerable to convective cooling (Gilchrist 1990; Pivnick and McNeil 1986).

Small pierid butterflies in western North America have been used as model insect ectotherms by Watt and Kingsolver in a series of comprehensive studies at all levels of organization. These sulphurs and whites (*Colias, Pieris, Pontia*) live in open areas and have high flight speeds, with vigorous flight being confined to a high and narrow range of T_b (35–39°C) which is common to species in a wide range of habitats and thermal conditions (Watt 1968). Flight occurs only in bright sunlight and at low wind speeds, and never outside the temperature range of 29–41°C, so may be limited to a few hours a day. The highest wingbeat frequencies are also measured in the T_b range of 35–39°C (Tsuji *et al.* 1986). (Note that this optimum range of T_b is similar to that of large endothermic insects.) The physical basis of behavioural thermoregulation by basking and heat avoidance, with wings oriented perpendicular or parallel to the sun, has been extensively investigated. Models of heat balance have been developed for basking (e.g. Kingsolver 1983) and flying pierids, in which testing is more difficult (Tsuji *et al.* 1986). Wing patterns and basking postures interact to determine body temperatures, and manipulation of wing pattern (butterfly wings are easily painted) causes changes in wing angle during basking (Kingsolver 1987). Heat gain in flight is primarily through solar radiation, not metabolism, and there is no regulation of heat loss by shunting excess heat to the abdomen.

Weather plays a major role in the population dynamics of insects, and we now briefly touch on two case studies where it has been possible to explore the mechanisms involved (Kingsolver 1989), and a third example involving predation and its effect on the evolution of butterfly wings and body shapes. Pierid butterflies have been a favourite subject of experimental approaches to phenotype–environment interactions, not often accomplished in physiological ecology.

6.4.1 Variation at the phosphoglucose isomerase locus

Phosphoglucose isomerase (PGI) (see also Section 5.2.2) is involved in fuel supply to the flight muscles, catalysing an intervening step in glycolysis. Genetic variation at the PGI locus of *Colias* butterflies has been exploited as a model system for examining the fitness implications of different phenotypes (for review see Watt 1991). There are four major PGI alleles, and *in vitro* study shows that their thermal stability is inversely related to kinetic effectiveness. This trade-off between kinetics and stability of the enzyme should maintain the polymorphism in a thermally fluctuating environment. Among genotypes in the wild, thermally stable allozymes are favoured under warm conditions as predicted, while the most kinetically effective genotypes fly over a wider temperature range and survive better in low-to-moderate temperature habitats than their relatives with the thermally more stable genotypes (Watt *et al.* 1983). Much of the selection on the PGI locus appears to operate at more extreme temperatures. There is, thus, good correspondence between whole-organism results and biochemical properties. The major fitness components of adult survival and male mating success vary predictably with PGI genotype, and egg-laying by female *Colias* also depends on flight performance throughout the day, as females lay eggs singly and fly between scattered host plants. As predicted, the thermally stable genotypes have the lowest fecundity in a cool habitat (Watt 1992). Adult pierids have lifespans of a few days in the field, and their populations may be severely limited by the inability of females to lay their full complement of eggs (Kingsolver 1989).

Climate-related variation in PGI genotypes and other components of the thermal response have also been explored in chrysomelid beetles (Dahlhoff and Rank 2000; Neargarder *et al.* 2003; Section 5.2.2).

6.4.2 Wing colour

Many examples of both morphological plasticity (which is fixed during development) and physiological plasticity (which is reversible, and has traditionally been termed acclimation) are thought to be adaptive responses to variation in the environment. The beneficial acclimation hypothesis proposes that acclimation has a fitness benefit in the environment that caused it, but the adaptive signficance of acclimation has often been questioned (Section 5.2.1). Field experiments can show whether plasticity is adaptive (Kingsolver and Huey 1998). Seasonal polyphenism is a form of phenotypic plasticity, and seasonal changes in the wing colour of butterflies are frequently cited as examples. The plasticity of wing melanization and its fitness-related effects have been investigated in detail in western white butterflies, *Pontia occidentalis* (for review see Kingsolver and Huey 1998), using measurement (and sometimes manipulation) of a variety of quantitative traits, combined with tests of flight performance in natural environments. Melanization of the hindwings (the ventral surface and the base of the dorsal surface) is sensitive to developmental photoperiod and increases in spring. Basal melanin directly affects T_b achieved during basking, and the beneficial acclimation hypothesis would argue that darker wings provide an advantage by increasing activity time in cool seasons, while paler wings will reduce the risk of overheating in summer.

Alternative wing phenotypes were generated either by rearing *P. occidentalis* under short-day and long-day conditions, or by painting the wings of wild-collected individuals. Butterflies were then released into the field for subsequent monitoring of the effects on mating success and survival (Kingsolver 1995a, 1996). In the first study, mark–recapture of the paler butterflies reared under long-day conditions showed better survival during summer. The second study, using butterflies with painted wings, served to eliminate any phenotypic

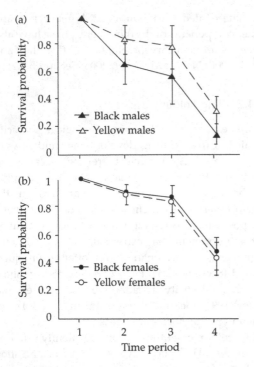

Figure 6.10 Survival of artificially blackened male butterflies (*Pontia occidentalis*) is less than for artificially yellowed butterflies (a), whereas for females paint colour has no effect (b).

Note: The four time periods are for different dates in July 1993. SEs for survival probablilities are indicated.

Source: *American Naturalist*, Kingsolver, **147**, 296–306. © 1996 by The University of Chicago. All rights reserved. 0003-0147/96/4702-0008$02.00

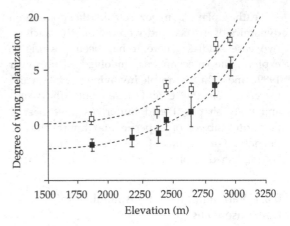

Figure 6.11 Wing melanization in *Colias philodice eriphyle* butterflies increases with elevation in seven populations.

Note: Mean ± SE, males shown by closed squares and females by open squares.

Source: Ellers and Boggs (2002).

differences other than colour which might have resulted from the rearing environment. Artificially blackened male *Colias* butterflies resembled the spring phenotype and showed lower survival probability under summer conditions than yellowed butterflies resembling summer phenotypes, but paint treatment had no effect on the less active and initially darker females (Fig. 6.10). Release of different cohorts of butterflies in different seasons showed that plasticity was sometimes, but not always, adaptive, providing partial support for the beneficial acclimation hypothesis. This is an important experimental study of natural selection. Short-term unpredictability in the weather probably prevents a more precise match between phenotype and environment (Kingsolver and Huey 1998). Watt (1991) has also stressed that habitat variability

is an important constraint on the precision of adaptation. The thermoregulatory phenotypes of *Colias* butterflies (colour and insulation) are less than optimal under average weather conditions because of the risk of occasional overheating.

There is, thus, strong selection for the thermal benefits of melanization in Pieridae, whether genetically or environmentally controlled, and wing pigmentation differs between the sexes, between spring and summer generations in a population, between populations of the same species, and between species. Figure 6.11 shows the increase in wing melanization in *Colias philodice eriphyle* butterflies with elevation in seven butterfly populations (Ellers and Boggs 2002). The other thermally relevant trait is insulating fur (modified scales) on the ventral thorax, which is thicker in *Colias* species living at higher altitudes: Together with wing colour, this results in identical flight temperatures and thermoregulatory behaviour despite habitat differences (Kingsolver 1983). Photoperiod is a cue not only for increased melanization of the wing bases, but in *C. eurytheme* it has been shown to control increases in insulation in cooler seasons (Jacobs and Watt 1994).

6.4.3 The influence of predation

Flight costs are low in butterflies owing to large wings and low wingbeat frequencies, but large wings also increase their visibility to aerial predators. Tropical butterflies are broadly divided into two groups where predator avoidance is concerned. The first group comprises unpalatable species. Defensive chemicals may be sequestered to make them unpalatable, and these species show warning colouration and fly slowly and regularly. Palatable species, on the other hand, fly fast and erratically, which can be energetically expensive (Dudley 2000). The impact of insectivorous birds and bats has thus selected for enhanced flight performance. This has been the subject of some elegant research involving neotropical butterflies, in which predation shapes morphology, flight patterns and thermal physiology (Chai and Srygley 1990; Srygley and Chai 1990). Palatability is assayed by means of the feeding response of an insectivorous bird, the rufous-tailed jacamar (illustrated in Dudley 2000): few butterfly species are intermediate in acceptability to this bird (Srygley and Chai 1990). A phylogeny of butterflies showing the distribution of palatability status is presented by Marden and Chai (1991).

Palatable butterfly species tend to be restricted to times and habitats with high solar radiation, while the reduced demand for rapid flight in unpalatable species favours cooler microhabitats and lower T_{th} (Srygley and Chai 1990). The proportion of palatable species increases in more open microhabitats, and they show higher activity during the middle of the day, resulting in significantly higher T_{th} (by 4°C). The lower T_{th} of unpalatable species is associated with broader spatial and temporal niches. Body mass itself is not correlated with palatability, but fast-flying palatable species have a greater thoracic mass, shorter bodies and shorter wings, and the centre of mass is more anterior, leading to better maneuverability in flight (Chai and Srygley 1990). This increase in flight muscle performance is associated with higher body temperatures. For 53 species of butterflies in the families Papilionidae, Pieridae, and Nymphalidae, Fig. 6.12 shows that palatable species have higher temperature excesses ($T_{th} - T_a$) and lower indices of body shape (= ratio of body length to thoracic width) (Chai and Srygley 1990). Body shape was a good predictor of T_{th}, flight speed, and the response of captive jacamars. In these neotropical butterfly assemblages, Srygley and Chai (1990) also

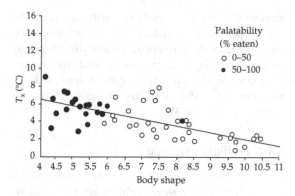

Figure 6.12 Relationship between temperature excess (T_x) and body shape (ratio of body length to thoracic width) for 53 species of butterflies of varying palatability.

Source: *American Naturalist*, Chai and Srygley, **135**, 748–765. © 1990 by The University of Chicago. All rights reserved. 003-0147/90/3506-0001$02.00

found that thermal characteristics of some mimics resembled those of their unpalatable models. Marden and Chai (1991) measured flight muscle ratios (the ratio of flight muscle to total mass) of 124 species of butterflies and diurnal moths in Costa Rica, and found significant effects of palatability, mimicry, and sex. Some unpalatable Mullerian mimics 'had so little flight muscle that they could just barely counteract gravity'. However, decreasing the flight muscle ratio in two pierid butterflies by adding tiny weights resulted in minor effects on survival, even for the more palatable species (Kingsolver and Srygley 2000).

6.5 Regulation by endothermy

Preflight warm-up rates and flight temperatures are strongly determined by body mass (Stone and Willmer 1989b). Heat loss is proportional to surface area and MHP is proportional to thoracic volume, with the result that at small body sizes convective heat loss predominates. In small insects endothermic abilities do not necessarily imply stable body temperatures. It is common for T_{th} to be elevated by endothermy during flight but not regulated, as in gypsy moths *Lymantria dispar* (Lymantriidae) in which T_{th} in free flight remains 6–7°C above T_a (Casey 1980). This species has a body mass of approximately 100 mg and the wings are large, so power requirements for flight are low and immediate flight is possible without preflight warm-up. From comparative data on moths of different families it is apparent that differences in metabolism, flight performance and the extent of thermoregulation are largely determined by morphometrics (for noctuids and geometrids see Casey and Joos 1983).

6.5.1 Preflight warm-up

Preflight thermogenesis involves simultaneous contractions of opposing flight muscles, so that all the energy is degraded to heat, and the metabolic power output during warm-up is as high as during flight (Bartholomew *et al.* 1981; Casey and Hegel-Little 1987). Stone and Willmer (1989*b*) investigated the importance of body size and phylogeny in the warm-up rates of a wide range of female bees from six families. They demonstrated a positive correlation between warm-up rates and body mass after controlling for phylogeny and thermal environment (characterized by the minimum T_a for foraging of each species). Bees from cooler environments fly at lower T_a and have greater endothermic abilities, shown by their higher flight temperatures and warm-up rates. An extreme example is the small solitary bee *Anthophora plumipes* (Anthophoridae) with a mean warm-up rate of $12°C\,min^{-1}$, and this species has been the subject of detailed study (Stone 1993, 1994*a*; Stone *et al.* 1995). Body mass ranges from 100 to 240 mg, females being larger, but males and females do not differ in warm-up rates after allowance is made for the effects of mass. Larger bees warm faster than small bees but also warm to higher T_{th} than small bees, and these two opposing influences may cancel out effects of mass on the duration of warm-up (Stone 1993). Correlations between body mass and warm-up rate have been demonstrated within this species (Stone 1993), among 19 species of the genus *Anthophora* (Stone 1994*b*), and across the Apoidea as a whole (Stone and Willmer 1989*b*). Even the smallest species investigated, a 10 mg halictid bee, elevated its T_{th} by 2–3°C before flight (Stone and Willmer 1989*b*).

Measurement of metabolic rate during preflight warm-up in *M. americanum* (Lasiocampidae) (Casey and Hegel-Little 1987) shows the importance of

T_a: heat production is related to T_{th}, passive heat loss is proportional to $T_{th} - T_a$ (the temperature excess), and so the cost of warm-up is inversely related to T_a. Stone (1993) calculated the energetic cost of warm-up in *A. plumipes* under different conditions. At T_a of 9 and 21°C and with a temperature excess of 6°C in each case, passive cooling accounts for 70 and 30 per cent, respectively, of the heat generated. It is not surprising that declining warm-up performance in *A. plumipes* in the laboratory is dramatically improved by the provision of sucrose solution. Endothermic warm-up is often supplemented by basking, especially at low T_a, leading to considerable savings in costs and time required. For example, a male *A. plumipes* basking at T_a of 9°C can achieve T_{th} of 20°C without metabolic cost (Stone *et al.* 1995). Male carpenter bees, *Xylocopa capitata* (Anthophoridae), use a combination of basking and wing buzzing to maintain T_{th} at a mean of 41°C during stationary intervals in territorial patrolling (Louw and Nicolson 1983). This buzzing with outstretched wings is not the same as preflight warm-up (in which the wings are motionless and folded dorsally) but serves to maintain readiness for immediate flight.

6.5.2 Regulation of heat gain

Do insects vary their flight performance for thermoregulatory as well as aerodynamic purposes? Metabolic rate in flight has been shown to be independent of T_a in various large endothermic insects: for example, sphinx moths *Hyles lineata* (Casey 1976), bumblebees *Bombus* spp. (Heinrich 1975), carpenter bees *Xylocopa capitata* (Nicolson and Louw 1982). If $\dot{V}O_2$ in flight is unchanged with T_a, heat production must be constant and regulation of T_{th} can only be achieved by heat loss. This generally accepted picture (Heinrich 1993) changed when Harrison *et al.* (1996) demonstrated that free-flying honeybees subjected to a rise in T_a from 20 to 40°C decrease their wingbeat frequency and MHP (calculated from CO_2 production). Variation in MHP accounts for most of the thermoregulation of flying honeybees between T_a of 21 and 33°C, whereas evaporative heat loss is also important between 33 and 45°C (see the negative water balance at high T_a in Fig. 4.5) (Roberts and Harrison 1999).

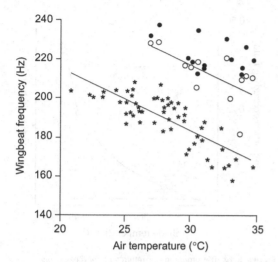

Figure 6.13 Wingbeat frequency as a function of air temperature in *Centris pallida* (Anthophoridae) flying in a respirometer (circles) and hovering in the field (stars).

Note: The negative relationship between wingbeat frequency and T_a was evident by the fourth minute of flight in the respirometer (open circles), but not during the first minute (closed circles).

Source: Roberts and Harrison (1999).

Similarly, field and laboratory measurements show that *Centris pallida* (Anthophoridae) decreases wingbeat frequency (Fig. 6.13) and MHP at high T_a, thereby achieving very precise thoracic thermo-regulation (Roberts and Harrison 1998; Roberts *et al.* 1998). Differences between field and respirometer measurements of wingbeat frequency are attributed by these authors to difficulties with maintaining continuous flight in closed-system respirometry. It seems likely that thermoregulation via modulation of MHP may be more widespread than is currently appreciated (Harrison and Roberts 2000). It is not clear whether the reduced metabolic rate at high T_a is a secondary consequence of greater efficiency of flight at high T_a, or a regulated response (Roberts and Harrison 1998).

The optimal temperature for force production during tethered flight of honeybees is 38°C (Coelho 1991), and flight metabolic rates of honeybees may decrease at T_{th} above or below 38°C, which can account for some of the discrepancies between different studies (Harrison and Fewell 2002). Sources of individual variation in metabolic rates of flying honeybees are illustrated in Fig. 3.19 (Chapter 3) and bees flying at high T_a (45°C) have among the

lowest measured values. It follows that high air temperatures must contribute to an increase in food collecting efficiency and colony performance (Harrison and Fewell 2002).

Power output is also altered during flight in the green darner dragonfly *Anax junius* (Odonata, Aeschnidae). Field measurements under various conditions show increased wingbeat frequency and inferred heat production at low T_a, together with a reduction in the proportion of time spent glid-ing (May 1995). Giant Palaeozoic dragonflies may have used gliding to reduce endogenous heat production (Dudley 2000).

6.5.3 Regulation of heat loss

Essentially there are three ways in which endothermic insects can regulate heat loss: by convective cooling, using the haemolymph as a heat-exchanging fluid, or evaporative cooling. At thermal equilibrium, MHP equals heat loss:

$$MHP = (T_{th} - T_a)C + RHL + EHL, \qquad (1)$$

where C is convective thermal conductance, RHL is radiative heat loss and EHL is evaporative heat loss (Harrison and Fewell 2002). Radiative heat gain is negligible in indoor respirometry experiments (Roberts and Harrison 1999) but can be substantial in the field (Cooper *et al.* 1985). Convective cooling from a flying insect depends on air speed and the difference between body and air temperatures, so tends to decrease at higher T_a when the temper-ature excess is less. Convective heat loss is dominant in small insects and most insects fly with T_{th} close to T_a. Increases in wingbeat frequency with T_a have been measured in small stingless bees (*Trigona jaty*, 2.5 mg) and in flies (Unwin and Corbet 1984), and this will have a net cooling effect because surface effects predominate in these small insects. Metabolic rate is independent of flight speed in free-flying bumblebees (Ellington *et al.* 1990), and increasing flight speed has been suggested as a mechanism of increasing forced convection at high T_a (Heinrich and Buchmann 1986).

Convective heat loss may also be varied by altering the distribution of heat within the insect: shifting heat to the abdomen or head increases the surface area for convective and radiative heat loss (Roberts and Harrison 1998). Physiological heat transfer uses the haemolymph as a heat-exchanging fluid between body compartments, and this transfer is minimized during warm-up and maximized during overheating. Countercurrent heat exchange in the bumblebee petiole between thorax and abdomen ensures that heat is sequestered in the thorax, except under conditions of heat stress when the exchanger is bypassed; the circulatory anatomy is described for *Bombus vosnesenskii* by Heinrich (1976). Active heat transfer to the abdomen is highly variable in bees and not explained by phylogeny (Roberts and Harrison 1998). Although best known for *Bombus* and possibly *Xylocopa* (Heinrich and Buchmann 1986), it also occurs at high T_a in the much smaller *A. plumipes* (Stone 1993), and has been found in other groups such as beetles (Chown and Scholtz 1993).

Evaporative cooling has traditionally been considered negligible, but is important in bees flying at high T_a such as *Apis mellifera* (Louw and Hadley 1985) and *Xylocopa varipuncta* (Heinrich and Buchmann 1986). It will be especially effective at the low RH of desert air, and honeybees returning to the hive with a fluid droplet on the tongue can cool themselves down and reduce the load carried at the same time. Foraging honeybees (*A. mellifera caucasica*) in the Sonoran Desert fly at air temperatures up to 46°C. Pollen foraging decreases at high T_a, and water and nectar foragers extrude droplets and have lower head and thoracic temperatures as a result (Cooper *et al.* 1985). Figure 6.14 shows that at T_a of 40°C, 40 per cent of nectar gatherers are carrying droplets. The resulting evaporative cooling enables honeybees to achieve head temperatures below T_a during flight at high T_a (Heinrich 1980; Roberts and Harrison 1999). Regurgitation of fluid has also been observed in wasps returning to the nest at high T_a (Coelho and Ross 1996). These hymenopterans can not use the abdomen as a heat radiator as do bumblebees and their abdominal temperature tracks T_a, although some evaporation may occur from the abdominal surface of honeybees at $T_a > 40°C$ (Roberts and Harrison 1999). Evaporative heat loss, while considerable, may be less important to the thermoregulation of flying *X. capitata* and *C. pallida* at

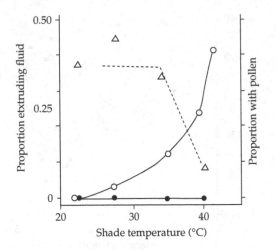

Figure 6.14 The proportion of returning honeybees (*Apis mellifera caucasica*) extruding fluid droplets increases with air temperature in the Sonoran Desert. Nectar and water collectors shown by hollow circles. Pollen gatherers (solid circles) did not extrude fluid, and the proportion of bees returning with pollen (triangles) declined at high T_a.

Source: Cooper *et al.* (1985).

more moderate T_a (up to 35°C) (Nicolson and Louw 1982; Roberts *et al.* 1998).

6.6 Endothermy: ecological and evolutionary aspects

Facultative endothermy is an option available only to adult insects with functional flight muscles. Other than preflight warm-up and flight, the flight muscles are used for endothermy during terrestrial activity (Bartholomew and Casey 1977), for stridulation in katydids (Stevens and Josephson 1977), and also for brood incubation and regulation of nest temperature in social insects (for review see Heinrich 1993). Colony homeostasis provides not only an incubator for the brood, but also a thermal refuge in which individuals can exploit temperature gradients to regulate their own T_b (Section 6.6.2). Both social and individual thermoregulation depends on flight muscle activity of adults.

In some beetles MHP can be independent of flight or preparation for flight, and examples of terrestrial endothermy have been recorded in large, nocturnally active beetles (Bartholomew and Casey 1977). Synchronized mating activity in male rain beetles (*Pleocoma* spp., Scarabaeidae) is characterized by large temperature excesses maintained during both flight and ground searching (Morgan 1987). Competition for the dung of mammalian herbivores has been an important factor in the evolution of endothermy in ball-rolling dung beetles, which maintain elevated T_b while making or rolling balls, and in which contests for possession of already constructed balls are usually won by the beetle with higher T_b which has most recently ceased flight (Heinrich and Bartholomew 1979). Also explicable in terms of competition for dung is the temporal partitioning of flight times in dung beetle communities, and 11 species of the genus *Onitis* show distinctive daily flight behaviour which is dependent on light intensity (Caveney *et al.* 1995). Smaller species of *Onitis* (<0.4 g) are less likely to fly at night or at dawn owing to excessive radiant heat loss. The unusually low conductance of another scarabaeid, the foliage-feeding *Sparrmannia flava* (0.8 g) makes its thermal physiology more like that of larger beetles, and its activity peaks at 3–4.30 AM in the Kalahari desert (Chown and Scholtz 1993), so maximizing predator avoidance. Endothermy enables nocturnal activity in moths and beetles, and this is a major difference from its significance in bees.

Constant body temperature is considered an advantage because biochemical systems evolve to function best at a single temperature. Endothermy in flight is an inevitable biophysical consequence of the relationship between body size and high rates of energy expenditure, and preflight warm-up is necessary because muscles become adapted to operate at high temperatures (Heinrich 1977). The evolution of metabolic architecture is complicated by temperature change, because the temperature dependence of enzyme reactions in a pathway will differ, and so precise regulation of the metabolic pathway will be possible only in a narrow temperature range (Watt 1991). For many insects there is a common T_{th} ceiling around 40°C, near the upper lethal temperature. When capacities and loads are compared at the biochemical level it is evident that, during flight, enzymes in the flight muscles of honeybees operate closer to V_{max} than in other muscles which have been studied (Suarez *et al.* 1996). When combined with high T_a, high

rates of MHP can lead to flight temperatures which come close to upper thermal limits. This is especially true of bees with mating systems involving patrolling or territorial behaviour by males, such as hovering *C. pallida* (Anthophoridae) in a Californian desert with T_{th} at 47°C, within 2–3°C of lethality (Chappell 1984).

6.6.1 Bees: body size and foraging

Even the smallest bees warm up endothermically before flight, although at relatively low rates (Stone and Willmer 1989*b*). Andrenid bees weighing only 29 mg, *Andrena bicolor*, bask inside flowers and fly at T_{th} of 22–31°C, with an average T_{th} during free flight of 27°C (Herrera 1995*a*). These values are about 10°C lower than those reported for many species of Apidae and Anthophoridae, so *A. bicolor* appears to be ectothermic; but many small solitary bees use their endothermic warm-up abilities rather infrequently (Willmer 1991*b*).

The phylogenetically based comparative study of Stone and Willmer (1989*b*) has already been discussed in the context of preflight warm-up in bees. The best endothermic abilities are seen in larger species and those active in cool climates. Bees from cooler environments fly at lower T_a and have higher flight temperatures and warm-up rates. For warm climate bees, tolerance of high T_{th} and T_a is more important, which means that warm-up at low T_a and activity at high T_a are not compatible. The superior thermoregulation of bumblebees compared to solitary bees is largely, but not entirely, explained by size differences (Bishop and Armbruster 1999). *Bombus* has a long history of specialization to cold climates and the richness of this group is greatest in the high latitudes of the Northern Hemisphere (Williams 1994).

Anthophora plumipes is active early in the European spring and can fly at low T_a by tolerating low T_{th} (Stone and Willmer 1989*b*). Males are smaller than females and both sexes vary considerably in size. Male–female differences in endothermic physiology, and the relationship to foraging and reproductive behaviour, have been considered in detail by Stone and colleagues (Stone 1994*a*; Stone *et al.* 1995). Provisioning activity in female *A. plumipes* involves long working days, and the mass of the nectar and pollen loads collected increases with T_a. At low T_a only the larger females are able to collect pollen, and thus complete provisioning of cells. They are also able to forage earlier in the morning, and these factors influence the body size of their offspring. Reproductive success in male *A. plumipes* is also dependent on weather and body size: larger males can forage for longer and have competitive advantages in conflicts with other males. There is no difference in endothermic abilities between males and females of the same mass, but the sexes show marked differences in thermal biology.

In *A. plumipes* the slope of the relationship between voluntary flight temperature (VFT—when a bee initiated tethered flight) and T_a is lower than that between stable flight temperature (SFT—measured after a period of continuous flight) and T_a, which indicates that there was significant cooling of the thorax during tethered flight at low T_a (Fig. 6.15) (Stone 1993). Decreases in T_{th} after the initiation of flight have been reported in many insects. Thermoregulation is thus more effective before than during flight, and comparison of T_{th}/T_a gradients before and during flight, where available, shows that this is true of bees, moths, and beetles (Stone 1993, 1994b). Apparently endothermic insects are unable to compensate fully for convective losses at low T_a or MHP at high T_a. Voluntary flight temperature and SFT are also significantly correlated

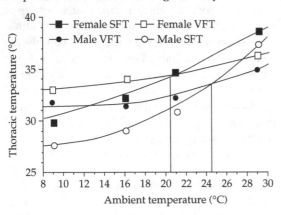

Figure 6.15 Voluntary (VFT) and stable flight temperatures (SFT) as functions of T_a for male and female *Anthophora plumipes*.

Note: VFT is the temperature at which the bee initiates tethered flight and SFT is the temperature after a period of continuous flight. Curves are fitted to the means for each sex.

Source: Stone (1993).

with body mass within the genus *Anthophora* and across the Apoidea (Stone 1994b).

Thermal constraints on flight activity limit the effectiveness of bees as pollinators (Corbet *et al.* 1993). Diurnal changes in microclimate define a thermal window suitable for flight, but floral resources also vary throughout the day. The activity patterns of female bees are determined by interactions between their physiology, changing floral resources, and surrounding microclimatic changes, as well as by nest construction in solitary species (Stone *et al.* 1999). Temperature and water relations of desert bees have been compared with those of temperate species by Willmer and Stone (1997), who make the interesting suggestion that endothermy in bees may have evolved in arid zones. Unlike ectothermic ants and beetles, bees of desert habitats show similar warm-up rates, cooling mechanisms, and upper critical temperatures to their temperate relatives. (Because bee activity is accompanied by heat generation, there has to be a safety margin.) Size strongly influences the activity of bees in arid environments, either on a daily basis (examples in Willmer and Stone 1997) or seasonally, as when mean body size of solitary bees decreases through spring and summer in Mediterranean scrublands of Israel (Shmida and Dukas 1990). Activity patterns of desert bees show avoidance of midday heat, and the endothermy needed for cold desert mornings and late afternoons could have pre-adapted bees for invasion of cool-temperate areas. Arid environments are centres of diversity for bees, including the genus *Anthophora*, and patterns of nectar secretion in these environments may have shaped the high levels of endothermy in this genus (Stone 1994b). Flowers pollinated by *Anthophora* species have deep, tubular corollas and peaks in nectar secretion early or late in the day, and activity patterns of the bees are strongly bimodal. Morning and evening peaks in nectar secretion may minimize water stress to desert plants, but may also be a response to the thermal tolerances of their pollinators.

Honeybees of the genus *Apis* are all of tropical origin and only *A. mellifera* occurs in northern climates. Dyer and Seeley (1987) measured flight temperatures in three Asian species of honeybees and compared them with data for *A. mellifera*.

Although there was a fivefold range of body mass in foragers of the four species, the temperature excesses did not increase with body size as expected. *Apis mellifera* and *A. cerana*, the two species of intermediate body mass, have disproportionately high wing loading, flight speed, and MHP (calculated from $T_{th} - T_a$ in flight and thoracic conductance). These two species build cavity nests. Slower physiology is apparent in *A. florea* and *A. dorsata*, the smallest and largest species respectively, which build exposed combs protected by curtains of bees which are effective through sheer numbers rather than high activity levels (Dyer and Seeley 1987). In a follow-up study, Underwood (1991) studied the thermoregulation of the largest honeybee, *A. laboriosa*, in Nepal, and it appears that this species can be grouped with *A. florea* and *A. dorsata* as a relatively low-powered, open-nesting honeybee.

6.6.2 Bees: food quality and body temperature

In bees, it appears that food quality modulates thermal behaviour. For example, workers of the Asian honeybees *A. cerana* and *A. dorsata* landing at highly congested feeders tend to have higher T_{th}, and congestion depends on food quality (Dyer and Seeley 1987). Similarly, the Himalayan honeybee *A. laboriosa* maintains high temperature excesses when arriving at feeders to collect concentrated sugar solution (Underwood 1991). Thermal imaging has proved a valuable technique in studying this variability in thermoregulation of honeybee foragers under field conditions. Honeybees regulate T_{th} at higher levels and more accurately for fast exploitation of profitable food sources (Schmaranzer and Stabentheiner 1988). Honeybees foraging on different plant species also have variable T_{th}, with dandelion foragers being 10°C warmer than those visiting sunflowers (Kovac and Schmaranzer 1996). Perhaps the observation that bumblebees allow their T_{th} to drop below the minimum for flight while foraging on dense inflorescences, especially if they contain low reward flowers (Heinrich and Heinrich 1983), is a result of motivational state, not simply an energy conservation mechanism. The T_{th} of vespine wasps also depends on food concentration (Kovac and Stabentheiner 1999), and the T_{th}

of water-collecting honeybees measured by thermography resembles that of bees feeding on 0.5 M sucrose, indicating similar motivation (Schmaranzer 2000). Similar results, showing T_{th} to be 3°C higher in bees feeding on high sucrose concentrations, have been obtained using thermocouples (Waddington 1990).

Not surprisingly, the metabolic rate of honeybees likewise varies with the reward rate at the food source and the motivational state of the bees. Direct effects of nectar load on metabolic rate (Wolf *et al.* 1989) can be eliminated by training bees to collect food in a respirometer so that they need not transport it (Moffat and Núñez 1997). The metabolic rate of free-flying bees collecting food in a much larger respirometer is also inversely proportional to T_a at constant sucrose flow rate (Moffatt 2001), supporting previous studies showing variation of heat production during flight (Roberts and Harrison 1999).

The beauty of infrared thermography is that it does not disturb social interactions such as the dancing behaviour of honeybees or the unloading of nectar by trophallaxis. Graduated thermal behaviour occurs during food unloading in the hive as well as at the feeding site. The temperature of dancing bees recruiting their nestmates increases with food quality and the number of brood cells, and decreases with distance of the food source from the hive and the amount of stored honey (Stabentheiner 1991, 2001). Foraging bees returning from feeders with high flow rates (8.2 μl min^{-1} of 50 per cent w/w sucrose solution) have high T_{th} during trophallaxis and transfer the food quickly. The receiver bees also raise their T_{th} during trophallaxis, and Farina and Wainselboim (2001) use a thermogram to show a 3.7°C increase in T_{th} of a receiver bee during 18 s of food transfer, equivalent to a heating rate of 12.3°C min^{-1}. During this time, T_{th} of the donor bee was relatively unchanged. Figure 6.16 illustrates linear increases in T_{th} of nine receiver bees during contact with donors that had fed at different flow rates (with the exception of one interaction at the lowest flow rate). Calculations show that the temperature of the exchanged solution can not account for the increase in T_{th} of receivers, and the linear increase also suggests active warm-up rather than passive heat transfer.

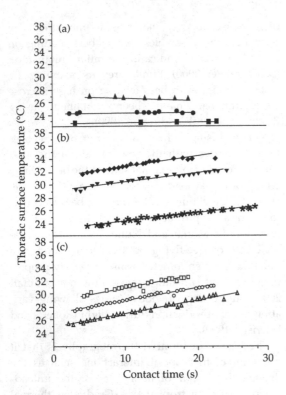

Figure 6.16 Thoracic surface temperatures of receiver honeybees during trophallaxis. Donor bees have returned from three feeders supplying 50% w/w sucrose solution at different flow rates, (a) 1.0 µl min⁻¹, (b) 2.4 µl min⁻¹, and (c) 8.2 µl min⁻¹.
Source: Farina and Wainselboim (2001).

It is not known whether receivers monitor the head temperature of the donor or the quality of the donation. From the colony point of view, increased T_{th} of the receiver bees ensures that their activity level is intensified and nectar from more profitable sources will be processed faster (Farina and Wainselboim 2001).

While the endothermic abilities of insects are determined by body size and shape, phylogeny, and environment (Stone and Willmer 1989*b*), sociality creates an entirely new dimension. Solitary bees show sophisticated thermoregulation compared to other insects, but they lack the thermal benefit and refuge of a stable nest microclimate (Willmer 1991*b*). Body temperatures of individual honeybees are much more than a consequence of flight metabolism and are finely regulated according to expected and actual foraging gains. As suggested by Waddington (1990), the thermal strategies of honeybees ensure that when profits are low the bees cool down and save energy, which has long-term benefits for the colony, but when profits are high the bees warm up and maximize short-term rates of energy gain. While a great deal might be known about the mechanisms of insect thermoregulation, the extent to which thermoregulation influences all facets of insect ecology is only beginning to be explored.

CHAPTER 7

Conclusion

It would seem, then, that the real life of animals (as opposed to the constricted and difficult life that they attempt to live in zoological textbooks) is the end-result of a number of fundamentally different processes.

Elton (1930)

In Chapters 2–6, the ways in which insects respond to different components of the external environment have been dealt with in relative isolation. Nonetheless, each of the chapters has revealed similarities and connections between these responses. In this final chapter, these associations and similarities are discussed more explicitly. In particular, four key themes are explored:

(1) Large-scale patterns, their identification, and ecological and evolutionary implications;
(2) The interdependence of body size and several physiological and life history traits;
(3) Interactions between physiological responses, and underlying covariation in the abiotic environment;
(4) The current and likely future responses of insects to global environmental change.

7.1 Spatial variation and its implications

The simultaneous consideration, in this book, of small-scale mechanisms and large-scale patterns has shown that, in the main, a relatively limited suite of cellular and molecular level mechanisms has given rise to a broad array of individual-level responses, which, nonetheless, show detectable and coherent patterns of taxonomic and environmental variation. In Chapter 1 we demonstrated that much of the taxonomic variation is partitioned at levels above the species in the genealogical

hierarchy. In the following chapters, we drew attention to coherent, though sometimes weak, large-scale spatial variation in traits, including metabolic rate (Chapter 3), water loss rate (Chapter 4) and thermal tolerances (Chapter 5). In our view, one of the most biologically significant results emerging from the large-scale perspective is that, at several spatial scales, there appears to be a general decoupling of upper and lower lethal temperature limits.

7.1.1 Decoupling of upper and lower lethal limits

The apparent decoupling of thermal limits in insects is of interest because it is unusual, at least by the standards of marine species. In many of these organisms, upper and lower lethal tolerances vary in synchrony. Before discussing this issue in detail it is worth pausing to determine just what is meant by coupled or correlated responses. Huey and Kingsolver (1993) argued that if selection results in a shift of the entire performance curve (Fig. 6.1), then performance at high and low temperatures is inversely correlated. That is, if performance improves at high temperature, it declines at low temperature. Similarly, Hercus *et al.* (2000) argue that a positive correlation between mean knockdown times at high temperature and greater cold resistance means an improvement in both cold resistance and heat resistance. Although this terminology appears to be straightforward, it is partly dependent on the way in which performance is measured. In both examples, the authors indicated that a positive correlation meant that an improvement in tolerance of one stress was associated with an increase in tolerance of the other. However, in some circumstances the term 'correlation' could be misleading, depending on the measure of tolerance used. Thus, if knockdown temperatures (critical minima and maxima) were simply plotted against each other, a negative correlation would imply that a decline in CT_{min} accompanies an increase in CT_{max}. This is precisely the converse of the terminology adopted by Huey and Kingsolver (1993). However, if the time to mortality for 50 per cent of the sample at each temperature was plotted, an inverse correlation would be perfectly in keeping with their terminology. In the present discussion, decoupling between high and low temperature responses is taken to mean that they are unrelated, or at best weakly inversely correlated such that an improvement in heat resistance means a decline in cold resistance (i.e. a shift of the performance curve).

In marine species it has long been known that performance at high and low temperatures is often inversely correlated among populations or species. That is, an improvement in high temperature tolerance leads to a decline in low temperature tolerance and vice versa (Prosser 1986; Cossins and Bowler 1987). Building on previous work (Ushakov 1964; Prosser 1986), Pörtner and his colleagues (Pörtner 2001, 2002; Pörtner et al. 1998, 2000) have provided the most extensive exploration of this relationship. In essence, they have argued that in complex metazoans, critical temperatures that affect fitness (i.e. survival and reproduction) are not set by cellular level responses (such as the stress protein response), but are rather set by a transition to anaerobic metabolism. These pejus (= deleterious) temperatures, which are less extreme than traditionally measured critical limits, result from insufficient aerobic capacity of mitochondria at low temperatures, and a mismatch between excessive oxygen demand by mitochondria and insufficient oxygen uptake and distribution by ventilation and circulation at high temperatures. In other words, whole-animal aerobic capacity is limited at both low and high temperatures, and this sets limits to animal performance. Adjustments to temperatures both seasonally, and over evolutionary time are made by altering mitochondrial densities, and this change has concomitant effects on both high and low deleterious temperatures. In consequence, there is an inverse correlation in performance at high and low temperatures when measured in either a population across seasons, or among populations from habitats differing in their thermal regimes. However, the nature of the change in deleterious temperature depends to some extent on whether stenothermal or eurythermal species are being examined. In addition, it appears that the width of the tolerance window between the upper and lower deleterious temperatures varies substantially with latitude, being broader in tropical and temperate than in polar species. That is, tropical and temperate species tend to have a broader tolerance range than do polar ones (Pörtner 2001).

Despite apparently limited information, Pörtner (2001) has argued that although terrestrial species may be more eurythermal, owing to a reduction in cost of ventilation (a result of higher oxygen levels), thermal tolerance limits in these species may be set in a way similar to those of marine taxa. However, in terrestrial insects, decoupling of upper and lower lethal limits at global and regional scales, between populations of the same species, and in selection experiments (see Chapter 5 and Chown 2001), suggests that this is not the case. This does not necessarily imply that oxygen delivery has no effect on thermal tolerance limits. Indeed, in several species anoxia and hypoxia have profound effects on tolerances, generally reducing them substantially (Yocum and Denlinger 1994; Denlinger and Yocum 1998, but see also Coulson and Bale 1991 who show that anoxia actually induces rapid cold hardening in house flies). Nonetheless, a major prediction of Pörtner's hypothesis—that a decline in critical temperature is expected to accompany hypoxia (Pörtner 2001)—which, to date, has been tested only once in insects, has not been supported. An assessment of the effects of both hyperoxia and hypoxia on critical thermal maxima, using a novel technique (thermolimit respirometry—Lighton and Turner 2004), on a tenebrionid beetle, revealed no effect of both treatments (Klok et al. 2004) (Fig. 7.1).

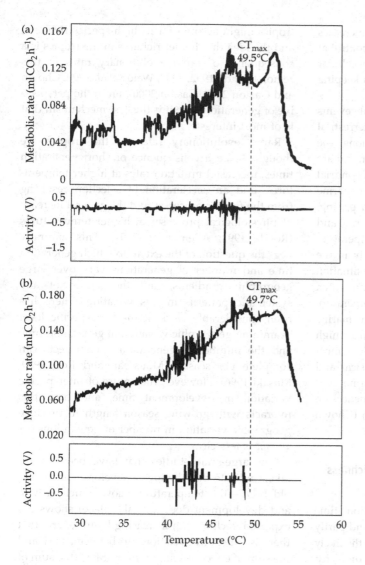

Figure 7.1 Representative data for thermolimit respirometry experiments on the tenebrionid beetle, *Gonocephalum simplex*. (a) 21% and (b) 10% oxygen.

Note: The upper trace shows $\dot{V}CO_2$, while the lower trace indicates activity in arbitrary units, where both negative and positive values indicate activity.

Source: Klok *et al.* (2004).

In other words, the critical thermal maximum is not set by a failure of oxygen delivery, but more likely as a consequence of cellular level damage (see also Barclay and Robertson 2000). This is perhaps not surprising given the highly efficient tracheal system that is responsible for gas exchange in insects (Chapter 3).

By contrast, it appears that insufficient aerobic capacity of mitochondria at low temperature might well be important in setting lower critical limits. In *Pringleophaga marioni* (Lepidoptera, Tineidae) caterpillars, there is a precipitous decline in metabolic rate at the critical thermal minimum (Sinclair

et al. 2004). Moreover, in both honeybees and *Drosophila*, decreasing temperature results in a steady decline in the resting potential of flight muscle neurons, and critical thermal minimum appears to be the temperature at which the Na^+/K^+-ATPase pump can no longer maintain nerve cell polarization to a level where action potentials could be produced (Hosler *et al.* 2000). Thus, lack of energy owing to insufficient aerobic capacity might well set lower limits in insects. In *P. marioni* caterpillars, cells continue to respire, indicated by no difference in metabolic rate in caterpillars that die from freezing and those that survive (this is a freezing tolerant species), but water loss increases rapidly in the former. This indicates that control at the organismal level, rather than at the cellular level, is lost (Sinclair *et al.* 2004). This is in keeping with Pörtner's (2001) hypothesis.

Therefore, although the physiological events needed to attain thermal tolerance in terrestrial insects probably require aerobic conditions, the high and low-temperature responses differ, and are probably not linked by alterations in mitochondrial density. Such profound differences between marine and terrestrial species should not be surprising, given the physical characteristics of the marine and terrestrial environments. For example, temperature change can take place much more rapidly in the terrestrial environment than in the marine situation because air has a lower heat capacity than water. Furthermore, latitudinal variation in temperature variability differs considerably between marine environments, which show low variation at high latitudes, and terrestrial systems, which show most variability at high latitudes. These physical and biological differences between the marine and terrestrial realms may have profound influences on the distribution of diversity within them (Chown *et al.* 2000).

7.1.2 Latitudinal variation in species richness and generation time

Large-scale patterns of insect generation time (partly a function of developmental rate and partly of diapause) can also provide insight into the likely determinants of broad-scale variation in diversity. Of the growing number of macroecological patterns that are exciting renewed interest among ecologists and conservation biologists, the latitudinal gradient in diversity is one of the most well known, but perplexing. As has long been acknowledged, overall taxonomic diversity increases from high to low latitudes (Gaston 2000). The perplexing nature of this general pattern has to do both with its exceptions, and with the plethora of mechanisms proposed to explain it (Rohde 1992).

Among these mechanisms, three have come to the fore in recent debate. These are the extent to which species richness is dictated by available energy, the extent to which the large area of the

tropics might account for its high species richness, and the idea that higher richness in the tropics is a consequence of rapid evolutionary rates in this region (Rohde 1992, 1998; Waide *et al.* 1999; Chown and Gaston 2000; Gaston 2000). From the perspective of generation time it is this last mechanism that is of most interest.

Rapid evolutionary rates in the tropics are thought to be a consequence of short generation times, increased mutation rates at higher temperatures, and an acceleration of selection resulting from the former processes and the general increase of physiological processes at higher temperatures (Rohde 1992; Allen *et al.* 2002). This naturally begs the question of the extent to which generation time and number of generations vary over large geographic gradients, and the nature of any systematic patterns in this variation. There has been considerable work at the intraspecific level examining geographic variation in generation time and the number of generations that insects can complete per season (Mousseau and Roff 1989; Masaki 1996). However, the nature of interspecific variation in development time, and how this interacts with growing season length to produce geographic variation in number of generations *per annum* is less clear.

The large-scale studies that have been undertaken indicate that the lower development threshold (LDT—the temperature below which growth and development does not take place) shows the expected decline with increasing latitude, and that there is a negative relationship between LDT and the sum of effective temperatures (SET) (the sum of day degrees above the LDT required for an insect stage to complete development) (Honěk 1996). These empirical relationships mean that SET should increase with latitude, and that at lower temperatures polar species should have more rapid development rates than their tropical counterparts and vice versa (Honěk and Kocourek 1990). The latter implies equivalent development rates across species from different latitudes, and this is indeed what is found (Fig. 7.2). For total development (egg to adult) the relationship is complicated by variation between species in size, higher taxonomic group membership, and dietary specialization (Honěk 1996, 1999), but considerable variation in

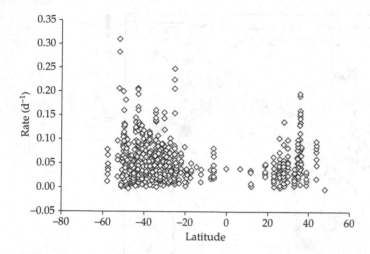

Figure 7.2 Geographic variation in development rate (egg to adult) (negative latitudes are in the North) for a variety of insect species.
Source: Data compiled by Addo-Bediako (2001).

development rates within a given area suggest that a consistent trend might be difficult to detect. Broad overviews of insect voltinism, which should provide greater insight into variation in generation time across space, unfortunately do little more to clarify the situation. Although multivoltinism in tropical species is probably common, some species behave as if the environment is highly seasonal, thus resulting in broad variety of strategies in the tropics (Wolda 1988).

Given these findings, and geographic variation in insect body size that regularly shows no consistent trends between groups and is often the consequence of latitudinal changes in the dominance of higher taxonomic groups (Hawkins and Lawton 1995), the actual form of the relationship between latitude and the number of generations seems remarkably difficult to determine. Thus, although these patterns do not resolve the question of the likelihood of declining generation times towards the tropics, especially owing to the complexities of interactions between size, generation time, voltinism, and feeding strategy (Chown *et al.* 2002*a*), they do suggest that the generation time assumption of the evolutionary rates hypothesis might be more complicated than previously thought. Intriguingly, Fig. 7.2 suggests that generation time actually increases towards the tropics. Moreover, metabolic rates show a decline towards the tropics (Addo-Bediako *et al.* 2002), suggesting that levels of DNA-damaging metabolites, which increase mutation rates, are unlikely to be higher in the tropics than elsewhere. Thus, more

rapid evolution in the tropics, via mutation rates elevated by metabolic damage (Martin and Palumbi 1993), is unlikely. These findings are broadly in keeping with a major study on latitudinal variation in the rates of molecular evolution in birds, which found no support for the evolutionary rates idea (Bromham and Cardillo 2003).

The investigation of insect development rates in this context raises several points that are of more general importance. Most significant among these are the interactions between the determinants of body size and the consequences for spatial variation in development rate and generation time (Chown *et al.* 2002*a*), and the spatial extent of the available data. In Fig. 7.2 the latter appears limited, especially for the tropics.

7.1.3 Spatial extent of the data

Many of the macrophysiological patterns we have explored and sought to understand from a mechanistic perspective have required data covering a range of latitudinal bands. This raises the important question of how geographically extensive are the data available for this purpose (Chown *et al.* 2002*a*).

Perhaps unsurprisingly, the majority of the work on insect physiological ecology appears to have been undertaken in the Holarctic (Fig. 7.3). This clearly has to do with both the geographic distribution of scientists in the field and the restrictions of the investigation undertaken by

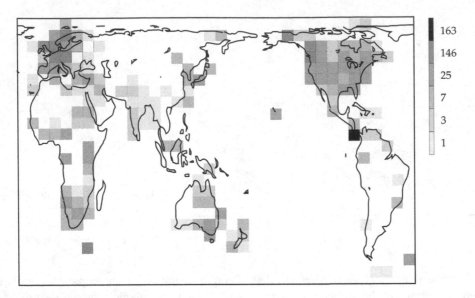

Figure 7.3 Global geographic extent of investigations of metabolic rate, water loss, development rate, upper lethal temperature, lower lethal temperature, and body temperature regulation in insects.

Source: Reprinted from *Comparative Biochemistry and Physiology B*, **131**, Chown *et al.*, 587–602, © 2002, with permission from Elsevier.

Chown *et al.* (2002*a*) (e.g. the non-Anglophone literature, especially from Asia, was largely excluded). Nonetheless, spatial variation in investigations of tolerances, metabolism, development and thermoregulation highlights several fascinating trends. Foremost among these is the tendency for investigations of particular traits to be biased to certain geographic regions. Thus, studies of lower lethal temperatures (LLTs) of insects tend to be undertaken in cold regions, and those of upper lethal temperatures in warm, often arid regions (Addo-Bediako *et al.* 2000). Similarly, investigations of desiccation resistance tend to be most common in desert regions (Addo-Bediako *et al.* 2001), and investigations of development rates in tropical species tend to be scarce (Honěk 1996). These tendencies preclude firm conclusions regarding a range of issues, such as hemispheric variation in responses (Section 7.4), latitudinal variation in development rate, and covariation in environmental temperature and insect responses (Section 7.3) (Chown *et al.* 2002*a*). Clearly, if these questions are to be addressed in the future, then a carefully considered expansion of the geographic extent of insect macrophysiology will have to be undertaken.

7.2 Body size

A second recurrent theme in this book has been the significance of body size-related variation in physiological traits. Indeed, it is widely appreciated that body size and many physiological variables are highly correlated, and that interactions between the latter and life history variables produce the range in body sizes that has been documented for various assemblages (Peters 1983; Schmidt-Nielsen 1984; Kozłowski and Gawelczyk 2002). That individual body size varies through space, therefore, potentially has profound implications not only for physiological functioning, but also for biodiversity as a whole (Gaston and Blackburn 2000; Allen *et al.* 2002). However, we have paid scant attention to spatial variation in body size.

Before discussing patterns in size variation, the mechanisms thought to explain them, and the constraints that might be associated with restriction to a given size, it is worth reiterating that there is considerable feedback between life history and physiological traits, and body size, which determines optimal individual size, and, in consequence, the body size-frequency distributions so typical of animal assemblages. These interactions have been

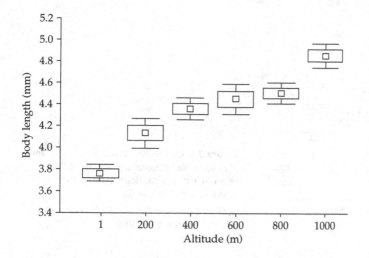

Figure 7.4 Body length variation (mean ± SE, and 2SE) across the altitudinal gradient (m) at the relatively aseasonal sub-Antarctic Marion Island for the weevil *Bothrometopus parvulus*.

Source: Chown and Klok (2003). *Ecography* **26**, 445–455, Blackwell Publishing.

explored in detail by Kozłowski and his colleagues (Kozłowski 1996; Kozłowski and Weiner 1997; Kozłowski and Gawelczyk 2002). Optimal size depends on mortality, assimilation, and respiration rates, all of which are size-dependent (Kozłowski and Gawelczyk 2002). Competitors can influence production rates by usurping or excluding access to resources, and predation and parasitism can alter mortality rates. These interactions also depend strongly on body size, so ultimately determining the optimal size. It is this variation in the size-dependence of mortality and production rates that ultimately produces the right-skewed body size frequency distributions which are so characteristic of assemblages (Gaston and Blackburn 2000).

Interspecific geographic variation in body size seems to be set in a major way by the spatial turnover of higher taxonomic groups (Hawkins and Lawton 1995; Chown and Klok 2003). In light of this finding, the arguments by Kozłowski and colleagues that interspecific patterns may well be an epiphenomenon of those at the intraspecific level, and the paucity of whole-assemblage species–body size frequency distributions (Chown *et al.* 2002*a*, but see Gaston *et al.* 2001), little attention will be given to geographic variation at the interspecific level (also known as Bergmann's Rule). Rather, most attention will be given to intraspecific geographic patterns in body size variation (also known as James' Rule—Blackburn *et al.* 1999).

Two robust, but contrary patterns in the intraspecific geographic variation of insect body sizes

have been identified. In some species, body size increases with increasing latitude (David and Bocquet 1975; Arnett and Gotelli 1999; Huey *et al.* 2000) or altitude (Smith *et al.* 2000; Chown and Klok 2003) (Fig. 7.4). The proximate cause of this variation is generally thought to be the negative relationship between rearing temperature and body size in ectothermic invertebrates (Atkinson 1994). In turn, increasing size at lower temperatures has ultimately been regarded either as adaptive (Partridge and French 1996; Atkinson and Sibly 1997; Fischer *et al.* 2003), or as a consequence of the differential sensitivity to temperature of growth and differentiation (van der Have and de Jong 1996). Although the causes of the latitudinal increase in body size continue to be the subject of some controversy, the general pattern appears to be well supported (Chown and Gaston 1999).

In contrast, a decrease in body size with latitude or with altitude has also been documented in several species (Mousseau and Roff 1989; Nylin and Svärd 1991; Chown and Klok 2003) (Fig. 7.5). This pattern has been ascribed to changes in growing season length, such that longer seasons mean a longer growing period, and hence a larger final body size. Evidence for this causal hypothesis has come from studies of species showing sawtooth clines, where extended growing seasons allow two or more generations, resulting in a substantial decline in adult body size of the bivoltine population relative to the adjacent univoltine one (Masaki 1996).

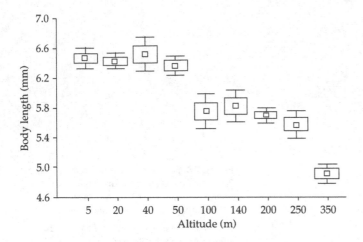

Figure 7.5 Body length variation (mean ± SE, and 2SE) across the altitudinal gradient (m) at the seasonal, sub-Antarctic Heard Island for the weevil *Ectemnorhinus viridis*.

Source: Chown and Klok (2003). *Ecography* **26**, 445–455, Blackwell Publishing.

The explanation for these contrary patterns lies in the relationship between growing season length and generation time (Roff 1980; Nylin and Gotthard 1998; Chown and Gaston 1999). If generation time is similar to or constitutes a significant proportion of the growing season length, then growing season length is likely to have a considerable influence on body size because of constraints on resource availability. Selection for differences in voltinism, via diapause propensity or development time, is also likely to be important. Hence, resource constraints, a consequence of the interaction between generation time and growing season length, mean that body size is likely to decline with declining temperature (and thus resource availability). In contrast, as generation time declines relative to growing season length, so resources effectively become available for longer, and selection for variation in voltinism is likely to be less important. Here, temperature influences on growth and differentiation, of the form widely seen in laboratory cultures and likely to be the physiological norm (Atkinson 1994; Ernsting and Isaaks 2000), are liable to become more pronounced, resulting in a negative relationship between temperature and body size.

These ideas were recently supported in an investigation of contrary altitudinal patterns in body size variation of sub-Antarctic weevils (Chown and Klok 2003). Species from the aseasonal Marion Island, which lies north of the cold Antarctic waters below the Antarctic Polar Frontal Zone (APFZ), can grow and develop year-round and show an increase in size with altitude (Fig. 7.4). By contrast, those from the more seasonal Heard Island, which lies to the south of the APFZ, show a decline in body size with altitude associated with shorter season-lengths at higher altitudes (Fig. 7.5).

While the interaction between season length and generation time provides a plausible explanation for both declines and increases in body size with altitude or latitude, the mechanistic basis underlying increases in size (e.g. increases in cell size or number) remains contentious. These mechanisms can vary depending on whether natural or laboratory populations are examined, with the trait in question, and with gender (Partridge and French 1996; Chown and Gaston 1999; Van'T Land *et al.* 1999). Moreover, other factors might also influence the size dependence of production or mortality. For example, it has been suggested that large body size enables better survival of both drought and food shortages (Lighton *et al.* 1994; Kaspari and Vargo 1995; Chown and Gaston 1999). Given that more than a single environmental variable is likely to influence insect survival (Chown 2002) and production (Slansky and Rodriguez 1987), it is obvious that body size is likely to respond to more than one variable. Therefore, without a careful consideration of the environment within which an insect finds itself, the reasons for variation in body size might be difficult to ascertain (Arnett and Gotelli 1999; Chown and Gaston 1999).

The feedback between physiological variables and body size also means that the requirements of body size might constrain the options open to insects for physiological regulation. In *Drosophila*, populations experiencing laboratory selection respond to dry conditions by increasing their water content—the canteen strategy (Gibbs *et al.* 1997). However, wild xeric and mesic species do not differ in water content, presumably because the large mass, associated with high water contents, reduces predator avoidance capabilities (Gibbs and Matzkin 2001). In sub-Antarctic weevils, it appears that a reduction in food quality, as a consequence of angiosperm extinction during the Pleistocene glaciations, resulted in a general decline in body size in species that managed to survive and reproduce under these conditions. In consequence, there was considerable selection for enhanced desiccation resistance because manipulation of body size, as a means to improve dehydration tolerance, was no longer open (Chown and Klok 2003b). Subsequent recolonization of the sub-Antarctic islands by angiosperms has meant an increase in body size of several of the weevil species that have once again started utilizing this nutritious resource. The increase in size has been accompanied by a decline in desiccation resistance (Fig. 4.16).

7.3 Interactions: internal and external

While separating thermal tolerances and regulation, water balance, metabolism, and growth and differentiation, is convenient, it in no way reflects reality. Insects must solve several environmental problems simultaneously (Park 1962; Scriber 2002). At any given time, they have to obtain sufficient resources for development and/or reproduction, avoid succumbing before they have reproduced, and cope with much environmental variation. For example, during dormancy (in an insect from seasonally dry subtropical South Africa) dehydration is likely to be prolonged and access to water and energy resources extremely limited. Therefore, there is likely to be strong selection for low metabolic rate to conserve energy resources, and for any mechanisms that might reduce water loss. Many arthropods also spend this time below ground,

where conditions are at least mildly hypoxic and hypercapnic, so there is also likely to be selection for improved gas exchange capabilities (Chown 2002). In consequence, a particular suite of characteristics, such as discontinuous gas exchange and low metabolic rate in this instance, might have evolved in response to more than a single variable. In this example, the need to conserve water, reduce internal resource depletion, and exchange gases efficiently, are all likely to have promoted discontinuous gas exchange. Environmental covariation (Loehle 1998; Hoffmann *et al.* 2003b) is also likely to promote cross-resistance, and to favour the co-option of mechanisms promoting survival of one set of circumstances to a host of other tasks. Unsurprisingly, environmental covariation also complicates adaptive inferences based on comparative studies (Davis *et al.* 2000).

7.3.1 Internal interactions

Cross-resistance (or cross-tolerance), where exposure to one kind of stress enhances resistance to others, is well known in insect physiological ecology (Hoffmann and Parsons 1991; Hoffmann *et al.* 2003b). In Section 5.4, the responses of upper or lower thermal tolerances to stress at the other end of the temperature spectrum were discussed in detail. While these effects are generally small and are probably the result of shared mechanisms, cross-resistance goes well beyond high and low-temperature interactions. It has been demonstrated in a wide range of laboratory selection experiments involving thermal tolerances, desiccation resistance, metabolic rate, and tolerance to ethanol in *Drosophila*. Stress protein expression is certainly characteristic of the response to many of these stressors (Feder and Hofmann 1999), and may well underlie the cross-tolerance here too. However, a reduction in metabolic rate is also thought to be a general response to stressful conditions, though this remains the subject of some dispute (Hoffmann and Parsons 1991; Chown and Gaston 1999).

That the response to one set of environmental stresses might have formed the basis for the response to another has been most widely explored in insect cold tolerance. It appears that the mechanisms that confer desiccation resistance have

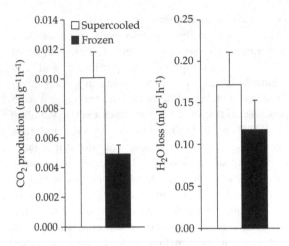

Figure 7.6 Metabolic rate (mean $\pm$ SE) and water loss rate of frozen and supercooled goldenrod gall fly larvae at $-5°C$ showing significant reductions in both rates in frozen insects.

Source: Irwin and Lee. *Journal of Experimental Zoology* **292**, © 2002. Reprinted by permission of Wiley-Liss, Inc., a subsidiary of John Wiley & Sons, Inc.

been co-opted to promote survival of subzero temperatures (Ring and Danks 1994). Indeed, the building blocks for both freeze intolerance and freezing tolerance are present in extant pathways of insect metabolism, and the biochemical responses may simply represent an upregulation of pathways already extant for protection against desiccation and other stresses (Pullin 1996; Sinclair *et al.* 2003c). This has been nicely illustrated for the springtail, *Folsomia candida*, where drought acclimation confers cold tolerance as a consequence of an increase in heat shock proteins and an increase in the molar percent of membrane fatty acids with a mid-chain double bond. The latter is thought to confer both drought and cold resistance (Bayley *et al.* 2001).

The cold hardiness strategy adopted by an insect can also influence its desiccation resistance and likely survival of prolonged starvation. Because the vapour pressure difference between ice and a supercooled insect is much higher than that between ice and a frozen one (Lundheim and Zachariassen 1993), freezing tolerance confers considerable desiccation resistance on the insects that adopt this strategy. Conversely, freeze intolerant insects have to have a greater desiccation resistance than those that can survive freezing (Klok and Chown 1998b). Moreover, frozen insects tend to have a much lower

metabolic rate than those that are supercooled. Therefore, freezing tolerance might also facilitate prolonged starvation resistance (Irwin and Lee 2002) (Fig. 7.6).

7.3.2 External interactions

It is also widely appreciated that variation in a given environmental variable might also influence responses to another. The simultaneous effects of two or more variables on survival, development, and egg production have been the subject of many studies, especially where insect pest species are concerned (Andrewartha and Birch 1954; Messenger 1959). Perhaps the most famous of these kinds of studies are those that were undertaken by Park, where the effect of environmental variation in insect performance was not only examined for single *Tribolium* (Coleoptera, Tenebrionidae) species, but also for the outcome of interactions between *T. confusum* and *T. castaneum* (Park 1962). Other studies of responses to two or more simultaneously varying parameters include several on diapause induction and termination (Denlinger 2002), and on the effects of temperature and hypoxia on development (e.g. Frazier *et al.* 2001).

Recently, investigations of the responses of species to concurrent changes in several environmental variables, especially in the presence of competitors and predators, have taken on a renewed urgency. This is mostly the consequence of the realization that species responses to climate change are likely to depend not only on their ability to overcome abiotic constraints, but also on the suite of species with which they either interact now, or with which they will interact under altered circumstances (Davis *et al.* 1998; Gaston 2003). Moreover, it is clear that the ability of a given species to overcome abiotic constraints is likely to have a considerable influence on community organization (Dunson and Travis 1991). This is particularly clear in the case of competing *Aedes* mosquitoes in Florida (Juliano *et al.* 2002). Egg survival in the indigenous *A. aegypti* is independent of temperature and humidity for at least 60 days, while eggs of the invasive *A. albopinctus* are sensitive to both high temperature and drought. Moreover, *A. albopinctus* is generally a superior competitor to *A. aegypti*. In consequence, in cool areas, with little or no dry season, *A. albopinctus* has largely replaced *A. aegypti*. In a similar way it has been suggested that rapid development times and considerable physiological flexibility in invasive arthropod species on sub-Antarctic Marion Island, compared to the slower, less flexible indigenous species is likely to mean replacement of the latter by the former as climates become warmer and drier on the island, in step with global climate change (Barendse and Chown 2001; Chown *et al.* 2002*b*).

7.3.3 Interactions: critical questions

Determining the likely abiotic thresholds for survival, development, or reproduction in the laboratory, or perhaps with limited caging experiments, is only the first step in explaining their relevance to population dynamics and ultimately to the abundance and distribution of a species (or suite of species) (Kingsolver 1989; Holt *et al.* 1997). Several critical questions remain, including which stage, gender, or age group is most likely to experience the critical population bottleneck (van der Have 2002), how physiological characteristics and fitness are actually related (Feder 1987; Kingsolver 1996), and how frequently environmental extremes might be encountered. In some cases, the immature stages might be most critical because development generally proceeds within a smaller range of environmental variables (e.g. temperature) than the adult organisms can survive (van der Have 2002). In consequence, both abundance and distribution might be determined by growth conditions faced by the immatures (Bryant *et al.* 1997). Alternatively, the time available for adult flight (and oviposition), and therefore realized fecundity of adults might explain population fluctuations (Kingsolver 1989). In *Drosophila pseudoobscura* survival of cold winter conditions by males is relatively insignificant because it is females and the sperm they carry from previous matings that are most important for subsequent population recovery (Collett and Jarman 2001). Surprisingly, investigations of ontogentic variation in physiological traits and its implications for survival, and consequently its role in determining

insect abundance and distribution, remain relatively scarce.

That year-to-year variation in environmental conditions has a large effect on insect populations has long been appreciated (Andrewartha and Birch 1954), and is widely supported by a range of studies (e.g. Roy *et al.* 2001). However, the importance of extreme events, and the way in which their likely recurrence and impact on populations should be investigated, have enjoyed less attention, particularly in the context of microclimates (Parmesan *et al.* 2000; Sinclair 2001*a*). For example, populations of *Euphydryas editha* (Lepidoptera, Nymphalidae) were driven to extinction as a consequence of three extreme weather events (and human landscape alteration). In 1989 minimal snow led to early April (rather than June) emergence of adults and their subsequent starvation owing to an absence of nectar. A year later emergence was once again early for the same reason, and a 'normal' snowstorm in May resulted in high mortalities. In 1992, unusually low temperatures killed most of the host plants, leaving caterpillars with no source of food (Thomas *et al.* 1996; Parmesan *et al.* 2000). Similarly, unseasonably warm temperatures in the Arctic lead to surface ice-formation and considerable mortality of several populations of soil-dwelling species (Coulson *et al.* 2000).

Many of the studies examining extreme environmental events have emphasized not only their extreme values, but also their duration, the rates at which they are approached, and their likely return times (Gaines and Denny 1993; Sinclair 2001*a*). These parameters might be more important than simple mean, variance, and absolute extremes of the climate of an area in determining the likely persistence of a population. Fortunately, there is a variety of techniques available for analysing extreme values (Ferguson and Messier 1996; Denny and Gaines 2000; Sinclair 2001*b*), and the importance of doing so at the microclimate level is being increasingly recognized (Sinclair 2001*a*; Williams *et al.* 2002).

For example, along the east coast of Australia, highest daily maximum temperature in the hottest month of the year does not vary with latitude. By contrast, mean daily maximum temperature declines with latitude, suggesting that the annual

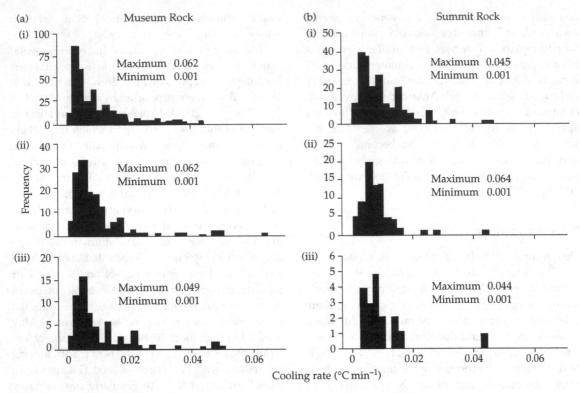

Figure 7.7 The frequency distribution of cooling rates recorded at an alpine site (a) Museum Rock and (b) Summit Rock in New Zealand. *Note*: (i) at 0°C, (ii) at −3.1°C and (iii) at −4.5°C.
Source: Sinclair (2001). *Oikos* **93**, 286–293, Blackwell Publishing.

number of warm days declines with latitude. It is this variation that is likely to be at the root of clinal variation of the 56H8 *hsp70* allele of *Drosophila melanogaster* (Bettencourt *et al.* 2002). In alpine New Zealand, measurements of the temperature in microhabitats of the cockroach *Celatoblatta quinquemaculata* and laboratory examination of critical temperatures indicated that there is substantial inter-annual variation in the risks of mortality (Sinclair 2001a). This variation is due mostly to the absence of well-developed snow cover in El Niño years. Moreover, cooling rates in field habitats differ substantially from those used in the laboratory, highlighting the need for more accurate information on microclimates to provide appropriate experimental conditions (Fig. 7.7). This work also highlights the substantial effect that increasing environmental variability, as a consequence of global climate change, might have on insects.

7.4 Climate change

From 1976 onwards, the Earth has been warming at a rate faster than it has done in the last 1000 years (Watson 2002). Both empirical studies and models have shown not only that global environmental change is happening, but also that its affects are heterogeneous. While some areas have experienced considerable and rapid warming, others have been less affected. Likewise, precipitation has largely increased in the mid- to high latitudes of the Northern Hemisphere, although several parts of Africa, Asia, and South America have experienced dramatic declines in rainfall. The Antarctic Peninsula has seen a striking increase in mean annual temperature and rain now falls there, while just to the north of the Antarctic Polar Frontal Zone, warming is generally being accompanied by a substantial decline in mean annual rainfall (as much as a 600 mm decrease in 50 years at Marion

Island—Bergstrom and Chown 1999). These global trends seem set to continue, accompanied by continuing increases in CO_2, tropospheric ozone, and trace gases of other kinds, and an ongoing increase in the frequency, persistence, and intensity of El Niño events (Watson 2002).

The likely and realized responses of insects to changes in temperature, water availability, elevated CO_2 levels and their interactions are the subject of a rapidly growing literature (Bazzaz 1990; Cammell and Knight 1992; Hoffmann and Parsons 1997; Cannon 1998; Coviella and Trumble 1999; Hill et al. 1999, 2002; Parmesan et al. 1999; Thomas et al. 2001; Bale et al. 2002; Erasmus et al. 2002). Indeed, it is now widely accepted that the signs of climate change are coherent and obvious for a range of both plant and animal taxa, and that these changes will continue (Walther et al. 2002; Parmesan and Yohe 2003; Root et al. 2003). For example, butterflies are expected to expand their northern range margins in the Northern Hemisphere, although this depends crucially on habitat availability, and the mobility of the species of concern. Highly mobile habitat generalists are likely to show much higher rates of change than less mobile specialists, which in some cases might be restricted by lack of habitat availability due to human land use (Hill et al. 1999, 2002). In two species of bush crickets, expansion of their range is being facilitated by a greater frequency of long-winged forms which can rapidly colonize new habitats (Thomas et al. 2001).

In the case of interactions between elevated temperature and CO_2 levels, plant nutritional status, and herbivore responses, it is widely expected that elevated CO_2 levels will result in a higher plant C : N ratio, prolonging herbivore larval growth rates, potentially reducing final adult size, and exposing larvae to higher predation rates (Buse et al. 1999; Coviella and Trumble 1999). Moreover, insects are expected to consume larger amounts of plant tissue and nutrient deficiencies are also likely to cascade up through higher trophic levels (Coviella and Trumble 1999). However, these generalizations belie complexities, such as those associated with the covariation of other gases, like O_3, and with the trophic or functional group of the insects concerned (leaf chewers, bark borers, or litter dwellers) (Karnosky et al. 2003).

Rather than provide a review of the realized and expected effects of climate change, which can be obtained from a host of recent works, we draw attention to several important areas in which physiological ecology might have a significant role to play in promoting an understanding of the responses of insects to current environmental change.

1. As in other aspects of physiological ecology (Chown et al. 2002a), the range of investigations of responses to change (apart from simulation models) is taxonomically biased relative to the distribution of species among higher taxa. Investigations on Lepidoptera and aphids are particularly common, while those on other taxa tend to be rarer (Coviella and Trumble 1999). Likewise, crop pests are much more likely to enjoy attention than wild species, even though there are considerable interactions between the two groups.

2. There is much support for the idea that the responses of insects in the Southern and Northern Hemispheres might differ considerably (Chown et al. 2002a; Sinclair et al. 2003c), and that the latitudinal richness patterns of the two hemispheres are dissimilar (Blackburn and Gaston 1996; Gaston 1996). However, these differences, and their implications for species responses to climate change, remain poorly explored. Indeed, several reviews make statements regarding the general responses of insects to climate change, when in reality only Northern Hemisphere species are concerned. Few studies are addressing the likely and realized responses of Southern Hemisphere insects to climate change, as is clear from recent reviews of empirical work (Parmesan and Yohe 2003; Root et al. 2003).

3. The responses of insects to elevated temperature and elevated CO_2 levels are more commonly investigated than responses to other variables. Nonetheless, not only is rainfall expected to decline in many areas, but extreme drought and extreme rainfall events are expected to increase in frequency (Watson 2002). Drought is known to have a substantial influence on a variety of insect species, particularly when the droughts are severe (Kindvall 1995; Pollard et al. 1997; Hawkins and Holyoak 1998), and prolonged hypoxia associated with submergence cannot be tolerated by most insect species. Inadequate understanding of the relationship between water availability and insect responses has been highlighted previously (Tauber *et al.* 1998). These responses can be quite subtle, though important. For example, the water status of host plants determines host quality, and therefore oviposition preference and larval performance, and so ultimately the population dynamics of the insect concerned (Scriber 2002).

4. While the differential responses of generalists versus specialists to climate change are becoming increasingly well known in butterflies in Europe (Hill *et al.* 2002), the relative importance of abiotic variables and biotic interactions to the responses of other species remain poorly known. In particular, determining whether there is a fast–slow life history dichotomy (Ricklefs and Wikelski 2002) which characterizes indigenous versus invasive insect species (Moller 1996) is of particular significance in the context of interactions between ongoing climate change and invasion (Sala *et al.* 2000).

7.5 To conclude

In this book we have explored the physiological responses of insects to a range of environmental variables, but mostly to variation in food quality, temperature, and water availability. The extent to which insects have to alter their physiology to cope with a changing environment depends, in large measure, on the extent of their behavioural flexibility, the microhabitats they select, and variation in the characteristics of these habitats. However, the physiological capabilities of insects also directly affect their behaviour. These capabilities determine under what conditions insects can survive, and therefore be active, and what habitats they can potentially occupy, so ultimately affecting not only their broad-scale distributions, but also community structure and coexistence at a local level. Because physiological tolerances play a large role in determining habitat selection, which in turn has an influence on survival, understanding the interactions between tolerances and alterations in habitat structure may also aid in understanding the effects of habitat alteration on the conservation of threatened species. There are many striking examples of these interdependencies, including interactions between thermal tolerance, overwintering, and deforestation threat (Anderson and Brower 1996), ant assemblage structure and the tolerances of the species comprising those communities (Cerdá *et al.* 1997, 1998; Bestelmeyer 2000; Cerdá and Retana 2000), and the ways in which tolerance might lead to assemblage nestedness (Worthen *et al.* 1998; Worthen and Haney 1999). Of course, physiological limits will not always be directly involved in determining variations in the population dynamics of species and consequently their abundances and distributions (Klok *et al.* 2003). Therefore, just as ecologists must pause to consider the physiological plausibility of the mechanisms they are proposing, so too must physiologists consider the ecological and evolutionary relevance of the mechanisms they are studying. An integrated physiological ecology has the advantage of promoting and integrating both perspectives. In consequence, it has much to offer biology.

References

Abisgold, J.D. and Simpson, S.J. (1987) The physiology of compensation by locusts for changes in dietary protein. *Journal of Experimental Biology* **129**, 329–346.

Abisgold, J.D., Simpson, S.J., and Douglas, A.E. (1994) Nutrient regulation in the pea aphid *Acyrthosiphon pisum*: applications of a novel geometric framework to sugar and amino acid consumption. *Physiological Entomology* **19**, 95–102.

Abril, A.B. and Bucher, E.H. (2002) Evidence that the fungus cultured by leaf-cutting ants does not metabolize cellulose. *Ecology Letters* **5**, 325–328.

Acar, E.B., Smith, B.N., Hansen, L.D., and Booth, G.M. (2001) Use of calorespirometry to determine effects of temperature on metabolic efficiency of an insect. *Environmental Entomology* **30**, 811–816.

Addo-Bediako, A. (2001) *Physiological Diversity in Insects: Large Scale Patterns.* Ph.D. Thesis, University of Pretoria, Pretoria.

Addo-Bediako, A., Chown, S.L., and Gaston, K.J. (2000) Thermal tolerance, climatic variability and latitude. *Proceedings of the Royal Society of London B* **267**, 739–745.

Addo-Bediako, A., Chown, S.L., and Gaston, K.J. (2001) Revisiting water loss in insects: a large scale view. *Journal of Insect Physiology* **47**, 1377–1388.

Addo-Bediako, A., Chown, S.L., and Gaston, K.J. (2002) Metabolic cold adaptation in insects: a large-scale perspective. *Functional Ecology* **16**, 332–338.

Ahearn, G.A. (1970) The control of water loss in desert tenebrionid beetles. *Journal of Experimental Biology* **53**, 573–595.

Aidley, D.J. (1976) Increase in respiratory rate during feeding in larvae of the armyworm, *Spodoptera exempta*. *Physiological Entomology* **1**, 73–75.

Alahiotis, S.N. (1983) Heat shock proteins. A new view on the temperature compensation. *Comparative Biochemistry and Physiology B* **75**, 379–387.

Alahiotis, S.N. and Stephanou, G. (1982) Temperature adaptation of *Drosophila* populations. The heat shock proteins system. *Comparative Biochemistry and Physiology B* **73**, 529–533.

Albaghdadi, L.F. (1987) Effects of starvation and dehydration on ionic balance in *Schistocerca gregaria*. *Journal of Insect Physiology* **33**, 269–277.

Allen, A.P., Brown, J.H., and Gillooly, J.F. (2002) Global biodiversity, biochemical kinetics, and the energetic-equivalence rule. *Science* **297**, 1545–1548.

Andersen, A.N. (1992) Regulation of 'momentary' diversity by dominant species in exceptionally rich ant communities of the Australian seasonal tropics. *American Naturalist* **140**, 401–420.

Anderson, A.R., Collinge, J.E., Hoffmann, A.A., Kellett, M., and McKechnie, S.W. (2003) Thermal tolerance trade-offs associated with the right arm of chromosome 3 and marked by the *hsr-omega* gene in *Drosophila melanogaster*. *Heredity* **90**, 195–202.

Anderson, J.B. and Brower, L.P. (1996) Freeze-protection of overwintering monarch butterflies in Mexico: critical role of the forest as a blanket and an umbrella. *Ecological Entomology* **21**, 107–116.

Andersson Escher, S. and Rasmuson-Lestander, A. (1999) The *Drosophila* glucose transporter gene: cDNA sequence, phylogenetic comparisons, analysis of functional sites and secondary structures. *Hereditas* **130**, 95–103.

Andersen, P.C., Brodbeck, B.V., and Mizell, R.F. (1992) Feeding by the leafhopper, *Homalodisca coagulata*, in relation to xylem fluid chemistry and tension. *Journal of Insect Physiology* **38**, 611–622.

Andrewartha, H.G. and Birch, L.C. (1954) *The Distribution and Abundance of Animals.* University of Chicago Press, Chicago.

Appel, H.M. (1994) The chewing herbivore gut lumen: physicochemical conditions and their impact on plant nutrients, allelochemicals, and insect pathogens. In *Insect-Plant Interactions* (ed. E.A. Bernays), Vol. 5, 209–223. CRC Press, Boca Raton.

Appel, H.M. and Joern, A. (1998) Gut physicochemistry of grassland grasshoppers. *Journal of Insect Physiology* **44**, 693–700.

Appel, H.M. and Martin, M.M. (1992) Significance of metabolic load in the evolution of host specificity of *Manduca sexta*. *Ecology* **73**, 216–228.

Applebaum, S.W. (1964) Physiological aspects of host specificity in the Bruchidae-I. General considerations of developmental compatibility. *Journal of Insect Physiology* 10, 783–788.

Applebaum, S.W. (1985) Biochemistry of digestion. In *Comprehensive Insect Physiology, Biochemistry and Pharmacology* (eds. G.A. Kerkut and L.I. Gilbert), Vol. 4, 279–311. Permagon Press, Oxford.

Arendt, J.D. (1997) Adaptive intrinsic growth rates: an integration across taxa. *Quarterly Review of Biology* 72, 149–177.

Arlian, L.G. (1979) Significance of passive sorption of atmospheric water vapor and feeding in water balance of the rice weevil, *Sitophilus oryzae*. *Comparative Biochemistry and Physiology A* 62, 725–733.

Arnett, A.E. and Gotelli, N.J. (1999) Bergmann's rule in the ant lion *Myrmeleon immaculatus* DeGeer (Neuroptera: Myrmeleontidae): geographic variation in body size and heterozygosity. *Journal of Biogeography* 26, 275–283.

Arrese, E.L., Canavoso, L.E., Jouni, Z.E., Pennington, J.E., Tsuchida, K., and Wells, M.A. (2001) Lipid storage and mobilization in insects: current status and future directions. *Insect Biochemistry and Molecular Biology* 31, 7–17.

Asahina, E. (1969) Frost resistance in insects. *Advances in Insect Physiology* 6, 1–49.

Ashby, P.D. (1997) Conservation of mass-specific metabolic rate among high- and low-elevation populations of the acridid grasshopper *Xanthippus corallipes*. *Physiological Zoology* 70, 701–711.

Ashby, P.D. (1998) The effect of standard metabolic rate on egg production in the acridid grasshopper, *Xanthippus corallipes*. *American Zoologist* 38, 561–567.

Ashford, D.A., Smith, W.A., and Douglas, A.E. (2000) Living on a high sugar diet: the fate of sucrose ingested by a phloem-feeding insect, the pea aphid *Acyrthosiphon pisum*. *Journal of Insect Physiology* 46, 335–341.

Atkinson, D. (1994) Temperature and organism size-A biological law for ectotherms? *Advances in Ecological Research* 25, 1–58.

Atkinson, D. and Sibly, R.M. (1997) Why are organisms usually bigger in colder environments? Making sense of a life history puzzle. *Trends in Ecology and Evolution* 12, 235–239.

Audsley, N., Coast, G.M., and Schooley, D.A. (1993) The effects of *Manduca sexta* diuretic hormone on fluid transport by the Malpighian tubules and cryptonephric complex of *Manduca sexta*. *Journal of Experimental Biology* 178, 231–243.

Auerswald, L., Schneider, P., and Gäde, G. (1998) Proline powers pre-flight warm-up in the African fruit beetle *Pachnoda sinuata* (Cetoniinae). *Journal of Experimental Biology* 201, 1651–1657.

Awmack, C.S. and Leather, S.R. (2002) Host plant quality and fecundity in herbivorous insects. *Annual Review of Entomology* 47, 817–844.

Ayres, M.P. (1993) Plant defense, herbivory, and climate change. In *Biotic Interactions and Global Change* (eds. P.M. Kareiva, J.G. Kingsolver, and R.B. Huey), 75–94. Sinauer, Sunderland, MA.

Bailey, E. (1975) Biochemistry of insect flight. Part 2—Fuel supply. In *Insect Biochemistry and Function* (eds. D.J. Candy and B.A. Kilby), 89–176. Chapman and Hall, London.

Bakken, G.S. (1992) Measurement and application of operative and standard operative temperatures in ecology. *American Zoologist* 32, 194–216.

Baldwin, W.F. (1954) Acclimation and lethal high temperatures for a parasitic insect. *Canadian Journal of Zoology* 32, 157–171.

Bale, J.S. (1987) Insect cold hardiness: freezing and supercooling—an ecophysiological perspective. *Journal of Insect Physiology* 33, 899–908.

Bale, J.S. (1993) Classes of insect cold hardiness. *Functional Ecology* 7, 751–753.

Bale, J.S. (1996) Insect cold hardiness: A matter of life and death. *European Journal of Entomology* 93, 369–382.

Bale, J.S., Masters, G.J., Hodkinson, I.D., Awmack, C., Bezemer, T.M., Brown, V.K. *et al.* (2002) Herbivory in global climate change research: direct effects of rising temperature on insect herbivores. *Global Change Biology* 8, 1–16.

Bale, J.S., Worland, M.R., and Block, W. (2001) Effects of summer frost exposures on the cold tolerance strategy of a sub-Antarctic beetle. *Journal of Insect Physiology* 47, 1161–1167.

Banavar, J.R., Damuth, J., Maritan, A., and Rinaldo, A. (2003) Allometric cascades. *Nature* 421, 713–714.

Barbehenn, R.V. (1992) Digestion of uncrushed leaf tissues by leaf-snipping larval Lepidoptera. *Oecologia* 89, 229–235.

Barbehenn, R.V., Bumgarner, S.L., Roosen, E.F., and Martin, M.M. (2001) Antioxidant defenses in caterpillars: role of the ascorbate-recycling system in the midgut lumen. *Journal of Insect Physiology* 47, 349–357.

Barbehenn, R.V. and Martin, M.M. (1995) Peritrophic envelope permeability in herbivorous insects. *Journal of Insect Physiology* 41, 303–311.

Barclay, J.W. and Robertson, R.M. (2000) Heat-shock induced thermoprotection of hindleg motor control in the locust. *Journal of Experimental Biology* 203, 941–950.

Barendse, J. and Chown, S.L. (2001) Abundance and seasonality of mid-altitude fellfield arthropods from Marion Island. *Polar Biology* 24, 73–82.

Barnby, M.A. (1987) Osmotic and ionic regulation of two brine fly species (Diptera: Ephydridae) from a saline hot spring. *Physiological Zoology* **60**, 327–338.

Bartholomew, G.A. and Casey, T.M. (1977) Body temperature and oxygen consumption during rest and activity in relation to body size in some tropical beetles. *Journal of Thermal Biology* **2**, 173–176.

Bartholomew, G.A. and Casey, T.M. (1978) Oxygen consumption of moths during rest, pre-flight warm-up, and flight in relation to body size and wing morphology. *Journal of Experimental Biology* **76**, 11–25.

Bartholomew, G.A. and Lighton, J.R.B. (1985) Ventilation and oxygen consumption during rest and locomotion in a tropical cockroach, *Blaberus giganteus*. *Journal of Experimental Biology* **118**, 449–454.

Bartholomew, G.A. and Lighton, J.R.B. (1986) Endothermy and energy metabolism of a giant tropical fly, *Pantophthalmus tabaninus* Thunberg. *Journal of Comparative Physiology B* **156**, 461–467.

Bartholomew, G.A., Lighton, J.R.B., and Feener, D.H. (1988) Energetics of trail running, load carriage, and emigration in the column-raiding army ant *Eciton hamatum*. *Physiological Zoology* **61**, 57–68.

Bartholomew, G.A., Vleck, D., and Vleck, C.M. (1981) Instantaneous measurements of oxygen consumption during pre-flight warm-up and post-flight cooling in sphingid and saturniid moths. *Journal of Experimental Biology* **90**, 17–32.

Barton Browne, L. (1968) Effects of altering the composition and volume of the haemolymph on water ingestion of the blowfly, *Lucilia cuprina*. *Journal of Insect Physiology* **14**, 1603–1620.

Barton Browne, L. and Raubenheimer, D. (2003) Ontogenetic changes in the rate of ingestion and estimates of food consumption in fourth and fifth instar *Helicoverpa armigera* caterpillars. *Journal of Insect Physiology* **49**, 63–71.

Baust, J.G. and Rojas, R.R. (1985) Insect cold hardiness: facts and fancy. *Journal of Insect Physiology* **31**, 755–759.

Bayley, M. and Holmstrup, M. (1999) Water vapor absorption in arthropods by accumulation of myo-inositol and glucose. *Science* **285**, 1909–1911.

Bayley, M., Petersen, S.O., Knigge, T., Köhler, H.-R., and Holmstrup, M. (2001) Drought acclimation confers cold tolerance in the soil collembolan *Folsomia candida*. *Journal of Insect Physiology* **47**, 1197–1204.

Bazzaz, F.A. (1990) The response of natural ecosystems to the rising global CO_2 levels. *Annual Review of Ecology and Systematics* **21**, 167–196.

Beaupre, S.J. and Dunham, A.E. (1995) A comparison of ratio-based and covariance analyses of a nutritional data set. *Functional Ecology* **9**, 876–880.

Becerra, J.X. (1997) Insects on plants: macroevolutionary chemical trends in host use. *Science* **276**, 253–256.

Bech, C., Langseth, I., and Gabrielsen, G.W. (1999) Repeatability of basal metabolism in breeding female kittiwakes *Rissa tridactyla*. *Proceedings of the Royal Society of London B* **266**, 2161–2167.

Beck, J., Mühlenberg, E., and Fiedler, K. (1999) Mud-puddling behavior in tropical butterflies: in search of proteins or minerals? *Oecologia* **119**, 140–148.

Behmer, S.T. and Joern, A. (1993) Diet choice by a grass-feeding grasshopper based on the need for a limiting nutrient. *Functional Ecology* **7**, 522–527.

Behmer, S.T. and Joern, A. (1994) The influence of proline on diet selection: sex-specific feeding preferences by the grasshoppers *Ageneotettix deorum* and *Phoetaliotes nebrascensis* (Orthoptera: Acrididae). *Oecologia* **98**, 76–82.

Behmer, S.T., Simpson, S.J., and Raubenheimer, D. (2002) Herbivore foraging in chemically heterogenous environments: nutrients and secondary metabolites. *Ecology* **83**, 2489–2501.

Bennett, A.F. (1987) The accomplishments of ecological physiology. In *New Directions in Ecological Physiology* (eds. M.E. Feder, A.F. Bennett, W.W. Burggren, and R.B. Huey), 1–8. Cambridge University Press, Cambridge.

Bennett, V.A., Kukal, O., and Lee, R.E. (1999) Metabolic opportunists: feeding and temperature influence the rate and pattern of respiration in the high Arctic woolly-bear caterpillar *Gynaephora groenlandica* (Lymantriidae). *Journal of Experimental Biology* **202**, 47–53.

Berenbaum, M. (1980) Adaptive significance of midgut pH in larval Lepidoptera. *American Naturalist* **115**, 138–146.

Bergstrom, D.M. and Chown, S.L. (1999) Life at the front: history, ecology and change on southern ocean islands. *Trends in Ecology and Evolution* **14**, 472–477.

Bernays, E.A. (1977) The physiological control of drinking behaviour in nymphs of *Locusta migratoria*. *Physiological Entomology* **2**, 261–273.

Bernays, E.A. (1981) Plant tannins and insect herbivores: an appraisal. *Ecological Entomology* **6**, 353–360.

Bernays, E.A. (1986*a*) Diet-induced head allometry among foliage-chewing insects and its importance for graminivores. *Science* **231**, 495–497.

Bernays, E.A. (1986*b*) Evolutionary contrasts in insects: nutritional advantages of holometabolous development. *Physiological Entomology* **11**, 377–382.

Bernays, E.A. (1997) Feeding by lepidopteran larvae is dangerous. *Ecological Entomology* **22**, 121–123.

Berrigan, D. (2000) Correlations between measures of thermal stress resistance within and between species. *Oikos* **89**, 301–304.

Berrigan, D. and Charnov, E.L. (1994) Reaction norms for age and size at maturity in response to temperature: a puzzle for life historians. *Oikos* **70**, 474–478.

Berrigan, D. and Hoffmann, A.A. (1998) Correlations between measures of heat resistance and acclimation in two species of *Drosophila* and their hybrids. *Biological Journal of the Linnean Society* **64**, 449–462.

Berrigan, D. and Lighton, J.R.B. (1993) Bioenergetic and kinematic consequences of limblessness in larval Diptera. *Journal of Experimental Biology* **179**, 245–259.

Berrigan, D. and Lighton, J.R.B. (1994) Energetics of pedestrian locomotion in adult male blowflies, *Protophormia terraenovae* (Diptera: Calliphoridae). *Physiological Zoology* **67**, 1140–1153.

Berrigan, D. and Partridge, L. (1997) Influence of temperature and activity on the metabolic rate of adult *Drosophila melanogaster*. *Comparative Biochemistry and Physiology A* **118**, 1301–1307.

Bertsch, A. (1984) Foraging in male bumblebees (*Bombus lucorum* L.): maximizing energy or minimizing water load? *Oecologia* **62**, 325–336.

Bestelmeyer, B.T. (2000) The trade-off between thermal tolerance and behavioural dominance in a subtropical South American ant community. *Journal of Animal Ecology* **69**, 998–1009.

Bettencourt, B.R. and Feder, M.E. (2002) Rapid concerted evolution via gene conversion at the *Drosophila hsp70* genes. *Journal of Molecular Evolution* **54**, 569–586.

Bettencourt, B.R., Feder, M.E., and Cavicchi, S. (1999) Experimental evolution of HSP70 expression and thermotolerance in *Drosophila melanogaster*. *Evolution* **53**, 484–492.

Bettencourt, B.R., Kim, I., Hoffmann, A.A., and Feder, M.E. (2002) Response to natural and laboratory selection at the *Drosophila HSP70* genes. *Evolution* **56**, 1796–1801.

Beyenbach, K.W. (1995) Mechanism and regulation of electrolyte transport in Malpighian tubules. *Journal of Insect Physiology* **41**, 197–207.

Beyenbach, K.W., Pannabecker, T.L., and Nagel, W. (2000) Central role of the apical membrane H^+-ATPase in electrogenesis and epithelial transport in Malpighian tubules. *Journal of Experimental Biology* **203**, 1459–1468.

Bezemer, T.M. and Jones, T.H. (1998) Plant–insect herbivore interactions in elevated atmospheric CO_2: quantitative analyses and guild effects. *Oikos* **82**, 212–222.

Birchard, G.F. (1991) Water vapor and oxygen exchange of praying mantis (*Tenodera aridifolia sinensis*) egg cases. *Physiological Zoology* **64**, 960–972.

Bishop, J.A. and Armbruster, W.S. (1999) Thermoregulatory abilities of Alaskan bees: effects of size, phylogeny and ecology. *Functional Ecology* **13**, 711–724.

Blackburn, T.M. and Gaston, K.J. (1996) A sideways look at patterns in species richness, or why there are so few species outside the tropics. *Biodiversity Letters* **3**, 44–53.

Blackburn, T.M. and Gaston, K.J. (1999) The relationship between animal abundance and body size: a review of the mechanisms. *Advances in Ecological Research* **28**, 181–210.

Blackburn, T.M., Gaston, K.J., and Loder, N. (1999) Geographic gradients in body size: a clarification of Bergmann's rule. *Diversity and Distributions* **5**, 165–174.

Blakemore, D., Williams, S., and Lehane, M.J. (1995) Protein stimulation of trypsin secretion from the opaque zone midgut cells of *Stomoxys calcitrans*. *Comparative Biochemistry and Physiology B* **110**, 301–307.

Blanckenhorn, W.U. and Hellriegel, B. (2002) Against Bergmann's rule: fly sperm size increases with temperature. *Ecology Letters* **5**, 7–10.

Blatt, J. and Roces, F. (2001) Haemolymph sugar levels in foraging honeybees (*Apis mellifera carnica*): dependence on metabolic rate and *in vivo* measurement of maximal rates of trehalose synthesis. *Journal of Experimental Biology* **204**, 2709–2716.

Blatt, J. and Roces, F. (2002) The control of the proventriculus in the honeybee (*Apis mellifera carnica* L.) I. A dynamic process influenced by food quality and quantity. *Journal of Insect Physiology* **48**, 643–654.

Block, W. (1982) Supercooling points of insects and mites on the Antarctic Peninsula. *Ecological Entomology* **7**, 1–8.

Block, W. (1990) Cold tolerance of insects and other arthropods. *Philosophical Transactions of the Royal Society of London B* **326**, 613–633.

Block, W. (1995) Insects and freezing. *Science Progress* **78**, 349–372.

Block, W. (1996) Cold or drought—the lesser of two evils for terrestrial arthropods? *European Journal of Entomology* **93**, 325–339.

Boggs, C.L. (1981) Nutritional and life-history determinants of resource allocation in holometabolous insects. *American Naturalist* **177**, 692–709.

Boggs, C.L. (1987) Ecology of nectar and pollen feeding in Lepidoptera. In *Nutritional Ecology of Insects, Mites, Spiders and Related Invertebrates* (eds. F. Slansky and J.G. Rodriguez), 369–391. John Wiley and Sons, New York.

Boggs, C.L. (1997) Dynamics of reproductive allocation from juvenile and adult feeding: radiotracer studies. *Ecology* **78**, 192–202.

Boggs, C.L. and Ross, C.L. (1993) The effect of adult food limitation on life history traits in *Speyeria mormonia* (Lepidoptera: Nymphalidae). *Ecology* **74**, 433–441.

Bolter, C.J. and Jongsma, M.A. (1995) Colorado potato beetles (*Leptinotarsa decemlineata*) adapt to proteinase inhibitors induced in potato leaves by methyl jasmonate. *Journal of Insect Physiology* **41**, 1071–1078.

Bosch, J. and Vicens, N. (2002) Body size as an estimator of production costs in a solitary bee. *Ecological Entomology* **27**, 129–137.

Bosch, M., Chown, S.L., and Scholtz, C.H. (2000) Discontinuous gas exchange and water loss in the keratin beetle *Omorgus radula*: further evidence against the water conservation hypothesis? *Physiological Entomology* **25**, 309–314.

Bowdan, E. (1988) Microstructure of feeding by tobacco hornworm caterpillars, *Manduca sexta*. *Entomologia Experimentalis et Applicata* **47**, 127–136.

Bradley, T.J. (1985) The excretory system: structure and physiology. In *Comprehensive Insect Physiology, Biochemistry and Pharmacology* (eds. G.A. Kerkut and L.I. Gilbert), Vol. 4, 421–465. Permagon Press, Oxford.

Bradley, T.J. (1994) The role of physiological capacity, morphology, and phylogeny in determining habitat use in mosquitoes. In *Ecological Morphology: Integrative Organismal Biology* (eds. P.C. Wainwright and S.M. Reilly), 303–318. University of Chicago Press, Chicago.

Bradley, T.J. (2000) The discontinuous gas exchange cycle in insects may serve to reduce oxygen supply to the tissues. *American Zoologist* **40**, 952.

Bradley, T.J., Brethorst, L., Robinson, S., and Hetz, S. (2003) Changes in the rate of CO_2 release following feeding in the insect *Rhodnius prolixus*. *Physiological and Biochemical Zoology* **76**, 302–309.

Bradley, T.J., Williams, A.E., and Rose, M.R. (1999) Physiological responses to selection for desiccation resistance in *Drosophila melanogaster*. *American Zoologist* **39**, 337–345.

Brakefield, P.M. and Willmer, P.G. (1985) The basis of thermal melanism in the ladybird *Adalia bipunctata*: differences in reflectance and thermal properties between the morphs. *Heredity* **54**, 9–14.

Breznak, J.A. (2000) Ecology of prokaryotic microbes in the guts of wood- and litter-feeding termites. In *Termites: Evolution, Sociality, Symbioses, Ecology* (eds. T. Abe, D.E. Bignell, and M. Higashi), 209–231. Kluwer Academic, Dordrecht, Netherlands.

Bridges, C.R. and Scheid, P. (1982) Buffering and CO_2 dissociation of body fluids in the pupa of the silkworm moth, *Hyalophora cecropia*. *Respiration Physiology* **48**, 183–197.

Broadway, R.M. and Duffey, S.S. (1986) Plant proteinase inhibitors: mechanism of action and effect on the growth and digestive physiology of larval *Heliothis zea* and *Spodoptera exigua*. *Journal of Insect Physiology* **32**, 827–833.

Brockway, A.P. and Schneiderman, H.A. (1967) Strain-gauge transducer studies on intratracheal pressure and pupal length during discontinuous respiration in diapausing silkworm pupae. *Journal of Insect Physiology* **13**, 1413–1451.

Brodbeck, B.V., Mizell, R.F., and Andersen, P.C. (1993) Physiological and behavioural adaptations of three species of leafhoppers in response to the dilute nutrient content of xylem fluid. *Journal of Insect Physiology* **39**, 73–81.

Bromham, L. and Cardillo, M. (2003) Testing the link between the latitudinal gradient in species richness and rates of molecular evolution. *Journal of Evolutionary Biology* **16**, 200–207.

Brooks, D.R. and Wiley, E.O. (1988) *Evolution as Entropy. Toward a Unified Theory of Biology*, 2nd edn. University of Chicago Press, Chicago.

Brown, J.H. and Maurer, B.A. (1989) Macroecology: The division of food and space among species on continents. *Science* **243**, 1145–1150.

Broza, M. (1979) Dew, fog and hygroscopic food as a source of water for desert arthropods. *Journal of Arid Environments* **2**, 43–49.

Bryant, J.P., Chapin, F.S., and Klein, D.R. (1983) Carbon/nutrient balance of boreal plants in relation to vertebrate herbivory. *Oikos* **40**, 357–368.

Bryant, S.R., Thomas, C.D., and Bale, J.S. (1997) Nettle-feeding nymphalid butterflies: temperature, development and distribution. *Ecological Entomology* **22**, 390–398.

Bryant, S.R., Thomas, C.D., and Bale, J.S. (2000) Thermal ecology of gregarious and solitary nettle-feeding nymphalid butterfly larvae. *Oecologia* **122**, 1–10.

Bryant, S.R., Thomas, C.D., and Bale, J.S. (2002) The influence of thermal ecology on the distribution of three nymphalid butterflies. *Journal of Applied Ecology* **39**, 43–55.

Bubli, O.A., Imasheva, A.G., and Loeschcke, V. (1998) Selection for knockdown resistance to heat in *Drosophila melanogaster* at high and low larval densities. *Evolution* **52**, 619–625.

Buck, J. (1962) Some physical aspects of insect respiration. *Annual Review of Entomology* **7**, 27–56.

Buck, J. and Keister, M. (1955) Cyclic CO_2 release in diapausing *Agapema* pupae. *Biological Bulletin* **109**, 144–163.

Burkett, B.N. and Schneiderman, H.A. (1974) Roles of oxygen and carbon dioxide in the control of spiracular function in cecropia pupae. *Biological Bulletin* **147**, 274–293.

Burks, C.S. and Hagstrum, D.W. (1999) Rapid cold hardening capacity in five species of coleopteran pests of stored grain. *Journal of Stored Products Research* **35**, 65–75.

Burmester, T. and Hankeln, T. (1999) A globin gene of *Drosophila melanogaster*. *Molecular Biology and Evolution* **16**, 1809–1811.

Burmester, T., Massey, H.C., Zakharkin, S.O., and Benes, H. (1998) The evolution of hexamerins and the phylogeny of insects. *Journal of Molecular Evolution* **47**, 93–108.

Bursell, E. (1957) Spiracular control of water loss in the tsetse fly. *Proceedings of the Royal Entomological Society of London A* **32**, 21–29.

Burton, V., Mitchell, H.K., Young, P., and Petersen, N.S. (1988) Heat shock protection against cold stress of *Drosophila melanogaster*. *Molecular and Cellular Biology* **8**, 3550–3552.

Buse, A., Dury, S.J., Woodburn, R.J.W., Perrins, C.M., and Good, J.E.G. (1999) Effects of elevated temperature on multi-species interactions: the case of Pedunculate Oak, Winter Moth and Tits. *Functional Ecology (Supplement)* **13**, 74–82.

Bustami, H.P., Harrison, J.F., and Hustert, R. (2002) Evidence for oxygen and carbon dioxide receptors in insect CNS influencing ventilation. *Comparative Biochemistry and Physiology A* **133**, 595–604.

Bustami, H.P. and Hustert, R. (2000) Typical ventilatory pattern of the intact locust is produced by the isolated CNS. *Journal of Insect Physiology* **46**, 1285–1293.

Butts, R.A., Howling, G.G., Bone, W., Bale, J.S., and Harrington, R. (1997) Contact with the host plant enhances aphid survival at low temperatures. *Ecological Entomology* **22**, 26–31.

Byrne, D.N. and Hadley, N.F. (1988) Particulate surface waxes of whiteflies: morphology, composition and waxing behaviour. *Physiological Entomology* **13**, 267–276.

Byrne, M.J. and Duncan, F.D. (2003) The role of the subelytral spiracles in respiration in the flightless dung beetle *Circellium bacchus*. *Journal of Experimental Biology* **206**, 1309–1318.

Cambefort, Y. (1991) From saprophagy to coprophagy. In *Dung Beetle Ecology* (eds. I. Hanski and Y. Cambefort), 22–35. Princeton University Press, Princeton.

Cammell, M.E. and Knight, J.D. (1992) Effects of climatic change on the population dynamics of crop pests. *Advances in Ecological Research* **22**, 117–162.

Candy, D.J., Becker, A., and Wegener, G. (1997) Coordination and integration of metabolism in insect flight. *Comparative Biochemistry and Physiology B* **117**, 497–512.

Cannon, R.J.C. (1998) The implications of predicted climate change for insect pests in the UK, with emphasis on non-indigenous species. *Global Change Biology* **4**, 785–796.

Cannon, R.J.C. and Block, W. (1988) Cold tolerance of microarthropods. *Biological Reviews* **63**, 23–77.

Casartelli, M., Leonardi, M.G., Fiandra, L., Parenti, P., and Giordana, B. (2001) Multiple transport pathways for dibasic amino acids in the larval midgut of the silkworm *Bombyx mori*. *Insect Biochemistry and Molecular Biology* **31**, 621–632.

Case, T.J. and Taper, M.L. (2000) Interspecific competition, environmental gradients, gene flow, and the coevolution of species' borders. *American Naturalist* **155**, 583–605.

Casey, T.M. (1976a) Activity patterns, body temperature and thermal ecology in two desert caterpillars (Lepidoptera: Sphingidae). *Ecology* **57**, 485–497.

Casey, T.M. (1976b) Flight energetics in sphinx moths: heat production and heat loss in *Hyles lineata* during free flight. *Journal of Experimental Biology* **64**, 545–560.

Casey, T.M. (1980) Flight energetics and heat exchange of gypsy moths in relation to air temperature. *Journal of Experimental Biology* **88**, 133–145.

Casey, T.M. (1981) Insect flight energetics. In *Locomotion and Energetics in Arthropods* (eds. C.F. Herreid and C.R. Fourtner), 419–452. Plenum Press, New York.

Casey, T.M. (1988) Thermoregulation and heat exchange. *Advances in Insect Physiology* **20**, 119–146.

Casey, T.M. (1989) Oxygen consumption during flight. In *Insect Flight* (eds. G.J. Goldsworthy and C.H. Wheeler), 257–272. CRC, Boca Raton.

Casey, T.M. (1991) Energetics of caterpillar locomotion: biomechanical constraints of a hydraulic skeleton. *Science* **252**, 112–114.

Casey, T.M. (1992) Biophysical ecology and heat exchange in insects. *American Zoologist* **32**, 225–237.

Casey, T.M. (1993) Effects of temperature on foraging of caterpillars. In *Caterpillars: Ecological and Evolutionary Constraints on Foraging* (eds. N.E. Stamp and T.M. Casey), 5–28. Chapman and Hall, New York.

Casey, T.M. and Hegel, J.R. (1981) Caterpillar setae: insulation for an ectotherm. *Science* **214**, 1131–1133.

Casey, T.M. and Hegel-Little, J.R. (1987) Instantaneous oxygen consumption and muscle stroke work in *Malacosoma americanum* during pre-flight warm-up. *Journal of Experimental Biology* **127**, 389–400.

Casey, T.M., Joos, B., Fitzgerald, T.D., Yurlina, M.E., and Young, P.A. (1988) Synchronized group foraging, thermoregulation, and growth of eastern tent caterpillars in relation to microclimate. *Physiological Zoology* **61**, 372–377.

Casey, T.M. and Joos, B.A. (1983) Morphometrics, conductance, thoracic temperature, and flight energetics of noctuid and geometrid moths. *Physiological Zoology* **56**, 160–173.

Casey, T.M., May, M.L., and Morgan, K.R. (1985) Flight energetics of euglossine bees in relation to morphology and wing stroke frequency. *Journal of Experimental Biology* **116**, 271–289.

Castagna, M., Shayakul, C., Trotti, D., Sacchi, V.F., Harvey, W.R., and Hediger, M.A. (1998) Cloning and characterization of a potassium-coupled amino acid transporter. *Proceedings of the National Academy of Sciences of the USA* **95**, 5395–5400.

Caveney, S., Scholtz, C.H., and McIntyre, P. (1995) Patterns of daily flight activity in onitine dung beetles (Scarabaeinae: Onitini). *Oecologia* **103**, 444–452.

Cavicchi, S., Guerra, D., La Torre, V., and Huey, R.B. (1995) Chromosomal analysis of heat-shock tolerance in *Drosophila melanogaster* evolving at different temperatures in the laboratory. *Evolution* **49**, 676–684.

Cerdá, X. and Retana, J. (2000) Alternative strategies by thermophilic ants to cope with extreme heat: individual versus colony level traits. *Oikos* **89**, 155–163.

Cerdá, X., Retana, J., and Cros, S. (1997) Thermal disruption of transitive hierarchies in Mediterranean ant communities. *Journal of Animal Ecology* **66**, 363–374.

Cerdá, X., Retana, J., and Cros, S. (1998) Critical thermal limits in Mediterranean ant species: trade-off between mortality risk and foraging performance. *Functional Ecology* **12**, 45–55.

Chai, P. and Srygley, R.B. (1990) Predation and the flight, morphology, and temperature of Neotropical rainforest butterflies. *American Naturalist* **135**, 748–765.

Chapman, R.F. (1998) *The Insects. Structure and Function*, 4th edn. Cambridge University Press, Cambridge.

Chapman, R.F. and De Boer, G. (1995) *Regulatory Mechanisms in Insect Feeding*. Chapman and Hall, New York.

Chappell, M.A. (1984) Temperature regulation and energetics of the solitary bee *Centris pallida* during foraging and intermale mate competition. *Physiological Zoology* **57**, 215–225.

Chappell, M.A. and Rogowitz, G.L. (2000) Mass, temperature and metabolic effects on discontinuous gas exchange cycles in Eucalyptus-boring beetles (Coleoptera: Cerambycidae). *Journal of Experimental Biology* **203**, 3809–3820.

Chaui-Berlinck, J.G., Monteiro, L.H.A., Navas, C.A., and Bicudo, J.E.P.W. (2001) Temperature effects on energy metabolism: a dynamic system analysis. *Proceedings of the Royal Society of London B* **269**, 15–19.

Chauvin, G., Vannier, G., and Gueguen, A. (1979) Larval case and water balance in *Tinea pellionella*. *Journal of Insect Physiology* **25**, 615–619.

Chen, C.-P. and Denlinger, D.L. (1992) Reduction of cold injury in flies using an intermittent pulse of high temperature. *Cryobiology* **29**, 138–143.

Chen, C.-P., Denlinger, D.L., and Lee, R.E. (1987) Cold-shock injury and rapid cold hardening in the flesh fly *Sarcophaga crassipalpis*. *Physiological Zoology* **60**, 297–304.

Chen, C.-P., Lee, R.E., and Denlinger, D.L. (1990) A comparison of the responses of tropical and temperate flies (Diptera: Sarcophagidae) to cold and heat stress. *Journal of Comparative Physiology B* **160**, 543–547.

Chen, C.-P., Lee, R.E., and Denlinger, D.L. (1991) Cold shock and heat shock: a comparison of the protection generated by brief pretreatment at less severe temperatures. *Physiological Entomology* **16**, 19–26.

Chen, C.-P. and Walker, V.K. (1994) Cold-shock and chilling tolerance in *Drosophila*. *Journal of Insect Physiology* **40**, 661–669.

Chen, Y., Veenstra, J.A., Hagedorn, H., and Davis, N.T. (1994) Leucokinin and diuretic hormone immunoreactivity of neurons in the tobacco hornworm, *Manduca sexta*, and co-localization of this immunoreactivity in lateral neurosecretory cells of abdominal ganglia. *Cell and Tissue Research* **278**, 493–507.

Cheung, W.W.K. and Marshall, A.T. (1973) Water and ion regulation in cicadas in relation to xylem feeding. *Journal of Insect Physiology* **19**, 1801–1816.

Chippindale, A.K., Gibbs, A.G., Sheik, M., Yee, K.J., Djawdan, M., Bradley, T.J. *et al.* (1998) Resource acquisition and the evolution of stress resistance in *Drosophila melanogaster*. *Evolution* **52**, 1342–1352.

Chown, S.L. (1993) Desiccation resistance in six sub-Antarctic weevils (Coleoptera: Curculionidae): humidity as an abiotic factor influencing assemblage structure. *Functional Ecology* **7**, 318–325.

Chown, S.L. (2001) Physiological variation in insects: hierarchical levels and implications. *Journal of Insect Physiology* **47**, 649–660.

Chown, S.L. (2002) Respiratory water loss in insects. *Comparative Biochemistry and Physiology A* **133**, 791–804.

Chown, S.L., Addo-Bediako, A., and Gaston, K.J. (2002*a*) Physiological variation in insects: large-scale patterns and their implications. *Comparative Biochemistry and Physiology B* **131**, 587–602.

Chown, S.L., Addo-Bediako, A., and Gaston, K.J. (2003) Physiological diversity: listening to the large-scale signal. *Functional Ecology* **17**, 568–572.

Chown, S.L. and Block, W. (1997) Comparative nutritional ecology of grass-feeding in a sub-Antarctic beetle: the impact of introduced species on *Hydromedion sparsutum* from South Georgia. *Oecologia* **111**, 216–224.

Chown, S.L. and Davis, A.L.V. (2003) Discontinuous gas exchange and the significance of respiratory water loss in scarabaeine beetles. *Journal of Experimental Biology* **206**, 3547–3556.

Chown, S.L. and Gaston, K.J. (1999) Exploring links between physiology and ecology at macro-scales: the role of respiratory metabolism in insects. *Biological Reviews* **74**, 87–120.

Chown, S.L. and Gaston, K.J. (2000) Areas, cradles and museums: the latitudinal gradient in species richness. *Trends in Ecology and Evolution* 15, 311–315.

Chown, S.L., Gaston, K.J., and Gremmen, N.J.M. (2000) Including the Antarctic: Insights for ecologists everywhere. In *Antarctic Ecosystems: Models for Wider Ecological Understanding* (eds. W. Davison, C. Howard-Williams and P.A. Broady), 1–15. New Zealand Natural Sciences, Christchurch.

Chown, S.L., Gaston, K.J., and Robinson, D. (2004) Macrophysiology: large-scale patterns in physiological traits and their ecological implications. *Functional Ecology* 18, 159–167.

Chown, S.L. and Holter, P. (2000) Discontinuous gas exchange cycles in *Aphodius fossor* (Scarabaeidae): a test of hypotheses concerning origins and mechanisms. *Journal of Experimental Biology* 203, 397–403.

Chown, S.L. and Klok, C.J. (2003a) Altitudinal body size clines: latitudinal effects associated with changing seasonality. *Ecography* 26, 445–455.

Chown, S.L. and Klok, C.J. (2003b) Water balance characteristics respond to changes in body size in sub-Antarctic weevils. *Physiological and Biochemical Zoology* 76, 634–643.

Chown, S.L., Le Lagadec, M.D., and Scholtz, C.H. (1999) Partitioning variance in a physiological trait: desiccation resistance in keratin beetles (Coleoptera, Trogidae). *Functional Ecology* 13, 838–844.

Chown, S.L., McGeoch, M.A., and Marshall, D.J. (2002b) Diversity and conservation of invertebrates on the sub-Antarctic Prince Edward Islands. *African Entomology* 10, 67–82.

Chown, S.L., Pistorius, P., and Scholtz, C.H. (1998) Morphological correlates of flightlessness in southern African Scarabaeinae (Coleoptera: Scarabaeidae): Testing a condition of the water conservation hypothesis. *Canadian Journal of Zoology* 76, 1123–1133.

Chown, S.L. and Scholtz, C.H. (1993) Temperature regulation in the nocturnal melolonthine *Sparrmannia flava*. *Journal of Thermal Biology* 18, 25–33.

Chown, S.L., Scholtz, C.H., Klok, C.J., Joubert, F.J., and Coles, K.S. (1995) Ecophysiology, range contraction and survival of a geographically restricted African dung beetle (Coleoptera: Scarabaeidae). *Functional Ecology* 9, 30–39.

Christian, K.A. and Morton, S.R. (1992) Extreme thermophilia in a Central Australian ant, *Melophorus bagoti*. *Physiological Zoology* 65, 885–905.

Cipollini, M.L., Paulk, E., and Cipollini, D.F. (2002) Effect of nitrogen and water treatment on leaf chemistry in horsenettle (*Solanum carolinense*), and relationship to herbivory by flea beetles (*Epitrix* spp.) and tobacco

hornworm (*Manduca sexta*). *Journal of Chemical Ecology* 28, 2377–2398.

Clark, B.R. and Faeth, S.H. (1998) The evolution of egg clustering in butterflies: a test of the egg desiccation hypothesis. *Evolutionary Ecology* 12, 543–552.

Clark, T.M. (1999) Evolution and adaptive significance of larval midgut alkalinization in the insect superorder Mecopterida. *Journal of Chemical Ecology* 25, 1945–1960.

Clarke, A. (1993) Seasonal acclimatization and latitudinal compensation in metabolism: do they exist? *Functional Ecology* 7, 139–149.

Clarke, A. (2004) Is there a Universal Temperature Dependence of metabolism? *Functional Ecology* 18, 252–256.

Clarke, K.U. (1957) The relationship of oxygen consumption to age and weight during the post-embryonic growth of *Locusta migratoria* L. *Journal of Experimental Biology* 34, 29–41.

Cloudsley-Thompson, J.L. (1962) Lethal temperatures of some desert arthropods and the mechanism of heat death. *Entomologia Experimentalis et Applicata* 5, 270–280.

Coast, G.M. (1995) Synergism between diuretic peptides controlling ion and fluid transport in insect Malpighian tubules. *Regulatory Peptides* 57, 283–296.

Coast, G.M. (1996) Neuropeptides implicated in the control of diuresis in insects. *Peptides* 17, 327–336.

Coast, G.M., Meredith, J., and Phillips, J.E. (1999) Target organ specificity of major neuropeptide stimulants in locust excretory systems. *Journal of Experimental Biology* 202, 3195–3203.

Coast, G.M., Orchard, I., Phillips, J.E., and Schooley, D.A. (2002) Insect diuretic and antidiuretic hormones. *Advances in Insect Physiology* 29, 279–409.

Cockbain, A.J. (1961) Water relationships of *Aphis fabae* Scop. during tethered flight. *Journal of Experimental Biology* 38, 175–180.

Coelho, J.R. (1991) The effect of thorax temperature on force production during tethered flight in honeybee (*Apis mellifera*) drones, workers, and queens. *Physiological Zoology* 64, 823–835.

Coelho, J.R. and Ross, A.J. (1996) Body temperature and thermoregulation in two species of yellowjackets, *Vespula germanica* and *V. maculifrons*. *Journal of Comparative Physiology B* 166, 68–76.

Coenen-Stass, D. (1986) Investigations on the water balance in the red wood ant, *Formica polyctena* (Hymenoptera, Formicidae): workers, their larvae and pupae. *Comparative Biochemistry and Physiology A* 83, 141–147.

Coleman, J.S., Heckathorn, S.A., and Hallberg, R.L. (1995) Heat-shock proteins and thermotolerance: linking molecular and ecological perspectives. *Trends in Ecology and Evolution* 10, 305–306.

Collett, J.I. and Jarman, M.G. (2001) Adult female-*Drosophila pseudoobscura* survive and carry fertile sperm through long periods in the cold: Populations are unlikely to suffer substantial bottlenecks in overwintering. *Evolution* **55**, 840–845.

Connor, E.F. and Taverner, M.P. (1997) The evolution and adaptive significance of the leaf-mining habit. *Oikos* **79**, 6–25.

Conradi-Larson, E.-M., and Sømme, L. (1973) Anaerobiosis in the overwintering beetle *Pelophila borealis*. *Nature* **245**, 388–390.

Coope, G.R. (1979) Late Cenozoic fossil Coleoptera: Evolution, biogeography, and ecology. *Annual Review of Ecology and Systematics* **10**, 247–267.

Cooper, P.D. (1982) Water balance and osmoregulation in a free-ranging tenebrionid beetle, *Onymacris unguicularis*, of the Namib desert. *Journal of Insect Physiology* **28**, 737–742.

Cooper, P.D. (1985) Seasonal changes in water budgets in two free-ranging tenebrionid beetles, *Eleodes armata* and *Cryptoglossa verrucosa*. *Physiological Zoology* **58**, 458–472.

Cooper, P.D., Schaffer, W.M., and Buchmann, S.L. (1985) Temperature regulation of honey bees (*Apis mellifera*) foraging in the Sonoran Desert. *Journal of Experimental Biology* **114**, 1–15.

Corbet, S.A. (1990) Pollination and the weather. *Israel Journal of Botany* **39**, 13–30.

Corbet, S.A. (1991) A fresh look at the arousal syndrome of insects. *Advances in Insect Physiology* **23**, 81–116.

Corbet, S.A., Fussell, M., Ake, R., Fraser, A., Gunson, C., Savage, A. *et al.* (1993) Temperature and the pollinating activity of social bees. *Ecological Entomology* **18**, 17–30.

Cossins, A.R. and Bowler, K. (1987) *Temperature Biology of Animals*. Chapman and Hall, London.

Costanzo, J.P., Moore, J.B., Lee, R.E., Kaufman, P.E., and Wyman, J.A. (1997) Influence of soil hydric parameters on the winter cold hardiness of a burrowing beetle, *Leptinotarsa decemlineata* (Say). *Journal of Comparative Physiology B* **167**, 169–176.

Coulson, S.C. and Bale, J.S. (1992) Effect of rapid cold hardening on reproduction and survival of offspring in the housefly *Musca domestica*. *Journal of Insect Physiology* **38**, 421–424.

Coulson, S.J. and Bale, J.S. (1990) Characterisation and limitations of the rapid cold-hardening response in the housefly *Musca domestica* (Diptera: Muscidae). *Journal of Insect Physiology* **36**, 207–211.

Coulson, S.J. and Bale, J.S. (1991) Anoxia induces rapid cold hardening in the housefly *Musca domestica* (Diptera: Muscidae). *Journal of Insect Physiology* **37**, 497–501.

Coulson, S.J., Hodkinson, I.D., Webb, N.R., and Harrison, J.A. (2002) Survival of terrestrial soil-dwelling arthropods on and in seawater: implications for trans-oceanic dispersal. *Functional Ecology* **16**, 353–356.

Coulson, S.J., Leinaas, H.P., Ims, R.A., and Søvik, G. (2000) Experimental manipulation of the winter surface ice layer: the effects on a High Arctic soil microarthropod community. *Ecography* **23**, 299–306.

Courtney, S.P. (1984) The evolution of egg clustering by butterflies and other insects. *American Naturalist* **123**, 276–281.

Coutchié, P.A. and Crowe, J.H. (1979a) Transport of water vapor by tenebrionid beetles. I. Kinetics. *Physiological Zoology* **52**, 67–87.

Coutchié, P.A. and Crowe, J.H. (1979b) Transport of water vapor by tenebrionid beetles. II. Regulation of the osmolarity and composition of the hemolymph. *Physiological Zoology* **52**, 88–100.

Coviella, C.E., Stipanovic, R.D., and Trumble, J.T. (2002) Plant allocation to defensive compounds: interactions between elevated CO_2 and nitrogen in transgenic cotton plants. *Journal of Experimental Botany* **53**, 323–331.

Coviella, C.E. and Trumble, J.T. (1999) Effects of elevated atmospheric carbon dioxide on insect-plant interactions. *Conservation Biology* **13**, 700–712.

Coyne, J.A., Bundgaard, J., and Prout, T. (1983) Geographic variation of tolerance to environmental stress in *Drosophila pseudoobscura*. *American Naturalist* **122**, 474–488.

Crafts-Brandner, S.J. (2002) Plant nitrogen status rapidly alters amino acid metabolism and excretion in *Bemisia tabaci*. *Journal of Insect Physiology* **48**, 33–41.

Crailsheim, K. (1988) Intestinal transport of sugars in the honeybee (*Apis mellifera* L.). *Journal of Insect Physiology* **34**, 839–845.

Crill, W.D., Huey, R.B., and Gilchrist, G.W. (1996) Within- and between-generation effects of temperature on the morphology and physiology of *Drosophila melanogaster*. *Evolution* **50**, 1205–1218.

Cristofoletti, P.T., Ribeiro, A.F., Deraison, C., Rahbé, Y., and Terra, W.R. (2003) Midgut adaptation and digestive enzyme distribution in a phloem feeding insect, the pea aphid *Acyrthosiphon pisum*. *Journal of Insect Physiology* **49**, 11–24.

Croghan, P.C., Noble-Nesbitt, J., and Appel, A.G. (1995) Measurement of water and carbon dioxide loss from insects using radioactive isotopes. *Journal of Experimental Biology* **198**, 227–233.

Crozier, A.J.G. (1979) Diel oxygen uptake rhythms in diapausing pupae of *Pieris brassicae* and *Papilio machaon*. *Journal of Insect Physiology* **25**, 647–652.

Czajka, M.C. and Lee, R.E. (1990) A rapid cold-hardening response protecting against cold shock injury in *Drosophila melanogaster. Journal of Experimental Biology* **148**, 245–254.

Dadd, R.H. (1985) Nutrition: organisms. In *Comprehensive Insect Physiology, Biochemistry and Pharmacology* (eds. G.A. Kerkut and L.I. Gilbert), Vol. 4, 313–390. Permagon Press, Oxford.

Dahlgaard, J., Loeschcke, V., Michalak, P., and Justesen, J. (1998) Induced thermotolerance and associated expression of the heat-shock protein Hsp70 in adult *Drosophila melanogaster. Functional Ecology* **12**, 786–793.

Dahlhoff, E.P. and Rank, N.E. (2000) Functional and physiological consequences of genetic variation at phosphoglucose isomerase: Heat shock protein expression is related to enzyme genotype in a montane beetle. *Proceedings of the National Academy of Sciences of the USA* **97**, 10056–10061.

Danks, H.V. (1999) Life cycles in polar arthropods-flexible or programmed? *European Journal of Entomology* **96**, 83–102.

Danks, H.V. (2000) Dehydration in dormant insects. *Journal of Insect Physiology* **46**, 837–852.

Danks, H.V. (2002) Modification of adverse conditions by insects. *Oikos* **99**, 10–24.

Darveau, C.-A., Suarez, R.K., Andrews, R.D., and Hochachka, P.W. (2002) Allometric cascade as a unifying principle of body mass effects on metabolism. *Nature* **417**, 166–170.

Darveau, C.-A., Suarez, R.K., Andrews, R.D., and Hochachka, P.W. (2003) Darveau *et al.* reply. *Nature* **421**, 714.

David, J.R. and Bocquet, C. (1975) Similarities and differences in latitudinal adaptation of two *Drosophila* sibling species. *Nature* **257**, 588–590.

David, J.R., Gibert, P., Moreteau, B., Gilchrist, G.W., and Huey, R.B. (2003) The fly that came in from the cold: geographic variation of recovery time from low temperature exposure in *Drosophila subobscura. Functional Ecology* **17**, 425–430.

Davidson, E.W., Segura, B.J., Steele, T., and Hendrix, D.L. (1994) Microorganisms influence the composition of honeydew produced by the silverleaf whitefly, *Bemisia argentifolii. Journal of Insect Physiology* **40**, 1069–1076.

Davies, S.A., Huesmann, G.R., Maddrell, S.H.P., O'Donnell, M.J., Skaer, N.J.V., Dow, J.A.T., and Tublitz, N.J. (1995) CAP_{2b}, a cardioacceleratory peptide, is present in *Drosophila* and stimulates tubule fluid secretion via cGMP. *American Journal of Physiology* **269**, R1321–R1326.

Davis, A.J., Jenkinson, L.S., Lawton, J.H., Shorrocks, B., and Wood, S. (1998) Making mistakes when predicting shifts in species range in response to global warming. *Nature* **391**, 783–786.

Davis, A.L.V., Chown, S.L., McGeoch, M.A., and Scholtz, C.H. (2000) A comparative analysis of metabolic rate in six *Scarabaeus* species (Coleoptera: Scarabaeidae) from southern Africa: further caveats when inferring adaptation. *Journal of Insect Physiology* **46**, 553–562.

Davis, A.L.V., Chown, S.L., and Scholtz, C.H. (1999) Discontinuous gas-exchange cycles in *Scarabaeus* dung beetles (Coleoptera: Scarabaeidae): mass-scaling and temperature dependence. *Physiological and Biochemical Zoology* **72**, 555–565.

Davis, E.E. and Friend, W.G. (1995) Regulation of a meal: blood feeders. In *Regulatory Mechanisms in Insect Feeding* (eds. R.F. Chapman and G. De Boer), 157–189. Chapman and Hall, New York.

Davis, M.B. and Shaw, R.G. (2001) Range shifts and adaptive responses to Quaternary climate change. *Science* **292**, 673–679.

De Jong, P.W., Gussekloo, S.W.S., and Brakefield, P.M. (1996) Differences in thermal balance, body temperature and activity between non-melanic and melanic two-spot ladybird beetles (*Adalia bipunctata*) under controlled conditions. *Journal of Experimental Biology* **199**, 2655–2666.

Denlinger, D.L. (1991) Relationship between cold hardiness and diapause. In *Insects at Low Temperatures* (eds. R.E. Lee and D.L. Denlinger), 174–198. Chapman and Hall, New York.

Denlinger, D.L. (2002) Regulation of diapause. *Annual Review of Entomology* **47**, 93–122.

Denlinger, D.L., Giebultowicz, J.M., and Saunders, D.S. (eds.) (2001) *Insect Timing: Circadian Rhythmicity to Seasonality.* Elsevier, Amsterdam.

Denlinger, D.L., Joplin, K.H., Chen, C.-P., and Lee, R.E. (1991) Cold shock and heat shock. In *Insects at Low Temperatures* (eds. R.E. Lee and D.L. Denlinger), 131–148. Chapman and Hall, New York.

Denlinger, D.L. and Lee, R.E. (1998) Physiology of cold sensitivity. In *Temperature Sensitivity in Insects and Application in Integrated Pest Management* (eds. G.J. Hallman and D.L. Denlinger), 55–95. Westview Press, Boulder.

Denlinger, D.L., Lee, R.E., Yocum, G.D., and Kukal, O. (1992) Role of chilling in the acquisition of cold tolerance and the capacitation to express stress proteins in diapausing pharate larvae of the gypsy moth, *Lymantria dispar. Archives of Insect Biochemistry and Physiology* **21**, 271–280.

Denlinger, D.L., Rinehart, J.P., and Yocum, G.D. (2001) Stress proteins: a role in insect diapause? In *Insect Timing: Circadian Rhythmicity to Seasonality* (eds.

D.L. Denlinger, J. Giebultowicz and D.S. Saunders), 155–171. Elsevier, Amsterdam.

Denlinger, D.L., Willis, J.H., and Fraenkel, G. (1972) Rates and cycles of oxygen consumption during pupal diapause in *Sarcophaga* flesh flies. *Journal of Insect Physiology* **18**, 871–882.

Denlinger, D.L. and Yocum, G.D. (1998) Physiology of heat sensitivity. In *Temperature Sensitivity in Insects and Application in Integrated Pest Management* (eds. G.J. Hallman and D.L. Denlinger), 7–53. Westview Press, Boulder.

Denno, R.F. and Benrey, B. (1997) Aggregation facilitates larval growth in the neotropical nymphalid butterfly *Chlosyne janais*. *Ecological Entomology* **22**, 133–141.

Denny, M. and Gaines, S. (2000) *Chance in Biology. Using Probability to Explore Nature*. Princeton University Press, Princeton.

Denton, D. (1982) *The Hunger for Salt*. Springer, New York.

D'Ettorre, P., Mora, P., Dibangou, V., Rouland, C., and Errard, C. (2002) The role of the symbiotic fungus in the digestive metabolism of two species of fungus-growing ants. *Journal of Comparative Physiology B* **172**, 169–176.

Devine, T.L. (1978) The turnover of the gut contents (traced with inulin-carboxyl-^{14}C), tritiated water and ^{22}Na in three stored product insects. *Journal of Stored Product Research* **14**, 189–211.

Diamond, J. (1991) Evolutionary design of intestinal nutrient absorption: enough but not too much. *News in Physiological Sciences* **6**, 92–96.

Dodds, P.S., Rothman, D.H., and Weitz, J.S. (2001) Re-examination of the '3/4-law' of metabolism. *Journal of Theoretical Biology* **209**, 9–27.

Dohm, M.R. (2002) Repeatability estimates do not always set an upper limit to heritability. *Functional Ecology* **16**, 273–280.

Douglas, A.E. (1989) Mycetocyte symbiosis in insects. *Biological Reviews* **64**, 409–434.

Douglas, A.E. (1998) Nutritional interactions in insect-microbial symbioses: aphids and their symbiotic bacteria *Buchnera*. *Annual Review of Entomology* **43**, 17–37.

Douglas, A.E., Minto, L.B., and Wilkinson, T.L. (2001) Quantifying nutrient production by the microbial symbionts in an aphid. *Journal of Experimental Biology* **204**, 349–358.

Dow, J.A.T. (1981) Countercurrent flows, water movements and nutrient absorption in the locust midgut. *Journal of Insect Physiology* **27**, 579–585.

Dow, J.A.T. (1984) Extremely high pH in biological systems: a model for carbonate transport. *American Journal of Physiology* **246**, R633–R635.

Dow, J.A.T. (1986) Insect midgut function. *Advances in Insect Physiology* **19**, 187–328.

Dow, J.A.T. and Davies, S.A. (2001) The *Drosophila melanogaster* Malpighian tubule. *Advances in Insect Physiology* **28**, 1–83.

Dow, J.A.T., Davies, S.A., Guo, Y., Graham, S., Finbow, M.E., and Kaiser, K. (1997) Molecular genetic analysis of V-ATPase function in *Drosophila melanogaster*. *Journal of Experimental Biology* **200**, 237–245.

Dow, J.A.T., Davies, S.A., and Sözen, M.A. (1998) Fluid secretion by the *Drosophila* Malpighian tubule. *American Zoologist* **38**, 450–460.

Down, R.E., Gatehouse, A.M.R., Hamilton, W.D.O., and Gatehouse, J.A. (1996) Snowdrop lectin inhibits development and decreases fecundity of the glasshouse potato aphid (*Aulacorthum solani*) when administered *in vitro* and via transgenic plants both in laboratory and glasshouse trials. *Journal of Insect Physiology* **42**, 1035–1045.

Downing, N. (1980) The regulation of sodium, potassium, and chloride in an aphid subjected to ionic stress. *Journal of Experimental Biology* **87**, 343–349.

Dreisig, H. (1980) Daily activity, thermoregulation and water loss in the tiger beetle *Cicindela hybrida*. *Oecologia* **44**, 376–389.

Dreisig, H. (1995) Thermoregulation and flight activity in territorial male graylings, *Hipparchia semele* (Satyridae), and large skippers, *Ochlodes venata* (Hesperiidae). *Oecologia* **101**, 169–176.

Dudley, R. (1998) Atmospheric oxygen, giant Paleozoic insects and the evolution of aerial locomotor performance. *Journal of Experimental Biology* **201**, 1043–1050.

Dudley, R. (2000) *The Biomechanics of Insect Flight. Form, Function, Evolution*. Princeton University Press, Princeton.

Dudley, R. and Vermeij, G.J. (1992) Do the power requirements of flapping flight constrain folivory in flying animals? *Functional Ecology* **6**, 101–104.

Duman, J.G. (2001) Antifreeze and ice nucleator proteins in terrestrial arthropods. *Annual Review of Physiology* **63**, 327–357.

Duman, J.G., Wu, D.W., Xu, L., Tursman, D., and Olsen, T.M. (1991) Adaptations of insects to subzero temperatures. *Quarterly Review of Biology* **66**, 387–410.

Duncan, F.D. and Byrne, M.J. (2000) Discontinuous gas exchange in dung beetles: patterns and ecological implications. *Oecologia* **122**, 452–458.

Duncan, F.D. and Byrne, M.J. (2002) Respiratory airflow in a wingless dung beetle. *Journal of Experimental Biology* **205**, 2489–2497.

Duncan, F.D. and Dickman, C.R. (2001) Respiratory patterns and metabolism in tenebrionid and carabid beetles from the Simpson Desert, Australia. *Oecologia* **129**, 509–517.

Duncan, F.D., Krasnov, B., and McMaster, M. (2002*a*) Metabolic rate and respiratory gas-exchange patterns in tenebrionid beetles from the Negev Highlands, Israel. *Journal of Experimental Biology* **205**, 791–798.

Duncan, F.D., Krasnov, B., and McMaster, M. (2002*b*) Novel case of a tenebrionid beetle using discontinuous gas exchange cycle when dehydrated. *Physiological Entomology* **27**, 79–83.

Duncan, F.D. and Lighton, J.R.B. (1994) Water relations in nocturnal and diurnal foragers of the desert honeypot ant *Myrmecocystus*: implications for colony-level selection. *Journal of Experimental Zoology* **270**, 350–359.

Duncan, F.D. and Lighton, J.R.B. (1997) Discontinuous ventilation and energetics of locomotion in the desert-dwelling female mutillid wasp, *Dasymutilla gloriosa*. *Physiological Entomology* **22**, 310–315.

Dunson, W.A. and Travis, J. (1991) The role of abiotic factors in community organisation. *American Naturalist* **138**, 1067–1091.

Dutton, A., Klein, H., Romeis, J., and Bigler, F. (2002) Uptake of Bt-toxin by herbivores feeding on transgenic maize and consequences for the predator *Chrysoperla carnea*. *Ecological Entomology* **27**, 441–447.

Dyer, F.C. and Seeley, T.D. (1987) Interspecific comparisons of endothermy in honey-bees (*Apis*): deviations from the expected size-related patterns. *Journal of Experimental Biology* **127**, 1–26.

Edgecomb, R.S., Harth, C.E., and Schneiderman, A.M. (1994) Regulation of feeding behavior in adult *Drosophila melanogaster* varies with feeding regime and nutritional state. *Journal of Experimental Biology* **197**, 215–235.

Edney, E.B. (1977) *Water Balance in Land Arthropods*. Springer, Berlin.

Eigenheer, R.A., Nicolson, S.W., Schegg, K.M., Hull, J.J., and Schooley, D.A. (2002) Identification of a potent antidiuretic factor acting on beetle Malpighian tubules. *Proceedings of the National Academy of Sciences of the USA* **99**, 84–89.

Ellers, J. and Boggs, C.L. (2002) The evolution of wing color in *Colias* butterflies: heritability, sex linkage, and population divergence. *Evolution* **56**, 836–840.

Ellington, C.P., Machin, K.E., and Casey, T.M. (1990) Oxygen consumption of bumblebees in forward flight. *Nature* **347**, 472–473.

Elser, J.J., Fagan, W.F., Denno, R.F., Dobberfuhl, D.R., Folarin, A., Huberty, A. *et al.* (2000) Nutritional constraints in terrestrial and freshwater food webs. *Nature* **408**, 578–580.

Elton, C. (1930) *Animal Ecology and Evolution*. Clarendon Press, Oxford.

Emlen, D.J. (1997) Diet alters male horn allometry in the beetle *Onthophagus acuminatus* (Coleoptera: Scarabaeidae). *Proceedings of the Royal Society of London B* **264**, 567–574.

Endler, J.A. (1986) *Natural Selection in the Wild*. Princeton University Press, Princeton.

Erasmus, B.F.N., van Jaarsveld, A.S., Chown, S.L., Kshatriya, M., and Wessels, K.J. (2002) Vulnerability of South African animal taxa to climate change. *Global Change Biology* **8**, 679–693.

Ernsting, G. and Isaaks, A. (2000) Ectotherms, temperature, and trade-offs: size and number of eggs in a carabid beetle. *American Naturalist* **155**, 804–813.

Escriche, B., De Decker, N., Van Rie, J., Jansens, S., and Van Kerkhove, E. (1998) Changes in permeability of brush border membrane vesicles from *Spodoptera littoralis* midgut induced by insecticidal crystal proteins from *Bacillus thuringiensis*. *Applied and Environmental Microbiology* **64**, 1563–1565.

Fagan, W.F., Siemann, E., Mitter, C., Denno, R.F., Huberty, A.F., Woods, H.A. *et al.* (2002) Nitrogen in insects: implications for trophic complexity and species diversification. *American Naturalist* **160**, 784–802.

Falconer, D.S. and Mackay, T.F.C. (1996) *Introduction to Quantitative Genetics*, 4th ed. Prentice Hall, Harlow.

Farhi, L.E. and Tenney, S.M. (1987) *Handbook of Physiology. Section 3: The Respiratory System. Volume IV. Gas Exchange*. American Physiological Society, Bethesda.

Farina, W.M. and Wainselboim, A.J. (2001) Changes in the thoracic temperature of honeybees while receiving nectar from foragers collecting at different reward rates. *Journal of Experimental Biology* **204**, 1653–1658.

Farrell, B.D. (1998) 'Inordinate fondness' explained: why are there so many beetles? *Science* **281**, 555–559.

Febvay, G., Rahbé, Y., Rynkiewicz, M., Guillaud, J., and Bonnot, G. (1999) Fate of dietary sucrose and neosynthesis of amino acids in the pea aphid, *Acyrthosiphon pisum*, reared on different diets. *Journal of Experimental Biology* **202**, 2639–2652.

Fedak, M.A. and Seeherman, H.J. (1979) Reappraisal of energetics of locomotion shows identical cost in bipeds and quadrupeds including ostrich and horse. *Nature* **282**, 713–716.

Feder, M.E. (1987) The analysis of physiological diversity: the prospects for pattern documentation and general questions in ecological physiology. In *New Directions in Ecological Physiology* (eds. M.E. Feder, A.F. Bennett, W.W. Burggren and R.B. Huey), 38–75. Cambridge University Press, Cambridge.

Feder, M.E. (1996) Ecological and evolutionary physiology of stress proteins and the stress response: the *Drosophila melanogaster* model. In *Animals and*

Temperature. Phenotypic and Evolutionary Adaptation (eds. I.A. Johnston and A.F. Bennett), 79–102. Cambridge University Press, Cambridge.

Feder, M.E. (1997) Necrotic fruit: a novel model system for thermal ecologists. *Journal of Thermal Biology* 22, 1–9.

Feder, M.E. (1999) Engineering candidate genes in studies of adaptation: the heat-shock protein Hsp70 in *Drosophila melanogaster*. *American Naturalist Supplement* 154, 55–66.

Feder, M.E., Bennett, A.F., and Huey, R.B. (2000a) Evolutionary physiology. *Annual Review of Ecology and Systematics* 31, 315–341.

Feder, M.E., Blair, N., and Figueras, H. (1997a) Natural thermal stress and heat-shock protein expression in *Drosophila* larvae and pupae. *Functional Ecology* 11, 90–100.

Feder, M.E., Blair, N., and Figueras, H. (1997b) Oviposition site selection: unresponsiveness of *Drosophila* to cues of potential thermal stress. *Animal Behaviour* 53, 585–588.

Feder, M.E., Cartaño, N.V., Milos, L., Krebs, R.A., and Lindquist, S.L. (1996) Effect of engineering *Hsp70* copy number on Hsp70 expression and tolerance of ecologically relevant heat shock in larvae and pupae of *Drosophila melanogaster*. *Journal of Experimental Biology* 199, 1837–1844.

Feder, M.E. and Hofmann, G.E. (1999) Heat-shock proteins, molecular chaperones, and the stress response: evolutionary and ecological physiology. *Annual Review of Physiology* 61, 243–282.

Feder, M.E. and Krebs, R.A. (1998) Natural and genetic engineering of the heat-shock protein Hsp70 in *Drosophila melanogaster*: consequences for thermotolerance. *American Zoologist* 38, 503–517.

Feder, M.E. and Mitchell-Olds, T. (2003) Evolutionary and ecological functional genomics. *Nature Reviews Genetics* 4, 649–655.

Feder, M.E., Roberts, S.P., and Bordelon, A.C. (2000b) Molecular thermal telemetry of free-ranging adult *Drosophila melanogaster*. *Oecologia* 123, 460–465.

Feener, D.H., Lighton, J.R.B., and Bartholomew, G.A. (1988) Curvilinear allometry, energetics and foraging ecology: a comparison of leaf-cutting ants and army ants. *Functional Ecology* 2, 509–520.

Felton, G.W. and Gatehouse, J.A. (1996) Antinutritive plant defence mechanisms. In *Biology of the Insect Midgut* (eds. M.J. Lehane and P.F. Billingsley), 373–416. Chapman and Hall, London.

Ferguson, S.H. and Messier, F. (1996) Ecological implications of a latitudinal gradient in inter-annual climatic varibility: a test using fractal and chaos theories. *Ecography* 19, 382–392.

Ferro, D.N. and Southwick, E.E. (1984) Microclimates of small arthropods: estimating humidity within the leaf boundary layer. *Environmental Entomology* 13, 926–929.

Feuerbacher, E., Fewell, J.H., Roberts, S.P., Smith, E.F., and Harrison, J.F. (2003) Effects of load type (pollen or nectar) and load mass on hovering metabolic rate and mechanical power output in the honey bee *Apis mellifera*. *Journal of Experimental Biology* 206, 1855–1865.

Fewell, J.H., Harrison, J.F., Lighton, J.R.B., and Breed, M.D. (1996) Foraging energetics of the ant, *Paraponera clavata*. *Oecologia* 105, 419–427.

Field, J. (1992) Patterns of nest provisioning and parental investment in the solitary digger wasp *Ammophila sabulosa*. *Ecological Entomology* 17, 43–51.

Field, L.H. and Matheson, T. (1998) Chordotonal organs of insects. *Advances in Insect Physiology* 27, 1–228.

Fischer, K., Bott, A.N.M., Brakefield, P.M., and Zwaan, B.J. (2003) Fitness consequences of temperature-mediated egg size plasticity in a butterfly. *Functional Ecology* 17, 803–810.

Fischer, K. and Fiedler, K. (2000) Response of the copper butterfly *Lycaena tityrus* to increased leaf nitrogen in natural food plants: evidence against the nitrogen limitation hypothesis. *Oecologia* 124, 235–241.

Fisher, D.B., Wright, J.P., and Mittler, T.E. (1984) Osmoregulation by the aphid *Myzus persicae*: a physiological role for honeydew oligosaccharides. *Journal of Insect Physiology* 30, 387–393.

Fitches, E., Audsley, N., Gatehouse, J.A., and Edwards, J.P. (2002) Fusion proteins containing neuropeptides as novel insect control agents: snowdrop lectin delivers fused allatostatin to insect haemolymph following oral ingestion. *Insect Biochemistry and Molecular Biology* 32, 1653–1661.

Fitzgerald, T.D. (1993) Sociality in caterpillars. In *Caterpillars. Ecological and Evolutionary Constraints on Foraging* (eds. N.E. Stamp and T.M. Casey), 372–403. Chapman and Hall, London.

Fitzgerald, T.D., Casey, T., and Joos, B. (1988) Daily foraging schedule of field colonies of the eastern tent caterpillar, *Malacosoma americanum*. *Oecologia* 76, 574–578.

Flannagan, R.D., Tammariello, S.P., Joplin, K.H., Cikra-Ireland, R.A., Yocum, G.D., and Denlinger, D.L. (1998) Diapause-specific gene expression in pupae of the flesh fly *Sarcophaga crassipalpis*. *Proceedings of the National Academy of Sciences of the USA* 95, 5616–5620.

Fogleman, J.C. and Danielson, P.B. (2001) Chemical interactions in the cactus-microorganism-*Drosophila* model system of the Sonoran Desert. *American Zoologist* 41, 877–889.

Folk, D.G., Han, C., and Bradley, T.J. (2001) Water acquisition and partitioning in *Drosophila melanogaster*: effects of selection for desiccation-resistance. *Journal of Experimental Biology* 204, 3323–3331.

Fox, C.W. (1997) The ecology of body size in a seed beetle, *Stator limbatus*: persistence of environmental variation across generations? *Evolution* 51, 1005–1010.

Fox, L.R. and Macauley, B.J. (1977) Insect grazing on *Eucalyptus* in response to variation in leaf tannins and nitrogen. *Oecologia* 29, 145–162.

Frazier, M.R., Harrison, J.F., and Behmer, S.T. (2000) Effects of diet on titratable acid–base excretion in grasshoppers. *Physiological and Biochemical Zoology* 73, 66–76.

Frazier, M.R., Woods, H.A., and Harrison, J.F. (2001) Interactive effects of rearing temperature and oxygen on the development of *Drosophila melanogaster*. *Physiological and Biochemical Zoology* 74, 641–650.

Freckleton, R.P., Harvey, P.H., and Pagel, M. (2002) Phylogenetic analysis and comparative data: a test and review of evidence. *American Naturalist* 160, 712–726.

Friedman, S. (1985) Intermediary metabolism. In *Fundamentals of Insect Physiology* (ed. M.S. Blum), 467–505. Wiley-Interscience, New York.

Frisbie, M.P. and Dunson, W.A. (1988) The effect of food consumption on sodium and water balance in the predaceous diving beetle, *Dytiscus verticalis*. *Journal of Comparative Physiology B* 158, 91–98.

Furuya, K., Milchak, R.J., Schegg, K.M., Zhang, J., Tobe, S.S., Coast, G.M. *et al.* (2000) Cockroach diuretic hormones: characterization of a calcitonin-like peptide in insects. *Proceedings of the National Academy of Sciences of the USA* 97, 6469–6474.

Gäde, G. (1985) Anaerobic energy metabolism. In *Environmental Physiology and Biochemistry of Insects* (ed. K.H. Hoffmann), 119–136. Springer, Berlin.

Gäde, G. (1991) Hyperglycaemia or hypertrehalosaemia? The effect of insect neuropeptides on haemolymph sugars. *Journal of Insect Physiology* 37, 483–487.

Gäde, G. and Auerswald, L. (2002) Beetles' choice—proline for energy output: control by AKHs. *Comparative Biochemistry and Physiology B* 132, 117–129.

Gäde, G., Hoffmann, K.-H., and Spring, J.H. (1997) Hormonal regulation in insects: facts, gaps and future directions. *Physiological Reviews* 77, 963–1032.

Gaede, K. (1992) On the water balance of *Phytoseiulus persimilis* A.-H. and its ecological significance. *Experimental and Applied Acarology* 15, 181–198.

Gaines, S.D. and Denny, M.W. (1993) The largest, smallest, highest, lowest, longest, and shortest: extremes in ecology. *Ecology* 74, 1677–1692.

Gainey, L.F. (1984) Osmoregulation in the larvae of *Odontomyia cincta* (Diptera: Stratiomyidae). *Physiological Zoology* 57, 663–672.

Garland, T. and Adolph, S.C. (1994) Why not to do two-species comparative studies: limitations on inferring adaptation. *Physiological Zoology* 67, 797–828.

Garland, T., Harvey, P.H., and Ives, A.R. (1992) Procedures for the analysis of comparative data using phylogenetically independent contrasts. *Systematic Biology* 41, 18–32.

Garten, C.T. (1976) Correlations between concentrations of elements in plants. *Nature* 261, 686–688.

Gaston, K.J. (1991) The magnitude of global insect species richness. *Conservation Biology* 5, 283–296.

Gaston, K.J. (1994) *Rarity*. Chapman and Hall, London.

Gaston, K.J. (1996) Biodiversity—latitudinal gradients. *Progress in Physical Geography* 20, 466–476.

Gaston, K.J. (2000) Global patterns in biodiversity. *Nature* 405, 220–227.

Gaston, K.J. (2003) *The Structure and Dynamics of Geographic Ranges*. Oxford University Press, Oxford.

Gaston, K.J. and Blackburn, T.M. (2000) *Pattern and Process in Macroecology*. Blackwell Science, Oxford.

Gaston, K.J., Blackburn, T.M., and Spicer, J.I. (1998) Rapoport's rule: time for an epitaph? *Trends in Ecology and Evolution* 13, 70–74.

Gaston, K.J. and Chown, S.L. (1999a) Elevation and climatic tolerance: a test using dung beetles. *Oikos* 86, 584–590.

Gaston, K.J. and Chown, S.L. (1999b) Geographic range size and speciation. In *Evolution of Biological Diversity* (eds. A.E. Magurran and R.M. May), 236–259. Oxford University Press, Oxford.

Gaston, K.J. and Chown, S.L. (1999c) Why Rapoport's rule does not generalise. *Oikos* 84, 309–312.

Gaston, K.J., Chown, S.L., and Mercer, R.D. (2001) The animal species-body size distribution of Marion Island. *Proceedings of the National Academy of Sciences of the USA* 98, 14493–14496.

Gaston, K.J., Chown, S.L., and Styles, C.V. (1997) Changing size and changing enemies: the case of the mopane worm. *Acta Oecologica* 18, 21–26.

Gatehouse, A.M.R., Norton, E., Davison, G.M., Babbé, S.M., Newell, C.A., and Gatehouse, J.A. (1999) Digestive proteolytic activity in larvae of tomato moth, *Lacanobia oleracea*; effects of plant protease inhibitors *in vitro* and *in vivo*. *Journal of Insect Physiology* 45, 545–558.

Gehring, W.J. and Wehner, R. (1995) Heat shock protein synthesis and thermotolerance in *Cataglyphis*, an ant from the Sahara desert. *Proceedings of the National Academy of Sciences of the USA* 92, 2994–2998.

Gehrken, U. (1984) Winter survival of an adult bark beetle *Ips acuminatus* Gyll. *Journal of Insect Physiology* **30**, 421–429.

Gibbs, A.G. (1998) Water-proofing properties of cuticular lipids. *American Zoologist* **38**, 471–482.

Gibbs, A.G. (1999) Laboratory selection for the comparative physiologist. *Journal of Experimental Biology* **202**, 2709–2718.

Gibbs, A.G. (2002a) Water balance in desert *Drosophila*: lessons from non-charismatic microfauna. *Comparative Biochemistry and Physiology A* **133**, 781–789.

Gibbs, A.G. (2002b) Lipid melting and cuticular permeability: new insights into an old problem. *Journal of Insect Physiology* **48**, 391–400.

Gibbs, A.G., Chippindale, A.K., and Rose, M.R. (1997) Physiological mechanisms of evolved desiccation resistance in *Drosophila melanogaster*. *Journal of Experimental Biology* **200**, 1821–1832.

Gibbs, A.G., Fukuzato, F., and Matzkin, L.M. (2003a) Evolution of water conservation mechanisms in *Drosophila*. *Journal of Experimental Biology* **206**, 1183–1192.

Gibbs, A.G. and Johnson, R.A. (2004) Discontinuous ventilation by insects: the chthonic hypothesis does not hold water. *Journal of Experimental Biology*, in press.

Gibbs, A.G., Louie, A.K., and Ayala, J.A. (1998) Effects of temperature on cuticular lipids and water balance in a desert *Drosophila*: is thermal acclimation beneficial? *Journal of Experimental Biology* **201**, 71–80.

Gibbs, A.G. and Matzkin, L.M. (2001) Evolution of water balance in the genus *Drosophila*. *Journal of Experimental Biology* **204**, 2331–2338.

Gibbs, A.G. and Mousseau, T.A. (1994) Thermal acclimation and genetic variation in cuticular lipids of the lesser migratory grasshopper (*Melanoplus sanguinipes*): effects of lipid composition on biophysical properties. *Physiological Zoology* **67**, 1523–1543.

Gibbs, A.G., Mousseau, T.A., and Crowe, J.H. (1991) Genetic and acclimatory variation in biophysical properties of insect cuticle lipids. *Proceedings of the National Academy of Sciences of the USA* **88**, 7257–7260.

Gibbs, A.G., Perkins, M.C., and Markow, T.A. (2003b) No place to hide: microclimates of Sonoran Desert *Drosophila*. *Journal of Thermal Biology* **28**, 353–362.

Gibert, P. and Huey, R.B. (2001) Chill-coma temperature in *Drosophila*: effects of developmental temperature, latitude, and phylogeny. *Physiological and Biochemical Zoology* **74**, 429–434.

Gibert, P., Huey, R.B., and Gilchrist, G.W. (2001) Locomotor performance of *Drosophila melanogaster*: interactions among developmental and adult temperatures, age, and geography. *Evolution* **55**, 205–209.

Giesel, J.T., Lanciani, C.A., and Anderson, J.F. (1989) Metabolic rate and sexual activity in *Drosophila simulans*. *Journal of Insect Physiology* **35**, 893–895.

Gilby, A.R. (1980) Transpiration, temperature and lipids in insect cuticle. *Advances in Insect Physiology* **15**, 1–33.

Gilchrist, G.W. (1990) The consequences of sexual dimorphism in body size for butterfly flight and thermoregulation. *Functional Ecology* **4**, 475–487.

Gilchrist, G.W. (1995) Specialists and generalists in changing environments. I. Fitness landscapes of thermal sensitivity. *American Naturalist* **146**, 252–270.

Gilchrist, G.W. (1996) A quantitative genetic analysis of thermal sensitivity in the locomotor performance curve of *Aphidius ervi*. *Evolution* **50**, 1560–1572.

Gilchrist, G.W., Huey, R.B., and Partridge, L. (1997) Thermal sensitivity of *Drosophila melanogaster*: evolutionary responses of adults and eggs to laboratory natural selection at different temperatures. *Physiological Zoology* **70**, 403–414.

Gillooly, J.F., Brown, J.H., West, G.B., Savage, V.M., and Charnov, E.L. (2001) Effects of size and temperature on metabolic rate. *Science* **293**, 2248–2251.

Giordana, B., Leonardi, M.G., Casartelli, M., Consonni, P., and Parenti, P. (1998) K+-neutral amino acid symport of *Bombyx mori* larval midgut: a system operative in extreme conditions. *American Journal of Physiology* **274**, R1361–R1371.

Giribet, G., Edgecombe, G.D., and Wheeler, W.C. (2001) Arthropod phylogeny based on eight molecular loci and morphology. *Nature* **413**, 157–161.

Giribet, G., Edgecombe, G.D., Wheeler, W.C., and Babbitt, C. (2002) Phylogeny and systematic position of Opiliones: a combined analysis of chelicerate relationships using morphological and molecular data. *Cladistics* **18**, 5–70.

Glass, L. and Mackey, M.C. (1988) *From Clocks to Chaos. The Rhythms of Life*. Princeton University Press, Princeton.

Gmeinbauer, R. and Crailsheim, K. (1993) Glucose utilization during flight of honeybee (*Apis mellifera*) workers, drones and queens. *Journal of Insect Physiology* **39**, 959–967.

Goodchild, A.J.P. (1966) Evolution of the alimentary canal in the Hemiptera. *Biological Reviews* **41**, 97–140.

Goto, S.G. and Kimura, M.T. (1998) Heat- and cold-shock responses and temperature adaptations in subtropical and temperate species of *Drosophila*. *Journal of Insect Physiology* **44**, 1233–1239.

Goto, S.G., Kitamura, H.W., and Kimura, M.T. (2000) Phylogenetic relationships and climatic adaptations in the *Drosophila takahashii* and *montium* species subgroups. *Molecular Phylogenetics and Evolution* **15**, 147–156.

Goto, S.G., Yoshida, K.M., and Kimura, M.T. (1998) Accumulation of *Hsp70* mRNA under environmental stresses in diapausing and non-diapausing adults of *Drosophila triauraria*. *Journal of Insect Physiology* **44**, 1009–1015.

Gotthard, K. (2000) Increased risk of predation as a cost of high growth rate: an experimental test in a butterfly. *Journal of Animal Ecology* **69**, 896–902.

Gotthard, K., Nylin, S., and Wiklund, C. (2000) Individual state controls temperature dependence in a butterfly (*Lassiommata maera*). *Proceedings of the Royal Society of London B* **267**, 589–593.

Gouveia, S.M., Simpson, S.J., Raubenheimer, D., and Zanotto, F.P. (2000) Patterns of respiration in *Locusta migratoria* nymphs when feeding. *Physiological Entomology* **25**, 88–93.

Gracey, A.Y., Logue, J., Tiku, P.E., and Cossins, A.R. (1996) Adaptation of biological membranes to temperature: biophysical perspectives and molecular mechanisms. In *Animals and Temperature. Phenotypic and Evolutionary Adaptation* (eds. I.A. Johnston and A.F. Bennett), 1–21. Cambridge University Press, Cambridge.

Graves, J.L., Toolson, E.C., Jeong, C., Vu, L.N., and Rose, M.R. (1992) Desiccation, flight, glycogen, and postponed senescence in *Drosophila melanogaster*. *Physiological Zoology* **65**, 268–286.

Gray, B.F. (1981) On the 'surface law' and basal metabolic rate. *Journal of Theoretical Biology* **93**, 757–767.

Gringorten, J.L. and Friend, W.G. (1979) Haemolymph-volume changes in *Rhodnius prolixus* during flight. *Journal of Experimental Biology* **83**, 325–333.

Grueber, W.B. and Bradley, T.J. (1994) The evolution of increased salinity tolerance in larvae of *Aedes* mosquitoes: a phylogenetic analysis. *Physiological Zoology* **67**, 566–579.

Gruntenko, N.E., Wilson, T.G., Monastirioti, M., and Rauschenbach, I.Y. (2000) Stress-reactivity and juvenile hormone degradation in *Drosophila melanogaster* strains having stress-related mutations. *Insect Biochemistry and Molecular Biology* **30**, 775–783.

Gulinson, S.L. and Harrison, J.F. (1996) Control of resting ventilation rate in grasshoppers. *Journal of Experimental Biology* **199**, 379–389.

Gupta, B.L., Wall, B.J., Oschman, J.L., and Hall, T.A. (1980) Direct microprobe evidence of local concentration gradients and recycling of electrolytes during fluid absorption in the rectal papillae of *Calliphora*. *Journal of Experimental Biology* **88**, 21–47.

Hack, M.A. (1997) The effects of mass and age on standard metabolic rate in house crickets. *Physiological Entomology* **22**, 325–331.

Hadley, N.F. (1989) Lipid water barriers in biological systems. *Progress in Lipid Research* **28**, 1–33.

Hadley, N.F. (1994a) *Water Relations of Terrestrial Arthropods*. Academic Press, San Diego.

Hadley, N.F. (1994b) Ventilatory patterns and respiratory transpiration in adult terrestrial insects. *Physiological Zoology* **67**, 175–189.

Hadley, N.F. and Quinlan, M.C. (1993) Discontinuous carbon dioxide release in the eastern lubber grasshopper *Romalea guttata* and its effect on respiratory transpiration. *Journal of Experimental Biology* **177**, 169–180.

Hadley, N.F., Quinlan, M.C., and Kennedy, M.L. (1991) Evaporative cooling in the desert cicada: thermal efficiency and water/metabolic costs. *Journal of Experimental Biology* **159**, 269–283.

Hadley, N.F., Stuart, J.L., and Quinlan, M. (1982) An air-flow system for measuring total transpiration and cuticular permeability in arthropods: studies on the centipede *Scolopendra polymorpha*. *Physiological Zoology* **55**, 393–404.

Hadley, N.F., Toolson, E.C., and Quinlan, M.C. (1989) Regional differences in cuticular permeability in the desert cicada *Diceroprocta apache*: implications for evaporative cooling. *Journal of Experimental Biology* **141**, 219–230.

Hagner-Holler, S., Schoen, A., Erker, W., Marden, J.H., Rupprecht, R., Decker, H., and Burmester, T. (2004) A respiratory hemocyanin from an insect. *Proceedings of the National Academy of Sciences of the USA* **101**, 871–874.

Hamilton, J.G., Zangerl, A.R., DeLucia, E.H., and Berenbaum, M.R. (2001) The carbon-nutrient balance hypothesis: its rise and fall. *Ecology Letters* **4**, 86–95.

Hamilton, J.V., Munks, R.J.L., Lehane, S.M., and Lehane, M.J. (2002) Association of midgut defensin with a novel serine protease in the blood-sucking fly *Stomoxys calcitrans*. *Insect Molecular Biology* **11**, 197–205.

Hankeln, T., Jaenicke, V., Kiger, L., Dewilde, S., Ungerechts, G., Schmidt, M. *et al.* (2002) Characterization of *Drosophila* hemoglobin. *Journal of Biological Chemistry* **277**, 29012–29017.

Hanrahan, J.W., Meredith, J., Phillips, J.E., and Brandys, D. (1984) Methods for the study of transport and control in insect hindgut. In *Measurement of Ion Transport and Metabolic Rate in Insects* (eds. T.J. Bradley and T.A. Miller), 19–67. Springer, New York.

Hanski, I. (1987) Nutritional ecology of dung- and carrion-feeding insects. In *Nutritional Ecology of Insects, Mites, Spiders and Related Invertebrates* (eds. F. Slansky and J.G. Rodriguez), 837–884. John Wiley and Sons, New York.

Harborne, J.B. (1993) *Introduction to Ecological Biochemistry*. 4th edn. Academic Press, London.

Harrison, J.F. (1997) Ventilatory mechanism and control in grasshoppers. *American Zoologist* **37**, 73–81.

Harrison, J.F. (2001) Insect acid–base physiology. *Annual Review of Entomology* **46**, 221–250.

Harrison, J.F., Camazine, S., Marden, J.H., Kirkton, S.D., Rozo, A., and Yang, X. (2001) Mite not make it home: tracheal mites reduce the safety margin for oxygen delivery of flying honeybees. *Journal of Experimental Biology* **204**, 805–814.

Harrison, J.F. and Fewell, J.H. (1995) Thermal effects on feeding behavior and net energy intake in a grasshopper experiencing large diurnal fluctuations in body temperature. *Physiological Zoology* **68**, 453–473.

Harrison, J.F. and Fewell, J.H. (2002) Environmental and genetic influences on flight metabolic rate in the honey bee, *Apis mellifera*. *Comparative Biochemistry and Physiology A* **133**, 323–333.

Harrison, J.F., Fewell, J.H., Roberts, S.P., and Hall, H.G. (1996) Achievement of thermal stability by varying metabolic heat production in flying honeybees. *Science* **274**, 88–90.

Harrison, J.F., Hadley, N.F., and Quinlan, M.C. (1995) Acid–base status and spiracular control during discontinuous ventilation in grasshoppers. *Journal of Experimental Biology* **198**, 1755–1763.

Harrison, J.F. and Kennedy, M.J. (1994) *In vivo* studies of the acid–base physiology of grasshoppers: the effect of feeding state on acid–base and nitrogen excretion. *Physiological Zoology* **67**, 120–141.

Harrison, J.F. and Lighton, J.R.B. (1998) Oxygen-sensitive flight metabolism in the dragonfly *Erythemis simplicicollis*. *Journal of Experimental Biology* **201**, 1739–1744.

Harrison, J.F., Phillips, J.E., and Gleeson, T.T. (1991) Activity physiology of the two-striped grasshopper, *Melanoplus bivittatus*: gas exchange, hemolymph acid–base status, lactate production, and the effect of temperature. *Physiological Zoology* **64**, 451–472.

Harrison, J.F. and Roberts, S.P. (2000) Flight respiration and energetics. *Annual Review of Physiology* **62**, 179–205.

Harshman, L.G. and Hoffmann, A.A. (2000) Laboratory selection experiments using *Drosophila*: what do they really tell us? *Trends in Ecology and Evolution* **15**, 32–36.

Harvey, P.H. and Pagel, M.D. (1991) *The Comparative Method in Evolutionary Biology*. Oxford University Press, Oxford.

Harvey, W.R., Maddrell, S.H.P., Telfer, W.H., and Wieczorek, H. (1998) H$^+$ V-ATPases energize animal plasma membranes for secretion and absorption of ions and fluids. *American Zoologist* **38**, 426–441.

Haukioja, E., Ossipov, V., Koricheva, J., Honkanen, T., Larsson, S., and Lempa, K. (1998) Biosynthetic origin of carbon-based secondary compounds: cause of variable responses of woody plants to fertilization? *Chemoecology* **8**, 133–139.

Hawkins, B.A. and Holyoak, M. (1998) Transcontinental crashes of insect populations? *American Naturalist* **152**, 480–484.

Hawkins, B.A. and Lawton, J.H. (1995) Latitudinal gradients in butterfly body sizes: is there a general pattern? *Oecologia* **102**, 31–36.

Heinrich, B. (1975) Thermoregulation in bumblebees II. Energetics of warm-up and free flight. *Journal of Comparative Physiology* **96**, 155–166.

Heinrich, B. (1976) Heat exchange in relation to blood flow between thorax and abdomen in bumblebees. *Journal of Experimental Biology* **64**, 561–585.

Heinrich, B. (1977) Why have some animals evolved to regulate a high body temperature? *American Naturalist* **111**, 623–640.

Heinrich, B. (1980) Mechanisms of body-temperature regulation in honeybees, *Apis mellifera*. I. Regulation of head temperature. *Journal of Experimental Biology* **85**, 61–72.

Heinrich, B. (1990) Is 'reflectance' basking real? *Journal of Experimental Biology* **154**, 31–43.

Heinrich, B. (1993) *The Hot-blooded Insects: Strategies and Mechanisms of Thermoregulation*. Harvard University Press, Cambridge, MA.

Heinrich, B. and Bartholomew, G.A. (1979) Roles of endothermy and size in inter- and intraspecific competition for elephant dung in an African dung beetle, *Scarabaeus laevistriatus*. *Physiological Zoology* **52**, 484–496.

Heinrich, B. and Buchmann, S.L. (1986) Thermoregulatory physiology of the carpenter bee, *Xylocopa varipuncta*. *Journal of Comparative Physiology B* **156**, 557–562.

Heinrich, B. and Casey, T.M. (1978) Heat transfer in dragonflies: 'fliers' and 'perchers'. *Journal of Experimental Biology* **74**, 17–36.

Heinrich, B. and Heinrich, M.J.E. (1983) Heterothermia in foraging workers and drones of the bumblebee *Bombus terricola*. *Physiological Zoology* **56**, 563–567.

Heinze, J., Foitzik, S., Kipyatkov, V.E., and Lopatina, E.B. (1998) Latitudinal variation in cold hardiness and body size in the boreal ant species *Leptothorax acervorum* (Hymenoptera: Formicidae). *Entomologia Generalis* **22**, 305–312.

Helversen, O. von, Volleth, M., and Núñez, J. (1986) A new method for obtaining blood from a small mammal without injuring the animal: use of triatomid bugs. *Experientia* **42**, 809–810.

Hendrichs, J., Cooley, S.S., and Prokopy, R.J. (1992) Post-feeding bubbling behaviour in fluid-feeding Diptera: concentration of crop contents by oral evaporation of excess water. *Physiological Entomology* **17**, 153–161.

Hendrick, J.P. and Hartl, F.-U. (1993) Molecular chaperone functions of heat-shock proteins. *Annual Review of Biochemistry* **62**, 349–384.

Herbst, D.B. and Bradley, T.J. (1989) A Malpigian tubule lime gland in an insect inhabiting alkaline salt lakes. *Journal of Experimental Biology* **145**, 63–78.

Herbst, D.B., Conte, F.P., and Brookes, V.J. (1988) Osmoregulation in an alkaline salt lake insect, *Ephydra* (*Hydropyrus*) *hians* Say (Diptera: Ephydridae) in relation to water chemistry. *Journal of Insect Physiology* **34**, 903–909.

Hercus, M.J., Berrigan, D., Blows, M.W., Magiafoglou, A., and Hoffmann, A.A. (2000) Resistance to temperature extremes between and within life cycle stages in *Drosophila serrata*, *D. birchii* and their hybrids: intraspecific and interspecific comparisons. *Biological Journal of the Linnean Society* **71**, 403–416.

Herford, G.M. (1938) Tracheal pulsation in the flea. *Journal of Experimental Biology* **15**, 327–338.

Herreid, C.F., Full, R.J., and Prawel, D.A. (1981) Energetics of cockroach locomotion. *Journal of Experimental Biology* **94**, 189–202.

Herrera, C.M. (1995*a*) Floral biology, microclimate, and pollination by ectothermic bees in an early-blooming herb. *Ecology* **76**, 218–228.

Herrera, C.M. (1995*b*) Microclimate and individual variation in pollinators: flowering plants are more than their flowers. *Ecology* **76**, 1516–1524.

Herrera, C.M. (1997) Thermal biology and foraging responses of insect pollinators to the forest floor irradiance mosaic. *Oikos* **78**, 601–611.

Hertel, W. and Pass, G. (2002) An evolutionary treatment of the morphology and physiology of circulatory organs in insects. *Comparative Biochemistry and Physiology A* **133**, 555–575.

Hertz, P.E., Huey, R.B., and Stevenson, R.D. (1993) Evaluating temperature regulation by field-active ectotherms: the fallacy of the inappropriate question. *American Naturalist* **142**, 796–818.

Hetz, S.K., Psota, E., and Wasserthal, L.T. (1999) Roles of aorta, ostia and tracheae in heartbeat and respiratory gas exchange in pupae of *Troides rhadamantus* Staudinger 1888 and *Ornithoptera priamus* L. 1758 (Lepidoptera, Papilionidae). *International Journal of Insect Morphology* **28**, 131–144.

Heusner, A.A. (1991) Size and power in mammals. *Journal of Experimental Biology* **160**, 25–54.

Hewitt, P.H., Nel, J.J.C., and Schoeman, I. (1971) Influence of group size on water imbibition by *Hodotermes mossambicus* alate termites. *Journal of Insect Physiology* **17**, 587–600.

Hill, J.K., Thomas, C.D., Fox, R., Telfer, M.G., Willis, S.G., Asher, J. *et al.* (2002) Responses of butterflies to twentieth century climate warming: implications for future ranges. *Proceedings of the Royal Society of London B* **269**, 2163–2171.

Hill, J.K., Thomas, C.D., and Huntley, B. (1999) Climate and habitat availability determine 20th century changes in a butterfly's range margin. *Proceedings of the Royal Society of London B* **266**, 1197–1206.

Hoback, W.W., Podrabsky, J.E., Higley, L.G., Stanley, D.W., and Hand, S.C. (2000) Anoxia tolerance of con-familial tiger beetle larvae is associated with differences in energy flow and anaerobiosis. *Journal of Comparative Physiology B* **170**, 307–314.

Hoback, W.W. and Stanley, D.W. (2001) Insects in hypoxia. *Journal of Insect Physiology* **47**, 533–542.

Hochachka, P.W., Buck, L.T., Doll, C.J., and Land, S.C. (1996) Unifying theory of hypoxia tolerance: molecular/metabolic defense and rescue mechanisms for surviving oxygen lack. *Proceedings of the National Academy of Sciences of the USA* **93**, 9493–9498.

Hochberg, M.E. and Ives, A.R. (1999) Can natural enemies enforce geographical range limits? *Ecography* **22**, 268–276.

Hodkinson, I.D. (2003) Metabolic cold adaptation in arthropods: a smaller-scale perspective. *Functional Ecology* **17**, 562–567.

Hoffmann, A.A. (1995) Acclimation: increasing survival at a cost. *Trends in Ecology and Evolution* **10**, 1–2.

Hoffmann, A.A., Anderson, A., and Hallas, R. (2002) Opposing clines for high and low temperature resistance in *Drosophila melanogaster*. *Ecology Letters* **5**, 614–618.

Hoffmann, A.A. and Blows, M.W. (1993) Evolutionary genetics and climate change: will animals adapt to global warming? In *Biotic Interactions and Global Change* (eds. P. Kareiva, J.G. Kingsolver, and R.B. Huey), 165–178. Sinauer, Sunderland, MA.

Hoffmann, A.A. and Blows, M.W. (1994) Species borders: ecological and evolutionary perspectives. *Trends in Ecology and Evolution* **9**, 223–227.

Hoffmann, A.A., Dagher, H., Hercus, M., and Berrigan, D. (1997) Comparing different measures of heat resistance in selected lines of *Drosophila melanogaster*. *Journal of Insect Physiology* **43**, 393–405.

Hoffmann, A.A., Hallas, R., Sinclair, C., and Partridge, L. (2001*a*) Rapid loss of stress resistance in *Drosophila melanogaster* under adaptation to laboratory culture. *Evolution* **55**, 436–438.

Hoffmann, A.A., Hallas, R., Sinclair, C., and Mitrovski, P. (2001*b*) Levels of variation in stress resistance in

Drosophila among strains, local populations, and geographic regions: patterns for desiccation, starvation, cold resistance, and associated traits. *Evolution* 55, 1621–1630.

Hoffmann, A.A., Hallas, R.J., Dean, J.A., and Schiffer, M. (2003*a*) Low potential for climatic stress adaptation in a rainforest *Drosophila* species. *Science* 301, 100–102.

Hoffmann, A.A., Sørensen, J.G., and Loeschcke, V. (2003*b*) Adaptation of *Drosophila* to temperature extremes: bringing together quantitative and molecular approaches. *Journal of Thermal Biology* 28, 175–216.

Hoffmann, A.A. and Hewa-Kapuge, S. (2000) Acclimation for heat resistance in *Trichogramma* nr. *brassicae*: can it occur without costs? *Functional Ecology* 14, 55–60.

Hoffmann, A.A. and Parsons, P.A. (1988) The analysis of quantitative variation in natural populations with isofemale strains. *Genetics, Selection, Evolution* 20, 87–98.

Hoffmann, A.A. and Parsons, P.A. (1991) *Evolutionary Genetics and Environmental Stress*. Oxford University Press, Oxford.

Hoffmann, A.A. and Parsons, P.A. (1997) *Extreme Environmental Change and Evolution*. Cambridge University Press, Cambridge.

Hoffmann, A.A. and Watson, M. (1993) Geographical variation in the acclimation responses of *Drosophila* to temperature extremes. *American Naturalist Supplement* 142, 93–113.

Hofmann, G.E., Buckley, B.A., Airaksinen, S., Keen, J.E., and Somero, G.N. (2000) Heat-shock protein expression is absent in the Antarctic fish *Trematomus bernacchii* (Family Nototheniidae). *Journal of Experimental Biology* 203, 2331–2339.

Holmstrup, M., Bayley, M., and Ramløv, H. (2002) Supercool or dehydrate? An experimental analysis of overwintering strategies in small permeable arctic invertebrates. *Proceedings of the National Academy of Sciences of the USA* 99, 5716–5720.

Holmstrup, M. and Sømme, L. (1998) Dehydration and cold hardiness in the Arctic Collembolan *Onychiurus arcticus* Tullberg 1876. *Journal of Comparative Physiology B* 168, 197–203.

Holt, R.D., Lawton, J.H., Gaston, K.J., and Blackburn, T.M. (1997) On the relationship between range size and local abundance: back to the basics. *Oikos* 78, 183–190.

Holter, P. (1991) Concentration of oxygen, carbon dioxide and methane in the air within dung pats. *Pedobiologia* 35, 381–386.

Holter, P., Scholtz, C.H., and Wardhaugh, K.G. (2002) Dung feeding in adult scarabaeines (tunnellers and endocoprids): even large dung beetles eat small particles. *Ecological Entomology* 27, 169–176.

Holter, P. and Spangenberg, A. (1997) Oxygen uptake in coprophilous beetles (*Aphodius Geotrupes, Sphaeridium*) at low oxygen and high carbon dioxide concentrations. *Physiological Entomology* 22, 339–343.

Honěk, A. (1996) Geographical variation in thermal requirements for insect development. *European Journal of Entomology* 93, 303–312.

Honěk, A. (1999) Constraints on thermal requirements for insect development. *Entomological Science* 2, 615–621.

Honěk, A. and Kocourek, F. (1990) Temperature and development time in insects: a general relationship between thermal constants. *Zoologische Jahrbuecher Systematik* 117, 401–439.

Hood, W.G. and Tschinkel, W.R. (1990) Desiccation resistance in arboreal and terrestrial ants. *Physiological Entomology* 15, 23–35.

Hori, Y. and Kimura, M.T. (1998) Relationship between cold stupor and cold tolerance in *Drosophila* (Diptera: Drosophilidae). *Environmental Entomology* 27, 1297–1302.

Hosler, J.S., Burns, J.E., and Esch, H.E. (2000) Flight muscle resting potential and species-specific differences in chill-coma. *Journal of Insect Physiology* 46, 621–627.

House, C.R. (1980) Physiology of invertebrate salivary glands. *Biological Reviews* 55, 417–473.

Huey, R.B. (1991) Physiological consequences of habitat selection. *American Naturalist Supplement* 137, 91–115.

Huey, R.B. and Berrigan, D. (1996) Testing evolutionary hypotheses of acclimation. In *Animals and Temperature. Phenotypic and Evolutionary Adaptation* (eds. I.A. Johnston and A.F. Bennett), 205–237. Cambridge University Press, Cambridge.

Huey, R.B., Berrigan, D., Gilchrist, G.W., and Herron, J.C. (1999) Testing the adaptive significance of acclimation: a strong inference approach. *American Zoologist* 39, 323–336.

Huey, R.B., Crill, W.D., Kingsolver, J.G., and Weber, K.E. (1992) A method for rapid measurement of heat or cold resistance of small insects. *Functional Ecology* 6, 489–494.

Huey, R.B., Gilchrist, G.W., Carlson, M.L., Berrigan, D., and Serra, L. (2000) Rapid evolution of a geographic cline in size in an introduced fly. *Science* 287, 308–309.

Huey, R.B., Hertz, P.E., and Sinervo, B. (2003) Behavioral drive versus behavioral inertia in evolution: a null model approach. *American Naturalist* 161, 357–366.

Huey, R.B. and Kingsolver, J.G. (1993) Evolution of resistance to high temperature in ectotherms. *American Naturalist Supplement* 142, 21–46.

Huey, R.B. and Stevenson, R.D. (1979) Integrating thermal physiology and ecology of ectotherms: a discussion of approaches. *American Zoologist* 19, 357–366.

Huey, R.B., Wakefield, T., Crill, W.D., and Gilchrist, G.W. (1995) Within- and between-generation effects of temperature on early fecundity of *Drosophila melanogaster*. *Heredity* **74**, 216–223.

Hunter, A.F. (1995) The ecology and evolution of reduced wings in forest macrolepidoptera. *Evolutionary Ecology* **9**, 275–287.

Hutchison, V.H. and Maness, J.D. (1979) The role of behavior in temperature acclimation and tolerance in ectotherms. *American Zoologist* **19**, 367–384.

Hyatt, A.D. and Marshall, A.T. (1985) Water and ion balance in the tissues of the dehydrated cockroach, *Periplaneta americana*. *Journal of Insect Physiology* **31**, 27–34.

Iaboni, A., Holman, G.M., Nachman, R.J., Orchard, I., and Coast, G.M. (1998) Immunocytochemical localisation and biological activity of diuretic peptides in the housefly, *Musca domestica*. *Cell and Tissue Research* **294**, 549–560.

Irvine, B., Audsley, N., Lechleitner, R., Meredith, J., Thomson, B., and Phillips, J. (1988) Transport properties of locust ileum *in vitro*: effects of cyclic AMP. *Journal of Experimental Biology* **137**, 361–385.

Irwin, J.T. and Lee, R.E. (2002) Energy and water conservation in frozen vs supercooled larvae of the goldenrod gall fly, *Eurosta solidaginis* (Fitch) (Diptera: Tephritidae). *Journal of Experimental Zoology* **292**, 345–350.

Isaacs, R., Byrne, D.N., and Hendrix, D.L. (1998) Feeding rates and carbohydrate metabolism by *Bemisia tabaci* (Homoptera: Aleyrodidae) on different quality phloem saps. *Physiological Entomology* **23**, 241–248.

Isaacson, L.C. and Nicolson, S.W. (1994) Concealed transepithelial potentials and current rectification in tsetse fly Malpighian tubules. *Journal of Experimental Biology* **186**, 199–213.

Jackson, R.B., Linder, C.R., Lynch, M., Purusganan, M., Somerville, S., and Thayer, S.S. (2002) Linking molecular insight and ecological research. *Trends in Ecology and Evolution* **17**, 409–414.

Jacobs, M.D. and Watt, W.B. (1994) Seasonal adaptation vs physiological constraint: photoperiod, thermoregulation and flight in *Colias* butterflies. *Functional Ecology* **8**, 366–376.

Jaenike, J. (1990) Host specialization in phytophagous insects. *Annual Review of Ecology and Systematics* **21**, 243–273.

Jaenike, J. and Markow, T.A. (2003) Comparative elemental stoichiometry of ecologically diverse *Drosophila*. *Functional Ecology* **17**, 115–120.

Jarecki, J., Johnson, E., and Krasnow, M.A. (1999) Oxygen regulation of airway branching in *Drosophila* is mediated by Branchless FGF. *Cell* **99**, 211–220.

Jenkins, N.L. and Hoffmann, A.A. (1999) Limits to the southern border of *Drosophila serrata*: cold resistance, heritable variation, and trade-offs. *Evolution* **53**, 1823–1834.

Joern, A. (1979) Feeding patterns in grasshoppers (Orthoptera: Acrididae): factors influencing diet specialization. *Oecologia* **38**, 325–347.

Johnson, K.S. and Rabosky, D. (2000) Phylogenetic distribution of cysteine proteinases in beetles: evidence for an evolutionary shift to an alkaline digestive strategy in Cerambycidae. *Comparative Biochemistry and Physiology B* **126**, 609–619.

Johnson, R.A., Thomas, R.J., Wood, T.G., and Swift, M.J. (1981) The inoculation of the fungus comb in newly founded colonies of some species of the Macrotermitinae (Isoptera) from Nigeria. *Journal of Natural History* **15**, 751–756.

Jones, C.G. and Firn, R.D. (1991) On the evolution of plant secondary chemical diversity. *Philosophical Transactions of the Royal Society of London B* **333**, 273–280.

Jones, S.A. and Raubenheimer, D. (2001) Nutritional regulation in nymphs of the German cockroach, *Blatella germanica*. *Journal of Insect Physiology* **47**, 1169–1180.

Jongsma, M.A. and Bolter, C. (1997) The adaptation of insects to plant protease inhibitors. *Journal of Insect Physiology* **43**, 885–895.

Joos, B. (1987) Carbohydrate use in the flight muscles of *Manduca sexta* during pre-flight warm-up. *Journal of Experimental Biology* **133**, 317–327.

Joos, B. (1992) Adaptations for locomotion at low body temperatures in Eastern Tent Caterpillars, *Malacosoma americanum*. *Physiological Zoology* **65**, 1148–1161.

Joos, B., Lighton, J.R.B., Harrison, J.F., Suarez, R.K., and Roberts, S.P. (1997) Effects of ambient oxygen tension on flight performance, metabolism, and water loss of the honeybee. *Physiological Zoology* **70**, 167–174.

Joos, B., Young, P.A., and Casey, T.M. (1991) Wingstroke frequency of foraging and hovering bumblebees in relation to morphology and temperature. *Physiological Entomology* **16**, 191–200.

Joplin, K.H. and Denlinger, D.L. (1990) Developmental and tissue specific control of the heat shock induced 70 kDa related proteins in the flesh fly, *Sarcophaga crassipalpis*. *Journal of Insect Physiology* **36**, 239–249.

Joplin, K.H., Yocum, G.D., and Denlinger, D.L. (1990) Cold shock elicits expression of heat shock proteins in the flesh fly, *Sarcophaga crassipalpis*. *Journal of Insect Physiology* **36**, 825–834.

Josens, R.B., Farina, W.M., and Roces, F. (1998) Nectar feeding by the ant *Camponotus mus*: intake rate and crop filling as a function of sucrose concentration. *Journal of Insect Physiology* **44**, 579–585.

Josephson, R.K. (1981) Temperature and the mechanical performance of insect muscle. In *Insect Thermoregulation* (ed. B. Heinrich), 19–44. John Wiley and Sons, New York.

Josephson, R.K., Malamud, J.G., and Stokes, D.R. (2000*a*) Asynchronous muscle: a primer. *Journal of Experimental Biology* **203**, 2713–2722.

Josephson, R.K., Malamud, J.G., and Stokes, D.R. (2000*b*) Power output by an asynchronous flight muscle from a beetle. *Journal of Experimental Biology* **203**, 2667–2689.

Josephson, R.K. and Young, D. (1979) Body temperature and singing in the bladder cicada, *Cystosoma saundersii*. *Journal of Experimental Biology* **80**, 69–81.

Juliano, S.A., O'Meara, G.F., Morrill, J.R., and Cutwa, M.M. (2002) Desiccation and thermal tolerance of eggs and the coexistence of competing mosquitoes. *Oecologia* **130**, 458–469.

Jungmann, R., Rothe, U., and Nachtigall, W. (1989) Flight of the honey bee. I. Thorax surface temperature and thermoregulation during tethered flight. *Journal of Comparative Physiology B* **158**, 711–718.

Jungreis, A.M., Ruhoy, M., and Cooper, P.D. (1982) Why don't tobacco hornworms (*Manduca sexta*) become dehydrated during larval-pupal and pupal-adult development. *Journal of Experimental Zoology* **222**, 265–276.

Kaars, C. (1981) Insects—spiracle control. In *Locomotion and Energetics in Arthropods* (eds. C.F. Herreid and C.R. Fourtner), 337–366. Plenum Press, New York.

Karley, A.J., Douglas, A.E., and Parker, W.E. (2002) Amino acid composition and nutritional quality of potato leaf phloem sap for aphids. *Journal of Experimental Biology* **205**, 3009–3018.

Karlsson, B. (1994) Feeding habits and change of body composition with age in three nymphalid butterfly species. *Oikos* **69**, 224–230.

Karnosky, D.F. *et al.* (2003) Tropospheric O_3 moderates responses of temperate hardwood forests to elevated CO_2: a synthesis of molecular to ecosystem results from the Aspen FACE project. *Functional Ecology* **17**, 289–304.

Kaspari, M. and Vargo, E.L. (1995) Colony size as a buffer against seasonality: Bergmann's rule in social insects. *American Naturalist* **145**, 610–632.

Kaspari, M. and Weiser, M.D. (1999) The size-grain hypothesis and interspecific scaling in ants. *Functional Ecology* **13**, 530–538.

Kataoka, H., Troetschler, R.G., Li, J.P., Kramer, S.J., Carney, R.L., and Schooley, D.A. (1989) Isolation and identification of a diuretic hormone from the tobacco hornworm, *Manduca sexta*. *Proceedings of the National Academy of Sciences of the USA* **86**, 2976–2980.

Kauffman, S.A. (1993) *The Origins of Order. Self-Organization and Selection in Evolution.* Oxford University Press, Oxford.

Keister, M. and Buck, J. (1961) Respiration of *Phormia regina* in relation to temperature and oxygen. *Journal of Insect Physiology* **7**, 51–72.

Keister, M. and Buck, J. (1964) Some endogenous and exogenous effects on rate of respiration. In *Physiology of Insecta* (ed. M. Rockstein), Vol. 3, 617–658. Academic Press, New York.

Kells, S.A., Vogt, J.T., Appel, A.G., and Bennett, G.W. (1999) Estimating nutritional status of German cockroaches, *Blatella germanica* (L.) (Dictyoptera: Blatellidae), in the field. *Journal of Insect Physiology* **45**, 709–717.

Kelty, J.D., Killian, K.A., and Lee, R.E. (1996) Cold shock and rapid cold-hardening of pharate adult flesh flies (*Sarcophaga crassipalpis*): effects on behaviour and neuromuscular function following eclosion. *Physiological Entomology* **21**, 283–288.

Kelty, J.D. and Lee, R.E. (1999) Induction of rapid cold hardening by cooling at ecologically relevant rates in *Drosophila melanogaster*. *Journal of Insect Physiology* **45**, 719–726.

Kelty, J.D. and Lee, R.E. (2001) Rapid cold-hardening of *Drosophila melanogaster* (Diptera: Drosophilidae) during ecologically based thermoperiodic cycles. *Journal of Experimental Biology* **204**, 1659–1666.

Kenagy, G.J. and Stevenson, R.D. (1982) Role of body temperature in the seasonality of daily activity in tenebrionid beetles of eastern Washington. *Ecology* **63**, 1491–1503.

Kerkut, G.A. and Gilbert, L.I. (eds.) (1985) *Comprehensive Insect Physiology, Biochemistry and Pharmacology.* Pergamon Press, Oxford.

Kestler, P. (1985) Respiration and respiratory water loss. In *Environmental Physiology and Biochemistry of Insects* (ed. K.H. Hoffmann), 137–183. Springer, Berlin.

Kevan, P.G. (1973) Flowers, insects and pollination ecology in the Canadian high arctic. *Polar Record* **16**, 667–674.

Kevan, P.G. (1975) Sun-tracking solar furnaces in High Arctic flowers: significance for pollination and insects. *Science* **189**, 723–726.

Kim, Y. and Kim, N. (1997) Cold hardiness in *Spodoptera exigua* (Lepidoptera: Noctuidae). *Environmental Entomology* **26**, 1117–1123.

Kimura, M.T., Ohtsu, T., Yoshida, T., Awasaki, T., and Lin, F.-J. (1994) Climatic adaptations and distributions in the *Drosophila takahashii* species subgroup (Dipera: Drosophilidae). *Journal of Natural History* **28**, 401–409.

Kindvall, O. (1995) The impact of extreme weather on habitat preference and survival in a metapopulation of the bush cricket *Metrioptera bicolor* in Sweden. *Biological Conservation* **73**, 51–58.

King, L.E., Steele, J.E., and Bajura, S.W. (1986) The effect of flight on the composition of haemolymph in the cockroach, *Periplaneta americana*. *Journal of Insect Physiology* **32**, 649–655.

Kingsolver, J.G. (1983) Thermoregulation and flight in *Colias* butterflies: elevational patterns and mechanistic limitations. *Ecology* **64**, 534–545.

Kingsolver, J.G. (1987) Evolution and coadaptation of thermoregulatory behavior and wing pigmentation pattern in pierid butterflies. *Evolution* **41**, 472–490.

Kingsolver, J.G. (1989) Weather and the population dynamics of insects: integrating physiological and population ecology. *Physiological Zoology* **62**, 314–334.

Kingsolver, J.G. (1995a) Fitness consequences of seasonal polyphenism in western white butterflies. *Evolution* **49**, 942–954.

Kingsolver, J.G. (1995b) Viability selection on seasonally polyphenic traits: wing melanin pattern in western white butterflies. *Evolution* **49**, 932–941.

Kingsolver, J.G. (1996) Experimental manipulation of wing pigment pattern and survival in western white butterflies. *American Naturalist* **147**, 296–306.

Kingsolver, J.G. (2000) Feeding, growth, and the thermal environment of cabbage white caterpillars, *Pieris rapae* L. *Physiological and Biochemical Zoology* **73**, 621–628.

Kingsolver, J.G. and Huey, R.B. (1998) Evolutionary analyses of morphological and physiological plasticity in thermally variable environments. *American Zoologist* **38**, 545–560.

Kingsolver, J.G. and Srygley, R.B. (2000) Experimental analyses of body size, flight and survival in pierid butterflies. *Evolutionary Ecology Research* **2**, 593–612.

Kingsolver, J.G. and Woods, H.A. (1997) Thermal sensitivity of growth and feeding in *Manduca sexta* catepillars. *Physiological Zoology* **70**, 631–638.

Kingsolver, J.G. and Woods, H.A. (1998) Interactions of temperature and dietary protein concentration in growth and feeding of *Manduca sexta* caterpillars. *Physiological Entomology* **23**, 354–359.

Klass, K.-D., Zompro, O., Kristensen, N.P., and Adis, J. (2002) Mantophasmatodea: a new insect order with extant members in the Afrotropics. *Science* **296**, 1456–1459.

Klein, U., Koch, A., and Moffet, D.F. (1996) Ion transport in Lepidoptera. In *Biology of the Insect Midgut* (eds. M.J. Lehane and P.F. Billingsley), 236–264. Chapman and Hall, London.

Klok, C.J. and Chown, S.L. (1997) Critical thermal limits, temperature tolerance and water balance of a sub-Antarctic caterpillar, *Pringleophaga marioni* (Lepidoptera: Tineidae). *Journal of Insect Physiology* **43**, 685–694.

Klok, C.J. and Chown, S.L. (1998a) Field thermal ecology and water relations of *gregaria* phase African army-worm caterpillars, *Spodoptera exempta* (Lepidoptera: Noctuidae). *Journal of Thermal Biology* **23**, 131–142.

Klok, C.J. and Chown, S.L. (1998b) Interactions between desiccation resistance, host-plant contact and the thermal biology of a leaf-dwelling sub-Antarctic caterpillar, *Embryonopsis halticella* (Lepidoptera: Yponomeutidae). *Journal of Insect Physiology* **44**, 615–628.

Klok, C.J. and Chown, S.L. (1999) Assessing the benefits of aggregation: thermal biology and water relations of Anomalous Emperor Moth caterpillars. *Functional Ecology* **13**, 417–427.

Klok, C.J. and Chown, S.L. (2001) Critical thermal limits, temperature tolerance and water balance of a sub-Antarctic kelp fly, *Paractora dreuxi* (Diptera: Helcomyzidae). *Journal of Insect Physiology* **47**, 95–109.

Klok, C.J. and Chown, S.L. (2003) Resistance to temperature extremes in sub-Antarctic weevils: interspecific variation, population differentiation and acclimation. *Biological Journal of the Linnean Society* **78**, 401–414.

Klok, C.J., Gaston, K.J., and Chown, S.L. (2003) The geographical range structure of the holly leaf-miner. III. Cold hardiness physiology. *Functional Ecology* **17**, 858–868.

Klok, C.J., Mercer, R.D., and Chown, S.L. (2002) Discontinuous gas-exchange in centipedes and its convergent evolution in tracheated arthropods. *Journal of Experimental Biology* **205**, 1019–1029.

Klok, C.J., Sinclair, B.J., and Chown, S.L. (2004) Upper thermal tolerance and oxygen limitation in terrestrial arthropods. *Journal of Experimental Biology* **207**, 2361–2370.

Klostermeyer, E.C., Mech, S.J., and Rasmussen, W.B. (1973) Sex and weight of *Megachile rotundata* (Hymenoptera: Megachilidae) progeny associated with provision weights. *Journal of the Kansas Entomological Society* **46**, 536–548.

Knapp, R. and Casey, T.M. (1986) Thermal ecology, behavior, and growth of gypsy moth and eastern tent caterpillars. *Ecology* **67**, 598–608.

Koiwa, H., Bressan, R.A., and Hasegawa, P.M. (1997) Regulation of protease inhibitors and plant defense. *Trends in Plant Sciences* **2**, 379–384.

Kölsch, G., Jakobi, K., Wegener, G., and Braune, H.J. (2002) Energy metabolism and metabolic rate of the alder leaf beetle *Agelastica alni* (L.) (Coleoptera, Chrysomelidae) under aerobic and anaerobic conditions: a microcalori-metric study. *Journal of Insect Physiology* **48**, 143–151.

Komai, Y. (1998) Augmented respiration in a flying insect. *Journal of Experimental Biology* **201**, 2359–2366.

Koptur, S. (1992) Extrafloral nectary-mediated inter-actions between insects and plants. In *Insect-Plant Interactions* (ed. E. Bernays), Vol. IV, 81–129. CRC Press, Boca Raton, Florida.

Kovac, H. and Schmaranzer, S. (1996) Thermoregulation of honeybees (*Apis mellifera*) foraging in spring and summer at different plants. *Journal of Insect Physiology* **42**, 1071–1076.

Kovac, H. and Stabentheiner, A. (1999) Effect of food quality on the body temperature of wasps (*Paravespula vulgaris*). *Journal of Insect Physiology* **45**, 183–190.

Kozłowski, J. (1996) Optimal initial size and adult size of animals: consequences for macroevolution and community structure. *American Naturalist* **147**, 101–114.

Kozłowski, J. and Gawelczyk, A.T. (2002) Why are spe-cies' body size distributions usually skewed to the right? *Functional Ecology* **16**, 419–432.

Kozłowski, J. and Weiner, J. (1997) Interspecific allometries are by-products of body size optimization. *American Naturalist* **149**, 352–380.

Krafsur, E.S. (1971) Behavior of thoracic spiracles of *Aedes* mosquitoes in controlled relative humidities. *Annals of the Entomological Society of America* **64**, 93–97.

Kram, R. (1996) Inexpensive load carrying by rhinoceros beetles. *Journal of Experimental Biology* **199**, 609–612.

Krebs, R.A. and Bettencourt, B.R. (1999) Evolution of thermotolerance and variation in the heat shock protein, Hsp70. *American Zoologist* **39**, 910–919.

Krebs, R.A. and Feder, M.E. (1997) Natural variation in the expression of the heat-shock protein Hsp70 in a population of *Drosophila melanogaster* and its correlation with tolerance of ecologically relevant thermal stress. *Evolution* **51**, 173–179.

Krebs, R.A. and Feder, M.E. (1998a) Hsp70 and larval thermotolerance in *Drosophila melanogaster*: how much is enough and when is more too much? *Journal of Insect Physiology* **44**, 1091–1101.

Krebs, R.A. and Feder, M.E. (1998b) Experimental manipulation of the cost of thermal acclimation in *Drosophila melanogaster*. *Biological Journal of the Linnean Society* **63**, 593–601.

Krebs, R.A., Feder, M.E., and Lee, J. (1998) Heritability of expression of the 70 kD heat-shock protein in *Drosophila melanogaster* and its relevance to the evolution of thermo-tolerance. *Evolution* **52**, 841–847.

Krebs, R.A. and Loeschcke, V. (1994) Costs and benefits of activation of the heat-shock response in *Drosophila melanogaster*. *Functional Ecology* **8**, 730–737.

Krebs, R.A. and Loeschcke, V. (1995a) Resistance to thermal stress in preadult *Drosophila buzzatii*: variation among populations and changes in relative resistance

across life stages. *Biological Journal of the Linnean Society* **56**, 517–531.

Krebs, R.A. and Loeschcke, V. (1995b) Resistance to thermal stress in adult *Drosophila buzzatii*: acclimation and variation among populations. *Biological Journal of the Linnean Society* **56**, 505–515.

Krebs, R.A. and Loeschcke, V. (1996) Acclimation and selection for increased resistance to thermal stress in *Drosophila buzzatii*. *Genetics* **142**, 471–479.

Krebs, R.A. and Loeschcke, V. (1997) Estimating herit-ability in a threshold trait: heat-shock tolerance in *Drosophila buzzatii*. *Heredity* **79**, 252–259.

Krolikowski, K. and Harrison, J.F. (1996) Haemolymph acid–base status, tracheal gas levels and the control of post-exercise ventilation rate in grasshoppers. *Journal of Experimental Biology* **199**, 391–399.

Kukal, O., Ayres, M.P., and Scriber, J.M. (1991) Cold tolerance of the pupae in relation to the distribution of swallowtail butterflies. *Canadian Journal of Zoology* **69**, 3028–3037.

Kukal, O. and Dawson, T.E. (1989) Temperature and food quality influences feeding behavior, assimilation effici-ency and growth rate af arctic woolly-bear caterpillars. *Oecologia* **79**, 526–532.

Kukal, O. and Duman, J.G. (1989) Switch in the over-wintering strategy of two insect species and latitudinal differences in cold hardiness. *Canadian Journal of Zoology* **67**, 825–827.

Kukal, O., Heinrich, B., and Duman, J.G. (1988) Beha-vioural thermoregulation in the freeze-tolerant Arctic caterpillar, *Gynaephora groenlandica*. *Journal of Experi-mental Biology* **138**, 181–193.

Kytö, M., Niemelä, P., and Larsson, S. (1996) Insects on trees: population and individual response to fertilization. *Oikos* **75**, 148–159.

Laird, T.B., Winston, P.W., and Braukman, M. (1972) Water storage in the cockroach *Leucophaea maderae* F. *Naturwissenschaften* **11**, 515–516.

Lanciani, C., Lipp, K.E., and Giesel, J.T. (1992) The effect of photoperiod on cold tolerance in *Drosophila melanogaster*. *Journal of Thermal Biology* **17**, 147–148.

Lansing, E., Justesen, J., and Loeschcke, V. (2000) Vari-ation in the expression of Hsp70, the major heat-shock protein, and thermotolerance in larval and adult selection lines in *Drosophila melanogaster*. *Journal of Thermal Biology* **25**, 443–450.

Larsen, K.J. and Lee, R.E. (1994) Cold tolerance including rapid cold-hardening and inoculative freezing of fall migrant monarch butterflies in Ohio. *Journal of Insect Physiology* **40**, 859–864.

Lawrence, P.K. and Koundal, K.R. (2002) Plant protease inhibitors in control of phytophagous insects. *Electronic*

Journal of Biotechnology [http://www.ejb.org/content/vol5/issue1/full/3] **5**, 93–109.

Lawton, J.H. (1991) From physiology to population dynamics and communities. *Functional Ecology* **5**, 155–161.

Lawton, J.H. (1992) There are not 10 million kinds of population dynamics. *Oikos* **63**, 337–338.

Lawton, J.H. (1999) Are there general laws in ecology? *Oikos* **84**, 177–192.

Le Lagadec, M.D., Chown, S.L., and Scholtz, C.H. (1998) Desiccation resistance and water balance in southern African keratin beetles (Coleoptera, Trogidae): the influence of body size and habitat. *Journal of Comparative Physiology B* **168**, 112–122.

Lederhouse, R.C., Finke, M.D., and Scriber, J.M. (1982) The contributions of larval growth and pupal duration to protandry in the black swallowtail butterfly, *Papilio polyxenes*. *Oecologia* **53**, 296–300.

Lee, R.E. (1989) Insect cold-hardiness: to freeze or not to freeze. *BioScience* **39**, 308–313.

Lee, R.E. (1991) Principles of insect low temperature tolerance. In *Insects at Low Temperatures* (eds. R.E. Lee and D.L. Denlinger), 17–46. Chapman and Hall, New York.

Lee, R.E., Chen, C.-P., and Denlinger, D.L. (1987) A rapid cold-hardening process in insects. *Science* **238**, 1415–1417.

Lee, R.E. and Costanzo, J.P. (1998) Biological ice nucleation and ice distribution in cold-hardy ectothermic animals. *Annual Review of Physiology* **60**, 55–72.

Lee, R.E., Costanzo, J.P., and Lee, M.R. (1998) Reducing cold-hardiness of insect pests using ice nucleating active microbes. In *Temperature Sensitivity in Insects and Application in Integrated Pest Management* (eds. G.J. Hallman and D.L. Denlinger), 97–124. Westview Press, Boulder.

Lee, R.E. and Denlinger, D.L. (1985) Cold tolerance in diapausing and non-diapausing stages of the flesh fly, *Sarcophaga crassipalpis*. *Physiological Entomology* **10**, 309–315.

Lee, R.E. and Denlinger, D.L. (eds.) (1991) *Insects at Low Temperatures*. Chapman and Hall, New York.

Lehane, M.J. and Billingsley, P.F. (eds.) (1996) *Biology of the Insect Midgut*. Chapman and Hall, London.

Lehane, M.J., Blakemore, D., Williams, S. and Moffatt, M.R. (1995) Regulation of digestive enzyme levels in insects. *Comparative Biochemistry and Physiology B* **110**, 285–289.

Lehmann, F.-O. (2001) Matching spiracle opening to metabolic need during flight in *Drosophila*. *Science* **294**, 1926–1929.

Lehmann, F.-O., Dickinson, M.H., and Staunton, J. (2000) The scaling of carbon dioxide release and respiratory

water loss in flying fruit flies (*Drosophila* spp.). *Journal of Experimental Biology* **203**, 1613–1624.

Leonardi, M.G., Casartelli, M., Parenti, P., and Giordana, B. (1998) Evidence for a low-affinity, high-capacity uniport for amino acids in *Bombyx mori* larval midgut. *American Journal of Physiology* **274**, R1372-R1375.

Leonardi, M.G., Fiandra, L., Casartelli, M., Cappellozza, S., and Giordana, B. (2001) Modulation of leucine absorption in the larval midgut of *Bombyx mori* (Lepidoptera, Bombycidae). *Comparative Biochemistry and Physiology A* **129**, 665–672.

Lerman, D.N. and Feder, M.E. (2001) Laboratory selection at different temperatures modifies heat-shock transcription factor (HSF) activation in *Drosophila melanogaster*. *Journal of Experimental Biology* **204**, 315–323.

Lerman, D.N., Michalak, P., Helin, A.B., Bettencourt, B.R., and Feder, M.E. (2003) Modification of heat-shock gene expression in *Drosophila melanogaster* populations via transposable elements. *Molecular Biology and Evolution* **20**, 135–144.

Lettau, J., Foster, W.A., Harker, J.E., and Treherne, J.E. (1977) Diel changes in potassium activity in the haemolymph of the cockroach *Leucophaea maderae*. *Journal of Experimental Biology* **71**, 171–186.

Levine, M. and Tjian, R. (2003) Transcription regulation and animal diversity. *Nature* **424**, 147–151.

Levins, R. (1969) Thermal acclimation and heat resistance in *Drosophila* species. *American Naturalist* **103**, 483–499.

Levy, R.I. and Schneiderman, H.A. (1966a) Discontinuous respiration in insects—II. The direct measurement and significance of changes in tracheal gas composition during the respiratory cycle of silkworm pupae. *Journal of Insect Physiology* **12**, 83–104.

Levy, R.I. and Schneiderman, H.A. (1966b) Discontinuous respiration in insects—III. The effect of temperature and ambient oxygen tension on the gaseous composition of the tracheal system of silkworm pupae. *Journal of Insect Physiology* **12**, 105–121.

Levy, R.I. and Schneiderman, H.A. (1966c) Discontinuous respiration in insects—IV. Changes in intratracheal pressure during the respiratory cycle of silkworm pupae. *Journal of Insect Physiology* **12**, 465–492.

Leyssens, A., Dijkstra, S., Van Kerkhove, E., and Steels, P. (1994) Mechanisms of K^+ uptake across the basal membrane of Malpighian tubules of *Formica polyctena*: the effect of ions and inhibitors. *Journal of Experimental Biology* **195**, 123–145.

Liebhold, A.M., Rossi, R.E., and Kemp, W.P. (1993) Geostatistics and geographic information systems in applied insect ecology. *Annual Review of Entomology* **38**, 303–327.

Lighton, J.R.B. (1985) Minimum cost of transport and ventilatory patterns in three African beetles. *Physiological Zoology* **58**, 390–399.

Lighton, J.R.B. (1987) Cost of tokking: the energetics of substrate communication in the tok-tok beetle, *Psammodes striatus*. *Journal of Comparative Physiology B* **157**, 11–20.

Lighton, J.R.B. (1988*a*) Discontinuous CO_2 emission in a small insect, the formicine ant *Camponotus vicinus*. *Journal of Experimental Biology* **134**, 363–376.

Lighton, J.R.B. (1988*b*) Simultaneous measurement of oxygen uptake and carbon dioxide emission during discontinuous ventilation in the tok-tok beetle, *Psammodes striatus*. *Journal of Insect Physiology* **34**, 361–367.

Lighton, J.R.B. (1989) Individual and whole-colony respiration in an African formicine ant. *Functional Ecology* **3**, 523–530.

Lighton, J.R.B. (1990) Slow discontinuous ventilation in the Namib dune-sea ant *Camponotus detritus* (Hymenoptera, Formicidae). *Journal of Experimental Biology* **151**, 71–82.

Lighton, J.R.B. (1991*a*) Ventilation in Namib desert tenebrionid beetles: mass scaling and evidence of a novel quantized flutter-phase. *Journal of Experimental Biology* **159**, 249–268.

Lighton, J.R.B. (1991*b*) Insects: measurements. In *Concise Encyclopedia on Biological and Biomedical Measurement Systems* (ed. P.A. Payne), 201–208. Pergamon Press, Oxford.

Lighton, J.R.B. (1992) Direct measurement of mass loss during discontinuous ventilation in two species of ants. *Journal of Experimental Biology* **173**, 289–293.

Lighton, J.R.B. (1994) Discontinuous ventilation in terrestrial insects. *Physiological Zoology* **67**, 142–162.

Lighton, J.R.B. (1996) Discontinuous gas exchange in insects. *Annual Review of Entomology* **41**, 309–324.

Lighton, J.R.B. (1998) Notes from underground: towards ultimate hypotheses of cyclic, discontinuous gas-exchange in tracheate arthropods. *American Zoologist* **38**, 483–491.

Lighton, J.R.B. (2002) Lack of discontinuous gas exchange in a tracheate arthropod, *Leiobunum townsendi* (Arachnida, Opiliones). *Physiological Entomology* **27**, 170–174.

Lighton, J.R.B. and Bartholomew, G.A. (1988) Standard energy metabolism of a desert harvester ant, *Pogonomyrmex rugosus*: effects of temperature, body mass, group size, and humidity. *Proceedings of the National Academy of Sciences of the USA* **85**, 4765–4769.

Lighton, J.R.B., Bartholomew, G.A., and Feener, D.H. (1987) Energetics of locomotion and load carriage and a model of the energy cost of foraging in the leaf-cutting ant *Atta colombica* Guer. *Physiological Zoology* **60**, 524–537.

Lighton, J.R.B. and Berrigan, D. (1995) Questioning paradigms: caste-specific ventilation in harvester ants, *Messor pergandei* and *M. julianus* (Hymenoptera: Formicidae). *Journal of Experimental Biology* **198**, 521–530.

Lighton, J.R.B., Brownell, P.H., Joos, B., and Turner, R.J. (2001) Low metabolic rate in scorpions: implications for population biomass and cannibalism. *Journal of Experimental Biology* **204**, 607–613.

Lighton, J.R.B. and Duncan, F.D. (1995) Standard and exercise metabolism and the dynamics of gas exchange in the giant red velvet mite, *Dinothrombium magnificum*. *Journal of Insect Physiology* **41**, 877–884.

Lighton, J.R.B. and Duncan, F.D. (2002) Energy cost of locomotion: validation of laboratory data by *in situ* respirometry. *Ecology* **83**, 3517–3522.

Lighton, J.R.B. and Feener, D.H. (1989) A comparison of energetics and ventilation of desert ants during voluntary and forced locomotion. *Nature* **342**, 174–175.

Lighton, J.R.B. and Fielden, L.J. (1995) Mass scaling of standard metabolism in ticks: a valid case of low metabolic rates in sit-and-wait strategists. *Physiological Zoology* **68**, 43–62.

Lighton, J.R.B. and Fielden, L.J. (1996) Gas exchange in wind spiders (Arachnida, Solphugidae): independent evolution of convergent control strategies in solphugids and insects. *Journal of Insect Physiology* **42**, 347–357.

Lighton, J.R.B., Fielden, L.J., and Rechav, Y. (1993*c*) Discontinuous ventilation in a non-insect, the tick *Amblyomma marmoreum* (Acari, Ixodidae): characterization and metabolic modulation. *Journal of Experimental Biology* **180**, 229–245.

Lighton, J.R.B., Fukushi, T., and Wehner, R. (1993*a*) Ventilation in *Cataglyphis bicolor*: regulation of carbon dioxide release from the thoracic and abdominal spiracles. *Journal of Insect Physiology* **39**, 687–699.

Lighton, J.R.B. and Garrigan, D. (1995) Ant breathing: testing regulation and mechanism hypotheses with hypoxia. *Journal of Experimental Biology* **198**, 1613–1620.

Lighton, J.R.B., Garrigan, D.A., Duncan, F.D., and Johnson, R.A. (1993*b*) Spiracular control of respiratory water loss in female alates of the harvester ant *Pogonomyrmex rugosus*. *Journal of Experimental Biology* **179**, 233–244.

Lighton, J.R.B. and Joos, B. (2002) Discontinuous gas exchange in the pseudoscorpion *Garypus californicus* is regulated by hypoxia, not hypercapnia. *Physiological and Biochemical Zoology* **75**, 345–349.

Lighton, J.R.B. and Lovegrove, B.G. (1990) A temperature-induced switch from diffusive to convective ventilation in the honeybee. *Journal of Experimental Biology* **154**, 509–516.

Lighton, J.R.B., Quinlan, M.C., and Feener, D.H. (1994) Is bigger better? Water balance in the polymorphic desert harvester ant *Messor pergandei*. *Physiological Entomology* **19**, 325–334.

Lighton, J.R.B. and Turner, R.J. (2004) Thermolimit respirometry: an objective assessment of critical thermal maxima in two sympatric desert harvester ants, *Pogonomyrmex rugosus* & *P. californicus*. *Journal of Experimental Biology* **207**, 1903–1913.

Lighton, J.R.B. and Wehner, R. (1993) Ventilation and respiratory metabolism in the thermophilic desert ant, *Cataglyphis bicolor* (Hymenoptera, Formicidae). *Journal of Comparative Physiology B* **163**, 11–17.

Lighton, J.R.B., Weier, J.A., and Feener, D.H. (1993*d*) The energetics of locomotion and load carriage in the desert harvester ant *Pogonomyrmex rugosus*. *Journal of Experimental Biology* **181**, 49–61.

Lincoln, D.E., Fajer, E.D., and Johnson, R.H. (1993) Plant–insect herbivore interactions in elevated CO_2 environments. *Trends in Ecology and Evolution* **8**, 64–68.

Lindquist, S. (1986) The heat-shock response. *Annual Review of Biochemistry* **55**, 1151–1191.

Lindroth, R.L., Klein, K.A., Hemming, J.D.C., and Feuker, A.M. (1997) Variation in temperature and dietary nitrogen affect performance of the gypsy moth (*Lymantria dispar* L.). *Physiological Entomology* **22**, 55–64.

Lindsay, K.L. and Marshall, A.T. (1981) The osmoregulatory role of the filter-chamber in relation to phloem-feeding in *Eurymela distincta* (Cicadelloidea, Homoptera). *Physiological Entomology* **6**, 413–419.

Linton, S.M. and O'Donnell, M.J. (1999) Contributions of K^+:Cl^- cotransport and Na^+/K^+-ATPase to basolateral ion transport in Malpighian tubules of *Drosophila melanogaster*. *Journal of Experimental Biology* **202**, 1561–1570.

Locke, M. (1998) Caterpillars have evolved lungs for hemocyte gas exchange. *Journal of Insect Physiology* **44**, 1–20.

Locke, M. (2001) The Wigglesworth Lecture: insects for studying fundamental problems in biology. *Journal of Insect Physiology* **47**, 495–507.

Lockey, K.H. (1991) Insect hydrocarbon classes: implications for chemotaxonomy. *Insect Biochemistry* **21**, 91–97.

Loehle, C. (1998) Height growth rate tradeoffs determine northern and southern range limits for trees. *Journal of Biogeography* **25**, 735–742.

Loeschcke, V. and Krebs, R.A. (1996) Selection for heat-shock resistance in larval and in adult *Drosophila buzzatii*:

comparing direct and indirect responses. *Evolution* **50**, 2354–2359.

Loeschcke, V., Krebs, R.A., and Barker, J.S.F. (1994) Genetic variation for resistance and acclimation to high temperature stress in *Drosophila buzzatii*. *Biological Journal of the Linnean Society* **52**, 83–92.

Loeschcke, V., Krebs, R.A., Dahlgaard, J., and Michalak, P. (1997) High-temperature stress and the evolution of thermal resistance in *Drosophila*. In *Environmental Stress, Adaptation and Evolution* (eds. R. Bijlsma and V. Loeschcke), 175–190. Birkhäuser, Basel.

Loudon, C. (1988) Development of *Tenebrio molitor* in low oxygen levels. *Journal of Insect Physiology* **34**, 97–103.

Louw, G.N. and Hadley, N.F. (1985) Water economy of the honeybee: a stoichiometric accounting. *Journal of Experimental Zoology* **235**, 147–150.

Louw, G.N. and Nicolson, S.W., (1983) Thermal, energetic and nutritional considerations in the foraging and reproduction of the carpenter bee *Xylocopa capitata*. *Journal of the Entomological Society of South Africa* **46**, 227–240.

Louw, G.N., Nicolson, S.W., and Seely, M.K. (1986) Respiration beneath desert sand: carbon dioxide diffusion and respiratory patterns in a tenebrionid beetle. *Journal of Experimental Biology* **120**, 443–447.

Lovegrove, B.G. (2000) The zoogeography of mammalian basal metabolic rate. *American Naturalist* **156**, 201–219.

Loveridge, J.P. (1968) The control of water loss in *Locusta migratoria migratorioides* R. & F. II. Water loss through the spiracles. *Journal of Experimental Biology* **49**, 15–29.

Loveridge, J.P. (1974) Studies on the water relations of adult locusts. II—Water gain in the food and loss in the faeces. *Transactions of the Rhodesia Scientific Association* **56**, 1–30.

Loveridge, J.P. (1980) Cuticular water relations techniques. In *Cuticle Techniques in Arthropods* (ed. T.A. Miller), 301–366. Springer, New York.

Luft, P.A., Paine, T.D., and Walker, G.P. (2001) Interactions of colonisation density and leaf environments on survival of *Trioza eugeniae* nymphs. *Ecological Entomology* **26**, 263–270.

Lundheim, R. and Zachariassen, K.E. (1993) Water balance of over-wintering beetles in relation to strategies for cold tolerance. *Journal of Comparative Physiology B* **163**, 1–4.

Lutterschmidt, W.I. and Hutchison, V.H. (1997) The critical thermal maximum: history and critique. *Canadian Journal of Zoology* **75**, 1561–1574.

MacArthur, R.H. (1972) *Geographical Ecology. Patterns in the Distribution of Species*. Harper and Row, New York.

Machin, J. (1981) Water compartmentalisation in insects. *Journal of Experimental Zoology* **215**, 327–333.

Machin, J. (1983) Water vapor absorption in insects. *American Journal of Physiology* **244**, R187-R192.

Machin, J., Kestler, P., and Lampert, G.J. (1991) Simultaneous measurements of spiracular and cuticular water losses in *Periplaneta americana*: implications for whole-animal mass loss studies. *Journal of Experimental Biology* **161**, 439–453.

Machin, J. and Lampert, G.J. (1987) An improved water content model for *Periplaneta* cuticle: effects of epidermis removal and cuticle damage. *Journal of Insect Physiology* **33**, 647–655.

Machin, J. and O'Donnell, M.J. (1991) Rectal complex ion activities and electrochemical gradients in larvae of the desert beetle, *Onymacris*: comparisons with *Tenebrio*. *Journal of Insect Physiology* **37**, 829–838.

Machin, K.E., Pringle, J.W.S., and Tamasige, M. (1962) The physiology of insect fibrillar muscle. IV. The effect of temperature on a beetle flight muscle. *Proceedings of the Royal Society of London B* **155**, 493–499.

Maddrell, S.H.P. (1991) The fastest fluid-secreting cell known: the upper Malpighian tubule cell of *Rhodnius*. *BioEssays* **13**, 357–362.

Maddrell, S.H.P., Herman, W.S., Farndale, R.W., and Riegel, J.A. (1993) Synergism of hormones controlling epithelial fluid transport in an insect. *Journal of Experimental Biology* **174**, 65–80.

Magiafoglou, A. and Hoffmann, A.A. (2003) Cross-generation effects due to cold exposure in *Drosophila serrata*. *Functional Ecology* **17**, 664–672.

Maisonhaute, C., Chihrane, J., and Lauge, G. (1999) Induction of thermotolerance in *Trichogramma brassicae* (Hymenoptera: Trichogrammatidae). *Environmental Entomology* **28**, 116–122.

Mangum, C.P. and Hochachka, P.W. (1998) New directions in comparative physiology and biochemistry: mechanisms, adaptations, and evolution. *Physiological Zoology* **71**, 471–484.

Marais, E. and Chown, S.L. (2003) Repeatability of standard metabolic rate and gas exchange characteristics in a highly variable cockroach, *Perisphaeria* sp. *Journal of Experimental Biology* **206**, 4565–4574.

Marden, J.H. (1989) Bodybuilding dragonflies: costs and benefits of maximizing flight muscle. *Physiological Zoology* **62**, 505–521.

Marden, J.H. (1995a) Evolutionary adaptation of contractile performance in muscle of ectothermic winter-flying moths. *Journal of Experimental Biology* **198**, 2087–2094.

Marden, J.H. (1995b) Large-scale changes in thermal sensitivity of flight performance during adult maturation in a dragonfly. *Journal of Experimental Biology* **198**, 2095–2102.

Marden, J.H. (2000) Variability in the size, composition, and function of insect flight muscles. *Annual Review of Physiology* **62**, 157–178.

Marden, J.H. and Chai, P. (1991) Aerial predation and butterfly design: how palatibility, mimicry, and the need for evasive flight constrain mass allocation. *American Naturalist* **138**, 15–36.

Marden, J.H., Fitzhugh, G.H., and Wolf, M.R. (1998) From molecules to mating success: integrative biology of muscle maturation in a dragonfly. *American Zoologist* **38**, 528–544.

Marden, J.H., Kramer, M.G., and Frisch, J. (1996) Age-related variation in body temperature, thermoregulation and activity in a thermally polymorphic dragonfly. *Journal of Experimental Biology* **199**, 529–535.

Margraf, N., Gotthard, K., and Rahier, M. (2003) The growth strategy of an alpine beetle: maximization or individual growth adjustment in relation to seasonal time horizons? *Functional Ecology* **17**, 605–610.

Marian, M.P., Pandian, T.J., and Muthukrishnan, J. (1982) Energy balance in *Speliphron violaceum* (Hymenoptera) and use of meconium weight as an index of bioenergetics components. *Oecologia* **55**, 264–267.

Marsh, A.C. (1985) Thermal responses and temperature tolerance in a diurnal desert ant, *Ocymyrmex barbiger*. *Physiological Zoology* **58**, 629–636.

Marshall, A.T., Kyriakou, P., Cooper, P.D., Coy, R., and Wright, A. (1995) Osmolality of rectal fluid from two species of osmoregulating brine fly larvae (Diptera: Ephydridae). *Journal of Insect Physiology* **41**, 413–418.

Martin, A.P. and Palumbi, S.R. (1993) Body size, metabolic rate, generation time, and the molecular clock. *Proceedings of the National Academy of Sciences of the USA* **90**, 4087–4091.

Martin, M.M. (1991) The evolution of cellulose digestion in insects. *Philosophical Transactions of the Royal Society of London B* **333**, 281–288.

Martin, M.M. and Van't Hof, H.M. (1988) The cause of reduced growth of *Manduca sexta* larvae on a low-water diet: increased metabolic processing costs or nutrient limitation? *Journal of Insect Physiology* **34**, 515–525.

Masaki, S. (1996) Geographical variation of life cycle in crickets (Ensifera: Grylloidea). *European Journal of Entomology* **93**, 281–302.

Mattson, W.J. (1980) Herbivory in relation to plant nitrogen content. *Annual Review of Ecology and Systematics* **11**, 119–161.

May, M.L. (1979a) Energy metabolism of dragonflies (Odonata: Anisoptera) at rest and during endothermic warm-up. *Journal of Experimental Biology* **83**, 79–94.

May, M.L. (1979b) Insect thermoregulation. *Annual Review of Entomology* **24**, 313–349.

May, M.L. (1995) Dependence of flight behavior and heat production on air temperature in the green darner dragonfly *Anax junius* (Odonata: Aeshnidae). *Journal of Experimental Biology* **198**, 2385–2392.

May, P.G. (1992) Flower selection and the dynamics of lipid reserves in two nectarivorous butterflies. *Ecology* **73**, 2181–2191.

May, R.M. (1986) The search for patterns in the balance of nature: advances and retreats. *Ecology* **67**, 1115–1126.

McCabe, J. and Partridge, L. (1997) An interaction between environmental temperature and genetic variation for body size for the fitness of adult female *Drosophila melanogaster*. *Evolution* **51**, 1164–1174.

McClain, E., Seely, M.K., Hadley, N.F., and Gray, V. (1985) Wax blooms in tenebrionid beetles of the Namib desert: correlations with environment. *Ecology* **66**, 112–118.

McColl, G., Hoffmann, A.A., and McKechnie, S.W. (1996) Response of two heat shock genes to selection for knockdown heat resistance in *Drosophila melanogaster*. *Genetics* **143**, 1615–1627.

McDonald, J.R., Bale, J.S., and Walters, K.F.A. (1997) Rapid cold hardening in the western flower thrips *Frankliniella occidentalis*. *Journal of Insect Physiology* **43**, 759–766.

McDonald, J.R., Head, J., Bale, J.S., and Walters, K.F.A. (2000) Cold tolerance, overwintering and establishment potential of *Thrips palmi*. *Physiological Entomology* **25**, 159–166.

McEvoy, P.B. (1984) Increase in respiratory rate during feeding in larvae of the cinnabar moth *Tyria jacobaeae*. *Physiological Entomology* **9**, 191–195.

McQuate, G.T. and Connor, E.F. (1990a) Insect responses to plant water deficits. I. Effect of water deficits in soybean plants on the feeding preference of Mexican bean beetle larvae. *Ecological Entomology* **15**, 419–431.

McQuate, G.T. and Connor, E.F. (1990b) Insect responses to plant water deficits. II. Effect of water deficits in soybean plants on the growth and survival of Mexican bean beetle larvae. *Ecological Entomology* **15**, 433–445.

Meats, A. (1973) Rapid acclimatization to low temperature in the Queensland fruit fly, *Dacus tryoni*. *Journal of Insect Physiology* **19**, 1903–1911.

Mellanby, K. (1932) The influence of atmospheric humidity on the thermal death point of a number of insects. *Journal of Experimental Biology* **9**, 222–231.

Meredith, J., Ring, M., Macins, A., Marschall, J., Cheng, N.N., Theilmann, D. *et al.* (1996) Locust ion transport peptide (ITP): primary structure, cDNA and expression in a baculovirus system. *Journal of Experimental Biology* **199**, 1053–1061.

Merzendorfer, H. and Zimoch, L. (2003) Chitin metabolism in insects: structure, function and regulation of chitin synthases and chitinases. *Journal of Experimental Biology* **206**, 4393–4412.

Messenger, P.S. (1959) Bioclimatic studies with insects. *Annual Review of Entomology* **4**, 183–206.

Miller, K. (1982) Cold-hardiness strategies of some adult and immature insects overwintering in interior Alaska. *Comparative Biochemistry and Physiology A* **73**, 595–604.

Miller, L.K. (1978) Freezing tolerance in relation to cooling rate in an adult insect. *Cryobiology* **15**, 345–349.

Miller, P.L. (1960) Respiration in the desert locust III. Ventilation and the spiracles during flight. *Journal of Experimental Biology* **37**, 264–278.

Miller, P.L. (1964) Factors altering spiracle control in adult dragonflies: water balance. *Journal of Experimental Biology* **41**, 331–343.

Miller, P.L. (1966) The supply of oxygen to the active flight muscles of some large beetles. *Journal of Experimental Biology* **45**, 285–304.

Miller, P.L. (1973) Spatial and temporal changes in the coupling of cockroach spiracles to ventilation. *Journal of Experimental Biology* **59**, 137–148.

Miller, P.L. (1974) Respiration—aerial gas transport. In *Physiology of Insecta, Volume VI* (ed. M. Rockstein), 345–402. Academic Press, New York.

Miller, P.L. (1981) Ventilation in active and inactive insects. In *Locomotion and Energetics in Arthropods* (eds. C.F. Herreid and C.R. Fourtner), 367–390. Plenum Press, New York.

Miller, W.E. (1996) Population behaviour and adult feeding capability in Lepidoptera. *Environmental Entomology* **25**, 213–226.

Minois, N. (2001) Resistance to stress as a function of age in transgenic *Drosophila melanogaster* overexpressing Hsp70. *Journal of Insect Physiology* **47**, 1007–1012.

Mira, A. (2000) Exuviae eating: a nitrogen meal? *Journal of Insect Physiology* **46**, 605–610.

Misener, S.R., Chen, C.-P., and Walker, V.K. (2001) Cold tolerance and proline metabolic gene expression in *Drosophila melanogaster*. *Journal of Insect Physiology* **47**, 393–400.

Moczek, A.P. (2002) Allometric plasticity in a polyphenic beetle. *Ecological Entomology* **27**, 58–67.

Moffatt, L. (2001) Metabolic rate and thermal stability during honeybee foraging at different reward rates. *Journal of Experimental Biology* **204**, 759–766.

Moffatt, L. and Núñez, J.A. (1997) Oxygen consumption in the foraging honeybee depends on the reward rate at the food source. *Journal of Comparative Physiology B* **167**, 36–42.

Moller, H. (1996) Lessons for invasion theory from social insects. *Biological Conservation* **78**, 125–142.

Montgomery, M.E. (1982) Life-cycle nitrogen budget for the gypsy moth, *Lymantria dispar*, reared on artificial diet. *Journal of Insect Physiology* **28**, 437–442.

Morgan, K.R. (1987) Temperature regulation, energy metabolism and mate-searching in rain beetles (*Pleocoma* spp.), winter-active, endothermic scarabs (Coleoptera). *Journal of Experimental Biology* **128**, 107–122.

Morgan, K.R. and Bartholomew, G.A. (1982) Home-othermic response to reduced ambient temperature in a scarab beetle. *Science* **216**, 1409–1410.

Morgan, K.R., Shelly, T.E., and Kimsey, L.S. (1985) Body temperature regulation, energy metabolism, and foraging in light-seeking and shade-seeking robber flies. *Journal of Comparative Physiology B* **155**, 561–570.

Morrow, P.A. and Fox, L.R. (1980) Effects of variation in *Eucalyptus* essential oil yield on insect growth and grazing damage. *Oecologia* **45**, 209–219.

Mousseau, T.A. and Roff, D.A. (1989) Adaptation to seasonality in a cricket: patterns of phenotypic and genotypic variation in body size and diapause expression along a cline in season length. *Evolution* **43**, 1483–1496.

Mueller, U.G. and Gerardo, N. (2002) Fungus-farming insects: multiple origins and diverse evolutionary histories. *Proceedings of the National Academy of Sciences of the USA* **99**, 15247–15249.

Muharsini, S., Dalrymple, B., Vuocolo, T., Hamilton, S., Willadsen, P., and Wijffels, G. (2001) Biochemical and molecular characterization of serine proteases from larvae of *Chrysomya bezziana*, the Old World screwworm fly. *Insect Biochemistry and Molecular Biology* **31**, 1029–1040.

Müller, C.B., Williams, I.S., and Hardie, J. (2001) The role of nutrition, crowding and interspecific interactions in the development of winged aphids. *Ecological Entomology* **26**, 330–340.

Müller, H.-M., Catteruccia, F., Vizioli, J., Della Torre, A., and Crisanti, A. (1995) Constitutive and blood meal-induced trypsin genes in *Anopheles gambiae*. *Experimental Parasitology* **81**, 371–385.

Mullins, D.E. (1985) Chemistry and physiology of the hemolymph. In *Comprehensive Insect Physiology, Biochemistry and Pharmacology* (eds. G.A. Kerkut and L.I. Gilbert), Vol. 3, 355–400. Pergamon Press, Oxford.

Nardi, J.B., Mackie, R.I., and Dawson, J.O. (2002) Could microbial symbionts of arthropod guts contribute significantly to nitrogen fixation in terrestrial ecosystems? *Journal of Insect Physiology* **48**, 751–763.

Nation, J.L. (1985) Respiratory systems. In *Fundamentals of Insect Physiology* (ed. M.S. Blum), 185–225. Wiley-Interscience, New York.

Nation, J.L. (2002) *Insect Physiology and Biochemistry.* CRC Press, Boca Raton.

Nayar, J.K. and Van Handel, E. (1971) The fuel for sustained mosquito flight. *Journal of Insect Physiology* **17**, 471–481.

Neargarder, G., Dahlhoff, E.P., and Rank, N.E. (2003) Variation in thermal tolerance is linked to phosphoglucose isomerase genotype in a montane leaf beetle. *Functional Ecology* **17**, 213–221.

Neufeld, D.S. and Leader, J.P. (1998) Freezing survival by isolated Malpighian tubules of the New Zealand alpine weta *Hemideina maori*. *Journal of Experimental Biology* **201**, 227–236.

Nichol, A.C. (2000) *Water Load: a Physiological Limitation to Bumblebee Foraging Behaviour?* Ph.D. thesis, University of Cambridge, Cambridge.

Nicolson, S. (1992) Excretory function in *Tenebrio molitor*: fast tubular secretion in a vapour-absorbing insect. *Journal of Insect Physiology* **38**, 139–146.

Nicolson, S.W. (1976) Diuresis in the cabbage white butterfly, *Pieris brassicae*: water and ion regulation and the role of the hindgut. *Journal of Insect Physiology* **22**, 1623–1630.

Nicolson, S.W. (1980) Water balance and osmoregulation in *Onymacris plana*, a tenebrionid beetle from the Namib Desert. *Journal of Insect Physiology* **26**, 315–320.

Nicolson, S.W. (1991) Diuresis or clearance: is there a physiological role for the 'diuretic hormone' of the desert beetle *Onymacris*? *Journal of Insect Physiology* **37**, 447–452.

Nicolson, S.W. (1993) The ionic basis of fluid secretion in insect Malpighian tubules: advances in the last ten years. *Journal of Insect Physiology* **39**, 451–458.

Nicolson, S.W. (1998) The importance of osmosis in nectar secretion and its consumption by insects. *American Zoologist* **38**, 418–425.

Nicolson, S.W. and Hanrahan, S.A. (1986) Diuresis in a desert beetle? Hormonal control of the Malpighian tubules of *Onymacris plana* (Coleoptera: Tenebrionidae). *Journal of Comparative Physiology B* **156**, 407–413.

Nicolson, S.W. and Louw, G.N. (1982) Simultaneous measurement of evaporative water loss, oxygen consumption, and thoracic temperature during flight in a carpenter bee. *Journal of Experimental Zoology* **222**, 287–296.

Nicolson, S.W., Louw, G.N., and Edney, E.B. (1984) Use of a ventilated capsule and tritiated water to measure evaporative water losses in a tenebrionid beetle. *Journal of Experimental Biology* **108**, 477–481.

Nielsen, M.G. (2001) Energetic cost of foraging in the ant *Rhytidoponera aurata* in tropical Australia. *Physiological Entomology* **26**, 248–253.

Nielsen, M.G., Jensen, T.F., and Holm-Jensen, I. (1982) Effect of load carriage on the respiratory metabolism of running worker ants of *Camponotus herculeanus* (Formicidae). *Oikos* **39**, 137–142.

Nijhout, H.F. (1994) *Insect Hormones*. Princeton University Press, Princeton.

Nijhout, H.F. (1999) Control mechanisms of polyphenic development in insects. *BioScience* **49**, 181–192.

Nijhout, H.F. and Emlen, D.J. (1998) Competition among body parts in the development and evolution of insect morphology. *Proceedings of the National Academy of Sciences of the USA* **95**, 3685–3689.

Noble-Nesbitt, J. (1991) Cuticular permeability and its control. In *The Physiology of the Insect Epidermis* (eds. K. Binnington and A. Retnakaran), 252–283. CSIRO Publications, Melbourne, Australia.

Noble-Nesbitt, J., Appel, A.G., and Croghan, P.C. (1995) Water and carbon dioxide loss from the cockroach *Periplaneta americana* (L.) measured using radioactive isotopes. *Journal of Experimental Biology* **198**, 235–240.

Novotný, V. and Wilson, M.R. (1997) Why are there no small species among xylem-sucking insects? *Evolutionary Ecology* **11**, 419–437.

Nowbahari, B. and Thibout, E. (1990) The cocoon and humidity in the development of *Acrolepiopsis assectella* (Lep.) pupae: consequences in adults. *Physiological Entomology* **15**, 363–368.

Nunamaker, R.A. (1993) Rapid cold-hardening in *Culicoides variipennis sonorensis* (Diptera: Ceratopogonidae). *Journal of Medical Entomology* **30**, 913–917.

Nylin, S. and Gotthard, K. (1998) Plasticity in life-history traits. *Annual Review of Entomology* **43**, 63–83.

Nylin, S. and Svärd, L. (1991) Latitudinal patterns in the size of European butterflies. *Holarctic Ecology* **14**, 192–202.

Nylund, L. (1991) Metabolic rates of *Calathus melanocephalus* (L.) (Coleoptera, Carabidae) from alpine and lowland habitats (Jeløy and Finse, Norway and Drenthe, The Netherlands). *Comparative Biochemistry and Physiology A* **100**, 853–862.

O'Brien, D.M. (1999) Fuel use in flight and its dependence on nectar feeding in the hawkmoth *Amphion floridensis*. *Journal of Experimental Biology* **202**, 441–451.

O'Brien, D.M., Fogel, M.L., and Boggs, C.L. (2002) Renewable and nonrenewable resources: amino acid turnover and allocation to reproduction in Lepidoptera. *Proceedings of the National Academy of Sciences of the USA* **99**, 4413–4418.

O'Brien, D.M., Schrag, D.P., and Martinez del Rio, C.M. (2000) Allocation to reproduction in a hawkmoth: a quantitative analysis using stable carbon isotopes. *Ecology* **81**, 2822–2831.

O'Connor, K.R. and Beyenbach, K.W. (2001) Chloride channels in apical membrane patches of stellate cells of Malpighian tubules of *Aedes aegypti*. *Journal of Experimental Biology* **204**, 367–378.

O'Donnell, M.J. (1997) Mechanisms of excretion and ion transport in invertebrates. In *Handbook of Physiology* (ed. W.H. Dantzler), Vol. 2, 1207–1289. Oxford University Press, Oxford.

O'Donnell, M.J., Dow, J.A.T., Huesmann, G.R., Tublitz, N.J., and Maddrell, S.H.P. (1996) Separate control of anion and cation transport in Malpighian tubules of *Drosophila melanogaster*. *Journal of Experimental Biology* **199**, 1163–1175.

O'Donnell, M.J. and Machin, J. (1988) Water vapor absorption by terrestrial organisms. *Advances in Comparative and Environmental Physiology* **2**, 47–90.

O'Donnell, M.J. and Machin, J. (1991) Ion activities and electrochemical gradients in the mealworm rectal complex. *Journal of Experimental Biology* **155**, 375–402.

O'Donnell, M.J., Rheault, M.R., Davies, S.A., Rosay, P., Harvey, B.J., Maddrell, S.H.P. *et al.* (1998) Hormonally controlled chloride movement across *Drosophila* tubules is via ion channels in stellate cells. *American Journal of Physiology* **274**, R1039–R1049.

O'Donnell, M.J. and Spring, J.H. (2000) Modes of control of insect Malpighian tubules: synergism, antagonism, cooperation and autonomous regulation. *Journal of Insect Physiology* **46**, 107–117.

Oertli, J.J. (1989) Relationship of wing beat frequency and temperature during take-off flight in temperate-zone beetles. *Journal of Experimental Biology* **145**, 321–338.

Ojeda-Avila, T., Woods, H.A., and Raguso, R.A. (2003) Effects of dietary variation on growth, composition, and maturation of *Manduca sexta* (Sphingidae: Lepidoptera). *Journal of Insect Physiology* **49**, 293–306.

Olckers, T. and Hulley, P.E. (1994) Host specificity tests on leaf-feeding insects: aberrations from the use of excised leaves. *African Entomology* **2**, 68–70.

O'Neill, K.M., Kemp, W.P., and Johnson, K.A. (1990) Behavioural thermoregulation in three species of robber flies (Diptera, Asilidae: *Efferia*). *Animal Behaviour* **39**, 181–191.

Ono, M., Igarashi, T., Ohno, E., and Sasaki, M. (1995) Unusual thermal defence by a honeybee against mass attack by hornets. *Nature* **377**, 334–336.

Packard, G.C. and Boardman, T.J. (1988) The misuse of ratios, indices, and percentages in ecophysiological research. *Physiological Zoology* **61**, 1–9.

Paim, U. and Beckel, W.E. (1963) Seasonal oxygen and carbon dioxide content of decaying wood as a component of the microenvironment of *Orthosoma brunneum* (Forster) (Coleoptera: Cerambycidae). *Canadian Journal of Zoology* **41**, 1133–1147.

Paim, U. and Beckel, W.E. (1964) The carbon dioxide related behavior of the adults of *Orthosoma brunneum* (Forster) (Coleoptera, Cerambycidae). *Canadian Journal of Zoology* **42**, 295–304.

Pannabecker, T. (1995) Physiology of the Malpighian tubule. *Annual Review of Entomology* **40**, 493–510.

Pannabecker, T.L., Hayes, T.K., and Beyenbach, K.W. (1993) Regulation of epithelial shunt conductance by the peptide leucokinin. *Journal of Membrane Biology* **132**, 63–76.

Pappenheimer, J.R. (1993) On the coupling of membrane digestion with intestinal absorption of sugars and amino acids. *American Journal of Physiology* **265**, G409–G417.

Parish, W.E.G. and Bale, J.S. (1990) The effect of feeding and gut contents on supercooling in larvae of *Pieris brassicae*. *Cryoletters* **11**, 67–74.

Park, T. (1962) Beetles, competition, and populations. *Science* **138**, 1369–1375.

Parmenter, R.R., Parmenter, C.A., and Cheney, C.D. (1989) Factors influencing microhabitat partitioning in arid-land darkling beetles (Tenebrionidae): temperature and water conservation. *Journal of Arid Environments* **17**, 57–67.

Parmesan, C., Root, T.L., and Willig, M.R. (2000) Impacts of extreme weather and climate on terrestrial biota. *Bulletin of the American Meteorological Society* **81**, 443–450.

Parmesan, C., Ryrholm, N., Stefanescu, C., Hill, J.K., Thomas, C.D., Descimon, H. *et al.* (1999) Poleward shifts in geographical ranges of butterfly species associated with regional warming. *Nature* **399**, 579–583.

Parmesan, C. and Yohe, G. (2003) A globally coherent fingerprint of climate change impacts across natural systems. *Nature* **421**, 37–42.

Parr, C.L. (2003) *Ant Assemblages in a Southern African Savanna: Local Processes and Conservation Implications.* Ph.D. Thesis, University of Pretoria, Pretoria.

Parr, Z.J.E., Parr, C.L., and Chown, S.L. (2003) The size-grain hypothesis: a phylogenetic and field test. *Ecological Entomology* **28**, 475–481.

Parsell, D.A. and Lindquist, S. (1993) The function of heat-shock proteins in stress tolerance: degradation and reactivation of damaged proteins. *Annual Review of Genetics* **27**, 437–496.

Parsons, P.A. (1977) Resistance to cold temperature stress in populations of *Drosophila melanogaster* and *D. simulans*. *Australian Journal of Zoology* **25**, 693–698.

Partridge, L., Barrie, B., Fowler, K., and French, V. (1994) Evolution and development of body size and cell size in *Drosophila melanogaster* in response to temperature. *Evolution* **48**, 1269–1276.

Partridge, L. and French, V. (1996) Thermal evolution of ecotherm body size: why get big in the cold? In *Animals and Temperature. Phenotypic and Evolutionary Adaptation* (eds. I.A. Johnston and A.F. Bennett), 265–292. Cambridge University Press, Cambridge.

Patrick, M.L. and Bradley, T.J. (2000a) The physiology of salinity tolerance in larvae of two species of *Culex* mosquitoes: the role of compatible solutes. *Journal of Experimental Biology* **203**, 821–830.

Patrick, M.L. and Bradley, T.J. (2000b) Regulation of compatible solute accumulation in larvae of the mosquito *Culex tarsalis*: osmolarity versus salinity. *Journal of Experimental Biology* **203**, 831–839.

Patrick, M.L., Gonzales, R.J., Wood, C.M., Wilson, R.W., Bradley, T.J., and Val, A.L. (2002) The characterization of ion regulation in Amazonian mosquito larvae: evidence of phenotypic plasticity, population-based disparity, and novel mechanisms of ion uptake. *Physiological and Biochemical Zoology* **75**, 223–236.

Peters, R.H. (1983) *The Ecological Implications of Body Size.* Cambridge University Press, Cambridge.

Petersen, C., Woods, H.A., and Kingsolver, J.G. (2000) Stage-specific effects of temperature and dietary protein on growth and survival of *Manduca sexta* caterpillars. *Physiological Entomology* **25**, 35–40.

Peumans, W.J. and Van Damme, E.J.M. (1995) Lectins as plant defense proteins. *Plant Physiology* **109**, 347–352.

Phillips, J. (1981) Comparative physiology of insect renal function. *American Journal of Physiology* **241**, R241–R257.

Phillips, J.E., Hanrahan, J., Chamberlin, M., and Thomson, B. (1986) Mechanisms and control of reabsorption in insect hindgut. *Advances in Insect Physiology* **19**, 329–422.

Phillips, J.E., Meredith, J., Audsley, N., Richardson, N., Macins, A., and Ring, M. (1998) Locust ion transport peptide (ITP): a putative hormone controlling water and ionic balance in terrestrial insects. *American Zoologist* **38**, 461–470.

Phillips, J.E., Thomson, R.B., Audsley, N., Peach, J.L., and Stagg, A.P. (1994) Mechanisms of acid–base transport and control in locust excretory system. *Physiological Zoology* **67**, 95–119.

Phillips, J.E., Wiens, C., Audsley, N., Jeffs, L., Bilgen, T., and Meredith, J. (1996a) Nature and control of chloride transport in insect absorptive epithelia. *Journal of Experimental Zoology* **275**, 292–299.

Phillips, S.A., Jusino-Atresino, R., and Thorvilson, H.G. (1996b) Desiccation resistance in populations of the red imported fire ant (Hymenoptera: Formicidae). *Environmental Entomology* **25**, 460–464.

Pietrantonio, P.V. and Gill, S.S. (1996) *Bacillus thuringiensis* endotoxins: action on the insect midgut. In *Biology of the Insect Midgut* (eds. M.J. Lehane and P.F. Billingsley), 345–372. Chapman and Hall, London.

Piiper, J., Dejours, P., Haab, P., and Rahn, H. (1971) Concepts and basic quantities in gas exchange physiology. *Respiration Physiology* **13**, 292–304.

Pivnick, K.A. and McNeil, J.N. (1986) Sexual differences in the thermoregulation of *Thymelicus lineola* adults (Lepidoptera: Hesperiidae). *Ecology* **67**, 1024–1035.

Pivnick, K.A. and McNeil, J.N. (1987) Puddling in butterflies: sodium affects reproductive success in *Thymelicus lineola*. *Physiological Entomology* **12**, 461–472.

Polcyn, D.M. (1994) Thermoregulation during summer activity in Mojave Desert dragonflies (Odonata: Anisoptera). *Functional Ecology* **8**, 441–449.

Pollard, E., Greatorex-Davies, J.N., and Thomas, J.A. (1997) Drought reduces breeding success of the butterfly *Aglais urticae*. *Ecological Entomology* **22**, 315–318.

Pörtner, H.O. (2001) Climate change and temperature-dependent biogeography: oxygen limitation of thermal tolerance in animals. *Naturwissenschaften* **88**, 137–146.

Pörtner, H.O. (2002) Climate variations and the physiological basis of temperature dependent biogeography: systemic to molecular hierarchy of thermal tolerance in animals. *Comparative Biochemistry and Physiology A* **132**, 739–761.

Pörtner, H.O., Hardewig, I., Sartoris, F.J., and van Dijk, P.L.M. (1998) Energetic aspects of cold adaptation: critical temperatures in metabolic, ionic and acid–base regulation? In *Cold Ocean Physiology* (eds. H.O. Pörtner and R. Playle), 88–120. Cambridge University Press, Cambridge.

Pörtner, H.O., van Dijk, P.L.M., Hardewig, I., and Sommer, A. (2000) Levels of metabolic cold adaptation: tradeoffs in eurythermal and stenothermal ectotherms. In *Antarctic Ecosystems: Models for Wider Ecological Understanding* (eds. W. Davison, C. Howard-Williams, and P.A. Broady), 109–122. New Zealand Natural Sciences, Christchurch.

Prange, H.D. (1990) Temperature regulation by respiratory evaporation in grasshoppers. *Journal of Experimental Biology* **154**, 463–474.

Prosser, C.L. (1986) *Adaptational Biology: Molecules to Organisms*. John Wiley & Sons, New York.

Pullin, A.S. (1996) Physiological relationships between insect diapause and cold tolerance: coevolution or coincidence? *European Journal of Entomology* **93**, 121–129.

Punt, A. (1950) The respiration of insects. *Physiologia Comparata et Oecologia* **2**, 59–74.

Punt, A., Parser, W.J., and Kuchlein, J. (1957) Oxygen uptake in insects with cyclic CO_2 release. *Biological Bulletin* **112**, 108–119.

Quinlan, M.C. and Hadley, N.F. (1993) Gas exchange, ventilatory patterns, and water loss in two lubber grasshoppers: quantifying cuticular and respiratory transpiration. *Physiological Zoology* **66**, 628–642.

Quinlan, M.C. and Lighton, J.R.B. (1999) Respiratory physiology and water relations of three species of *Pogonomyrmex* harvester ants (Hymenoptera: Formicidae). *Physiological Entomology* **24**, 293–302.

Quinn, R.M., Gaston, K.J., and Roy, D.B. (1997) Coincidence between consumer and host occurrence: macrolepidoptera in Britain. *Ecological Entomology* **22**, 197–208.

Ramirez, J.M. and Pearson, K.G. (1989) Distribution of intersegmental interneurones that can reset the respiratory rhythm of the locust. *Journal of Experimental Biology* **141**, 151–176.

Ramløv, H. (1999) Microclimate and variations in haemolymph composition in the freezing-tolerant New Zealand alpine weta *Hemideina maori* Hutton (Orthoptera: Stenopelmatidae). *Journal of Comparative Physiology B* **169**, 224–235.

Ramløv, H. (2000) Aspects of natural cold tolerance in ectothermic animals. *Human Reproduction* **15**, 26–46.

Ramløv, H. and Lee, R.E. (2000) Extreme resistance to desiccation in overwintering larvae of the gall fly *Eurosta solidaginis* (Diptera, Tephritidae). *Journal of Experimental Biology* **203**, 783–789.

Ramsay, J.A. (1964) The rectal complex of the mealworm *Tenebrio molitor*, L. (Coleoptera, Tenebrionidae). *Philosophical Transactions of the Royal Society of London B* **248**, 279–314.

Rasa, O.A.E. (1997) Aggregation in a desert tenebrionid beetle: a cost/benefit analysis. *Ethology* **103**, 466–487.

Ratte, H.T. (1985) Temperature and insect development. In *Environmental Physiology and Biochemistry of Insects* (ed. K.H. Hoffmann), 33–66. Springer, Berlin.

Raubenheimer, D. (1995) Problems with ratio analysis in nutritional studies. *Functional Ecology* **9**, 21–29.

Raubenheimer, D. and Bernays, E.A. (1993) Patterns of feeding in the polyphagous grasshopper *Taeniopoda eques*: a field study. *Animal Behaviour* **45**, 153–167.

Raubenheimer, D. and Gäde, G. (1993) Compensatory water intake by locusts (*Locusta migratoria*): implications for mechanisms regulating drink size. *Journal of Insect Physiology* **39**, 275–281.

Raubenheimer, D. and Gäde, G. (1994) Hunger-thirst interactions in the locust, *Locusta migratoria*. *Journal of Insect Physiology* **40**, 631–639.

Raubenheimer, D. and Gäde, G. (1996) Separating food and water deprivation in locusts: effects on the patterns of consumption, locomotion and growth. *Physiological Entomology* **21**, 76–84.

Raubenheimer, D. and Simpson, S.J. (1990) The effects of simultaneous variation in protein, digestible carbohydrate and tannic acid on the feeding behaviour of larval *Locusta migratoria* (L.) and *Schistocerca gregaria* (Forskal). I. Short-term studies. *Physiological Entomology* **15**, 219–233.

Raubenheimer, D. and Simpson, S.J. (1999) Integrating nutrition: a geometrical approach. *Entomologia Experimentalis et Applicata* **91**, 67–82.

Read, A.F. and Harvey, P.H. (1989) Life history differences among the eutherian radiations. *Journal of Zoology London* **219**, 329–353.

Reavey, D. (1993) Why body size matters to caterpillars. In *Caterpillars: Ecological and Evolutionary Constraints on Foraging* (eds. N.E. Stamp and T.M. Casey), 248–279. Chapman and Hall, New York.

Rees, C.J.C. (1986) Skeletal economy in certain herbivorous beetles as an adaptation to a poor dietary supply of nitrogen. *Ecological Entomology* **11**, 221–228.

Reineke, S., Wieczorek, H., and Merzendorfer, H. (2002) Expression of the *Manduca sexta* V-ATPase genes *mvB*, *mvG* and *mvd* is regulated by ecdysteroids. *Journal of Experimental Biology* **205**, 1059–1067.

Reinhold, K. (1999) Energetically costly behaviour and the evolution of resting metabolic rate in insects. *Functional Ecology* **13**, 217–224.

Reynolds, S.E. (1990) Feeding in caterpillars: maximizing or optimizing food acquisition? In *Animal Nutrition and Transport Processes. 1. Nutrition in Wild and Domestic Animals* (ed. J. Mellinger), Vol. 5, 106–118, Basel, Karger.

Reynolds, S.E. and Bellward, K. (1989) Water balance in *Manduca sexta* caterpillars: water recycling from the rectum. *Journal of Experimental Biology* **141**, 33–45.

Reynolds, S.E. and Nottingham, S.F. (1985) Effects of temperature on growth and efficiency of food utilization in fifth-instar caterpillars of the tobacco hornworm, *Manduca sexta*. *Journal of Insect Physiology* **31**, 129–134.

Reynolds, S.E., Nottingham, S.F., and Stephens, A.E. (1985) Food and water economy and its relation to growth in fifth-instar larvae of the tobacco hornworm, *Manduca sexta*. *Journal of Insect Physiology* **31**, 119–127.

Reynolds, S.E., Yeomans, M.R., and Timmins, W.A. (1986) The feeding behaviour of caterpillars (*Manduca sexta*) on tobacco and on artificial diet. *Physiological Entomology* **11**, 39–51.

Rhodes, J.D., Croghan, P.C., and Dixon, A.F.G. (1997) Dietary sucrose and oligosaccharide synthesis in relation to osmoregulation in the pea aphid *Acyrthosiphon pisum*. *Physiological Entomology* **22**, 373–379.

Rickards, J., Kelleher, M.J., and Storey, K.B. (1987) Strategies of freeze avoidance in larvae of the goldenrod gall moth, *Epiblema scudderiana*: winter profiles of a natural population. *Journal of Insect Physiology* **33**, 443–450.

Ricklefs, R.E. and Wikelski, M. (2002) The physiology/life-history nexus. *Trends in Ecology and Evolution* **17**, 462–468.

Rinehart, J.P. and Denlinger, D.L. (2000) Heat-shock protein 90 is down-regulated during pupal diapause in the flesh fly, *Sarcophaga crassipalpis*, but remains responsive to thermal stress. *Insect Molecular Biology* **9**, 641–645.

Rinehart, J.P., Yocum, G.D., and Denlinger, D.L. (2000*a*) Developmental upregulation of inducible hsp70 transcripts, but not the cognate form, during pupal diapause in the flesh fly, *Sarcophaga crassipalpis*. *Insect Biochemistry and Molecular Biology* **30**, 515–521.

Rinehart, J.P., Yocum, G.D., and Denlinger, D.L. (2000*b*) Thermotolerance and rapid cold hardening ameliorate the negative effects of brief exposures to high or low temperatures on fecundity in the flesh fly, *Sarcophaga crassipalpis*. *Physiological Entomology* **25**, 330–336.

Ring, R.A. (1982) Freezing-tolerant insects with low supercooling points. *Comparative Biochemistry and Physiology A* **73**, 605–612.

Ring, R.A. and Danks, H.V. (1994) Desiccation and cryoprotection: overlapping adaptations. *Cryoletters* **15**, 181–190.

Ring, R.A. and Danks, H.V. (1998) The role of trehalose in cold-hardiness and desiccation. *Cryoletters* **19**, 275–282.

Ring, R.A. and Riegert, P.W. (1991) A tribute to R.W. Salt. In *Insects at Low Temperatures* (eds. R.E. Lee and D.L. Denlinger), 3–16. Chapman and Hall, New York.

Roberts, C.S., Seely, M.K., Ward, D., Mitchell, D., and Campbell, J.D. (1991) Body temperatures of Namib Desert tenebrionid beetles: their relationship in laboratory and field. *Physiological Entomology* **16**, 463–475.

Roberts, S.P. and Feder, M.E. (1999) Natural hyperthermia and expression of the heat shock protein Hsp70 affect developmental abnormalities in *Drosophila melanogaster*. *Oecologia* **121**, 323–329.

Roberts, S.P. and Feder, M.E. (2000) Changing fitness consequences of *hsp70* copy number in transgenic *Drosophila* larvae undergoing natural thermal stress. *Functional Ecology* **14**, 353–357.

Roberts, S.P. and Harrison, J.F. (1998) Mechanisms of thermoregulation in flying bees. *American Zoologist* **38**, 492–502.

Roberts, S.P. and Harrison, J.F. (1999) Mechanisms of thermal stability during flight in the honeybee *Apis mellifera*. *Journal of Experimental Biology* **202**, 1523–1533.

Roberts, S.P., Harrison, J.F., and Hadley, N.F. (1998) Mechanisms of thermal balance in flying *Centris pallida* (Hymenoptera: Anthophoridae). *Journal of Experimental Biology* **201**, 2321–2331.

Robinson, T., Rogers, D., and Williams, B. (1997) Mapping tsetse habitat suitability in the common fly belt of Southern Africa using multivariate analysis of climate and remotely sensed vegetation data. *Medical and Veterinary Entomology* **11**, 235–245.

Robinson, W.R., Peters, R.H., and Zimmermann, J. (1983) The effects of body size and temperature on metabolic rate of organisms. *Canadian Journal of Zoology* **61**, 281–288.

Roces, F. and Blatt, J. (1999) Haemolymph sugars and the control of the proventriculus in the honey bee *Apis mellifera*. *Journal of Insect Physiology* **45**, 221–229.

Roces, F. and Lighton, J.R.B. (1995) Larger bites of leaf-cutting ants. *Nature* **373**, 392–393.

Roff, D.A. (1980) Optimizing development time in a seasonal environment: the 'ups and downs' of clinal variation. *Oecologia* **45**, 202–208.

Roff, D.A. (1990) The evolution of flightlessness in insects. *Ecological Monographs* **60**, 389–421.

Rogers, D.J. and Randolph, S.E. (1991) Mortality rates and population density of tsetse flies correlated with satellite imagery. *Nature* **351**, 739–741.

Rogowitz, G.L. and Chappell, M.A. (2000) Energy metabolism of eucalyptus-boring beetles at rest and during locomotion: gender makes a difference. *Journal of Experimental Biology* **203**, 1131–1139.

Rohde, K. (1992) Latitudinal gradients in species diversity: the search for the primary cause. *Oikos* **65**, 514–527.

Rohde, K. (1998) Latitudinal gradients in species diversity. Area matters, but how much? *Oikos* **82**, 184–190.

Rojas, R.R. and Leopold, R.A. (1996) Chilling injury in the house fly: evidence for the role of oxidative stress between pupariation and emergence. *Cryobiology* **33**, 447–458.

Root, T.L., Price, J.T., Hall, K.R., Schneider, S.H., Rosenzweig, C., and Pounds, J.A. (2003) Fingerprints of global warming on wild animals and plants. *Nature* **421**, 57–60.

Rosales, A.L., Krafsur, E.S., and Kim, Y. (1994) Cryobiology of the face fly and house fly (Diptera: Muscidae). *Journal of Medical Entomology* **31**, 671–680.

Rothery, P. and Block, W. (1992) Characterizing supercooling point distributions. *Cryoletters* **13**, 193–198.

Rourke, B.C. (2000) Geographic and altitudinal variation in water balance and metabolic rate in a California grasshopper, *Melanoplus sanguinipes*. *Journal of Experimental Biology* **203**, 2699–2712.

Roy, D.B., Rothery, P., Moss, D., Pollard, E., and Thomas, J.A. (2001) Butterfly numbers and weather: predicting historical trends in abundance and the future effects of climate change. *Journal of Animal Ecology* **70**, 201–217.

Rubin, G.M. *et al.* (2000) Comparative genomics of the eukaryotes. *Science* **287**, 2204–2215.

Rudolph, D. (1982) Occurrence, properties and biological implications of the active uptake of water vapour from the atmosphere in Psocoptera. *Journal of Insect Physiology* **28**, 111–121.

Ruf, C. and Fiedler, K. (2002) Plasticity in foraging patterns of larval colonies of the small Eggar moth, *Eriogaster lanestris* (Lepidoptera: Lasiocampidae). *Oecologia* **131**, 626–634.

Rutherford, S.L. and Lindquist, S. (1998) Hsp90 as a capacitor for morphological evolution. *Nature* **396**, 336–342.

Ryan, R.O. and van der Horst, D.J. (2000) Lipid transport biochemistry and its role in energy production. *Annual Review of Entomology* **45**, 233–260.

Sacchi, V.F. and Wolfersberger, M.G. (1996) Amino acid absorption. In *Biology of the Insect Midgut* (eds. M.J. Lehane and P.F. Billingsley), 265–292. Chapman and Hall, London.

Sala, O.E., Chapin, F.S., Armesto, J.J., Berlow, E., Bloomfield, J., Dirzo, R. *et al.* (2000) Global biodiversity scenarios for the year 2100. *Science* **287**, 1770–1774.

Salt, R.W. (1961) Principles of insect cold-hardiness. *Annual Review of Entomology* **6**, 55–74.

Salt, R.W. (1966) Effect of cooling rate on the freezing temperatures of supercooled insects. *Canadian Journal of Zoology* **44**, 655–659.

Salvucci, M.E. (2000) Sorbitol accumulation in whiteflies: evidence for a role in protecting proteins during heat stress. *Journal of Thermal Biology* **25**, 353–361.

Salvucci, M.E. and Crafts-Brandner, S.J. (2000) Effects of temperature and dietary sucrose concentration on respiration in the silverleaf whitefly, *Bemisia argentifolii*. *Journal of Insect Physiology* **46**, 1461–1467.

Salvucci, M.E., Stecher, D.S., and Henneberry, T.J. (2000) Heat shock proteins in whiteflies, an insect that accumulates sorbitol in response to heat stress. *Journal of Thermal Biology* **25**, 363–371.

Salvucci, M.E., Wolfe, G.R., and Hendrix, D.L. (1997) Effect of sucrose concentration on carbohydrate metabolism in *Bemisia argentifolii*: biochemical mechanism and physiological role for trehalulose synthesis in the silverleaf whitefly. *Journal of Insect Physiology* **43**, 457–464.

Sanborn, A.F. (2000) Comparative thermoregulation of sympatric endothermic and ectothermic cicadas (Homoptera: Cicadidae: *Tibicen winnemanna* and *Tibicen chloromerus*). *Journal of Comparative Physiology A* **186**, 551–556.

Sanborn, A.F., Heath, M.S., Heath, J.E., and Noriega, F.G. (1995) Diurnal activity, temperature responses and endothermy in three South American cicadas (Homoptera: Cicadidae: *Dorisiana bonaerensis*, *Quesada gigas* and *Fidicina mannifera*). *Journal of Thermal Biology* **20**, 451–460.

Sasaki, T., Kawamura, M., and Ishikawa, H. (1996) Nitrogen recycling in the brown planthopper, *Nilaparvata lugens*: involvement of yeast-like endosymbionts in uric acid metabolism. *Journal of Insect Physiology* **42**, 125–129.

Schmaranzer, S. (2000) Thermoregulation of water collecting honey bees (*Apis mellifera*). *Journal of Insect Physiology* **46**, 1187–1194.

Schmaranzer, S. and Stabentheiner, A. (1988) Variability of the thermal behavior of honeybees on a feeding place. *Journal of Comparative Physiology B* **158**, 135–141.

Schmidt, C. and Wägele, J.W. (2001) Morphology and evolution of respiratory structure in the pleopod exopodites of terrestrial Isopoda (Crustacea, Isopoda, Oniscidea). *Acta Zoologica* **82**, 315–330.

Schmidt, D.J. and Reese, J.C. (1986) Sources of error in nutritional index studies of insects on artificial diet. *Journal of Insect Physiology* **32**, 193–198.

Schmidt-Nielsen, K. (1972) Locomotion: energy cost of swimming, flying, and running. *Science* **177**, 222–228.

Schmidt-Nielsen, K. (1984) *Scaling. Why is Animal Size so Important?* Cambridge University Press, Cambridge.

Schmitz, A. and Perry, S.F. (1999) Stereological determination of tracheal volume and diffusing capacity of the tracheal walls in the stick insect *Carausius morosus* (Phasmatodea, Lonchodidae). *Physiological and Biochemical Zoology* **72**, 205–218.

Schmitz, A. and Perry, S.F. (2001) Bimodal breathing in jumping spiders: morphometric partitioning of the lungs and tracheae in *Salticus scenicus* (Arachnida, Araneae, Salticidae). *Journal of Experimental Biology* **204**, 4321–4334.

Schmitz, A. and Perry, S.F. (2002) Morphometric analysis of the tracheal walls of the harvestmen *Nemastoma lugubre* (Arachnida, Opiliones, Nemastomatidae). *Arthropod Structure and Development* **30**, 229–241.

Schmitz, H. (1994) Thermal characterization of butterfly wings. 1. Absorption in relation to different color, surface structure and basking type. *Journal of Thermal Biology* **19**, 403–412.

Schneiderman, H.A. (1960) Discontinous respiration in insects: role of the spiracles. *Biological Bulletin* **119**, 494–528.

Schneiderman, H.A. and Schechter, A.N. (1966) Discontinuous respiration in insects—V. Pressure and volume changes in the tracheal system of silkworm pupae. *Journal of Insect Physiology* **12**, 1143–1170.

Schneiderman, H.A. and Williams, C.M. (1955) An experimental analysis of the discontinuous respiration of the cecropia silkworm. *Biological Bulletin* **109**, 123–143.

Scholander, P.F., Flagg, W., Walters, V., and Irving, L. (1953) Climatic adaptation in Arctic and Tropical poikilotherms. *Physiological Zoology* **26**, 67–92.

Scholtz, C.H. and Caveney, S. (1992) Daily biphasic behaviour in keratin-feeding desert trogid beetles in relation to climate. *Ecological Entomology* **17**, 155–159.

Schoonhoven, L.M., Jermy, T., and van Loon, J.J.A. (1998) *Insect-Plant Biology*. Chapman and Hall, London.

Schultz, T.D. (1998) The utilization of patchy thermal microhabitats by the ectothermic insect predator, *Cicindela sexguttata*. *Ecological Entomology* **23**, 444–450.

Scott, M., Berrigan, D., and Hoffmann, A.A. (1997) Costs and benefits of acclimation to elevated temperature in *Trichogramma carverae*. *Entomologia Experimentalis et Applicata* **85**, 211–219.

Scriber, J.M. (2002) Latitudinal and local geographic mosaics in host plant preferences as shaped by thermal units and voltinism in *Papilio* spp. (Lepidoptera). *European Journal of Entomology* **99**, 225–239.

Scriber, J.M. and Feeny, P. (1979) Growth of herbivorous caterpillars in relation to feeding specialization and to the growth form of their food plants. *Ecology* **60**, 829–850.

Scriber, J.M. and Slansky, F. (1981) The nutritional ecology of immature insects. *Annual Review of Entomology* **26**, 183–211.

Scrivener, A.M. and Slaytor, M. (1994) Properties of the endogenous cellulase from *Panesthia cribrata* Saussure and purification of major endo-β-1,4-glucanase components. *Insect Biochemistry and Molecular Biology* **24**, 223–231.

Seely, M.K. (1979) Irregular fog as a water source for desert dune beetles. *Oecologia* **42**, 213–227.

Sernetz, M., Gelléri, B., and Hofmann, J. (1985) The organism as bioreactor. Interpretation of the reduction law of metabolism in terms of heterogeneous catalysis and fractal structure. *Journal of Theoretical Biology* **117**, 209–230.

Seymour, R.S., White, C.R., and Gibernau, M. (2003) Heat reward for insect pollinators. *Nature* **426**, 243–244.

Shelford, V.E. (1911) Physiological animal geography. *Journal of Morphology* **22**, 551–618.

Shelly, T.E. (1982) Comparative foraging behavior of light- versus shade-seeking adult damselflies in a lowland neotropical forest (Odonata: Zygoptera). *Physiological Zoology* **55**, 335–343.

Shelton, T.G. and Appel, A.G. (2001*a*) Cyclic CO_2 release and water loss in alates of the eastern subterranean termite (Isoptera: Rhinotermitidae). *Annals of the Entomological Society of America* **94**, 420–426.

Shelton, T.G. and Appel, A.G. (2001*b*) Cyclic CO_2 release in *Cryptotermes cavifrons* Banks, *Incisitermes tabogae*

(Snyder) and *I. minor* (Hagen) (Isoptera: Kalotermitidae). *Comparative Biochemistry and Physiology A* **129**, 681–693.

Shimada, K. and Riihimaa, A. (1990) Cold-induced freezing tolerance in diapausing and non-diapausing larvae of *Chymomyza costata* (Diptera: Drosophilidae) with accumulation of trehalose and proline. *Cryoletters* **11**, 243–250.

Shine, R. and Kearney, M. (2001) Field studies of reptile thermoregulation: how well do physical models predict operative temperatures? *Functional Ecology* **15**, 282–288.

Shmida, A. and Dukas, R. (1990) Progressive reduction in the mean body sizes of solitary bees active during the flowering season and its correlation with the sizes of bee flowers of the mint family (Lamiaceae). *Israel Journal of Botany* **39**, 133–141.

Sibly, R.M. and Calow, P. (1986) *Physiological Ecology of Animals: an Evolutionary Approach*. Blackwell Scientific Publications, Oxford.

Sieber, R. and Kokwaro, E.D. (1982) Water intake by the termite *Macrotermes michaelseni*. *Entomologia Experimentalis et Applicata* **31**, 147–153.

Silbermann, R. and Tatar, M. (2000) Reproductive costs of heat shock protein in transgenic *Drosophila melanogaster*. *Evolution* **54**, 2038–2045.

Sillén-Tullberg, B. (1988) Evolution of gregariousness in aposematic butterfly larvae: a phylogenetic analysis. *Evolution* **42**, 293–305.

Silva, A., Bacci, M., Siqueira, C.G., Bueno, O.C., Pagnocca, F.C., and Hebling, M.J.A. (2003) Survival of *Atta sexdens* workers on different food sources. *Journal of Insect Physiology* **49**, 307–313.

Simmonds, M.S.J. and Blaney, W.M. (1986) Effects of rearing density on development and feeding behaviour in larvae of *Spodoptera exempta*. *Journal of Insect Physiology* **32**, 1043–1053.

Simmons, F.H. and Bradley, T.J. (1997) An analysis of resource allocation in response to dietary yeast in *Drosophila melanogaster*. *Journal of Insect Physiology* **43**, 779–788.

Simpson, S.J. (1994) Experimental support for a model in which innate taste responses contribute to regulation of salt intake by nymphs of *Locusta migratoria*. *Journal of Insect Physiology* **40**, 555–559.

Simpson, S.J., Barton Browne, L., and van Gerwen, A.C.M. (1989) The patterning of compensatory sugar feeding in the Australian sheep blowfly. *Physiological Entomology* **14**, 91–105.

Simpson, S.J. and Raubenheimer, D. (1993a) A multi-level analysis of feeding behaviour: the geometry of nutritional decisions. *Philosophical Transactions of the Royal Society of London B* **342**, 381–402.

Simpson, S.J. and Raubenheimer, D. (1993b) The central role of the haemolymph in the regulation of nutrient intake in insects. *Physiological Entomology* **18**, 395–403.

Simpson, S.J., Raubenheimer, D., Behmer, S.T., Whitworth, A., and Wright, G.A. (2002) A comparison of nutritional regulation in solitarious- and gregarious-phase nymphs of the desert locust *Schistocerca gregaria*. *Journal of Experimental Biology* **205**, 121–129.

Simpson, S.J., Raubenheimer, D., and Chambers, P.G. (1995) The mechanisms of nutritional homeostasis. In *Regulatory Mechanisms in Insect Feeding* (eds. R.F. Chapman and G. de Boer), 251–278. Chapman and Hall, New York.

Simpson, S.J., Simmonds, S.J., Blaney, W.M., and Jones, J.P. (1990) Compensatory dietary selection occurs in larval *Locusta migratoria* but not *Spodoptera littoralis* after a single deficient meal during *ad libitum* feeding. *Physiological Entomology* **15**, 235–242.

Simpson, S.J. and Simpson, C.L. (1990) The mechanisms of nutritional compensation by phytophagous insects. In *Insect-Plant Interactions* (ed. E.A. Bernays), Vol. II, 111–160. CRC Press, Boca Raton.

Sinclair, B.J. (1999) Insect cold tolerance: how many kinds of frozen? *European Journal of Entomology* **96**, 157–164.

Sinclair, B.J. (2001a) Field ecology of freeze tolerance: interannual variation in cooling rates, freeze-thaw and thermal stress in the microhabitat of the alpine cockroach *Celatoblatta quinquemaculata*. *Oikos* **93**, 286–293.

Sinclair, B.J. (2001b) Biologically relevant environmental data: macros to make the most of microclimate recordings. *Cryoletters* **22**, 125–134.

Sinclair, B.J., Addo-Bediako, A., and Chown, S.L. (2003c) Climatic variability and the evolution of insect freeze tolerance. *Biological Reviews* **78**, 181–195.

Sinclair, B.J. and Chown, S.L. (2003) Rapid responses to high temperature and desiccation but not to low temperature in the freeze tolerant sub-Antarctic caterpillar *Pringleophaga marioni* (Lepidoptera, Tineidae). *Journal of Insect Physiology* **49**, 45–52.

Sinclair, B.J., Klok, C.J., and Chown, S.L. (2004) Metabolism of the sub-Antarctic caterpillar *Pringleophaga marioni* during cooling, freezing and thawing. *Journal of Experimental Biology* **207**, 1287–1294.

Sinclair, B.J., Klok, C.J., Scott, M.B., Terblanche, J.S., and Chown, S.L. (2003a) Diurnal variation in supercooling points of three species of Collembola from Cape Hallett, Antarctica. *Journal of Insect Physiology* **49**, 1049–1061.

Sinclair, B.J., Vernon, P., Klok, C.J., and Chown, S.L. (2003b) Insects at low temperatures: an ecological perspective. *Trends in Ecology and Evolution* **18**, 257–262.

Sinclair, B.J., Worland, M.R., and Wharton, D.A. (1999) Ice nucleation and freezing tolerance in New Zealand alpine and lowland weta, *Hemideina* spp. (Orthoptera: Stenopelmatidae). *Physiological Entomology* **24**, 56–63.

Singer, M.C. and Parmesan, C. (1993) Sources of variations in patterns of plant–insect association. *Nature* **361**, 251–253.

Singer, M.S. and Stireman, J.O. (2003) Does anti-parasitoid defense explain host-plant selection by a polyphagous caterpillar? *Oikos* **100**, 554–562.

Skalicki, N., Heran, H., and Crailsheim, K. (1988) Water budget of the honeybee during rest and flight. In *BIONA Report 3* (ed. W. Nachtigall), 103–118. Gustav Fischer, Stuttgart.

Sláma, K. (1984) Microrespirometry in small tissues and organs. In *Measurement of Ion Transport and Metabolic Rates in Insects* (eds. T.J. Bradley and T.A. Miller), 101–129. Springer, Berlin.

Sláma, K. (1988) A new look at insect respiration. *Biological Bulletin* **175**, 289–300.

Sláma, K. (1994) Regulation of respiratory acidemia by the autonomic nervous system (coelopulse) in insects and ticks. *Physiological Zoology* **67**, 163–174.

Sláma, K. (1999) Active regulation of insect respiration. *Annals of the Entomological Society of America* **92**, 916–929.

Sláma, K. and Coquillaud, M.-S. (1992) Homeostatic control of respiratory metabolism in beetles. *Journal of Insect Physiology* **38**, 783–791.

Slansky, F. (1993) Nutritional ecology: the fundamental quest for nutrients. In *Caterpillars: Ecological and Evolutionary Constraints on Foraging* (eds. N.E. Stamp and T.M. Casey), 29–91. Chapman and Hall, New York.

Slansky, F. and Feeny, P. (1977) Stabilization of the rate of nitrogen accumulation by larvae of the cabbage butterfly on wild and cultivated food plants. *Ecological Monographs* **47**, 209–228.

Slansky, F. and Rodriguez, J.G. (eds.) (1987) *Nutritional Ecology of Insects, Mites, Spiders and Related Invertebrates*. John Wiley and Sons, New York.

Slansky, F. and Scriber, J.M. (1985) Food consumption and utilization. In *Comprehensive Insect Physiology, Biochemistry and Pharmacology* (eds. G.A. Kerkut and L.I. Gilbert), Vol. 4, 87–163. Permagon Press, Oxford.

Slansky, F. and Wheeler, G.S. (1991) Food consumption and utilization responses to dietary dilution with cellulose and water by velvetbean caterpillars, *Anticarsia gemmatalis*. *Physiological Entomology* **16**, 99–116.

Slaytor, M. (1992) Cellulose digestion in termites and cockroaches: what role do symbionts play? *Comparative Biochemistry and Physiology B* **103**, 775–784.

Smedley, S.R. and Eisner, T. (1995) Sodium uptake by puddling in a moth. *Science* **270**, 1816–1818.

Smedley, S.R. and Eisner, T. (1996) Sodium: a male moth's gift to its offspring. *Proceedings of the National Academy of Sciences of the USA* **93**, 809–813.

Smith, R.J., Hines, A., Richmond, S., Merrick, M., Drew, A., and Fargo, R. (2000) Altitudinal variation in body size and population density of *Nicrophorus investigator* (Coleoptera: Silphidae). *Environmental Entomology* **29**, 290–298.

Smith, S.M., Turnball, D.A., and Taylor, P.D. (1994) Assembly, mating, and energetics of *Hybomitra arpadi* (Diptera: Tabanidae) at Churchill, Manitoba. *Journal of Insect Behavior* **7**, 355–383.

Snodgrass, R.E. (1935) *Principles of Insect Morphology*. McGraw-Hill, New York.

Snyder, M.J., Walding, J.K., and Feyereisen, R. (1994) Metabolic fate of the allelochemical nicotine in the tobacco hornworm *Manduca sexta*. *Insect Biochemistry and Molecular Biology* **24**, 837–846.

Somerfield, P.J., Clarke, K.R., and Olsgard, F. (2002) A comparison of the power of categorical and correlational tests applied to community ecology data from gradient studies. *Journal of Animal Ecology* **71**, 581–593.

Somero, G.N. (1995) Proteins and temperature. *Annual Review of Physiology* **57**, 43–68.

Somero, G.N., Dahlhoff, E., and Lin, J.J. (1996) Stenotherms and eurytherms: mechanisms establishing thermal optima and tolerance ranges. In *Animals and Temperature. Phenotypic and Evolutionary Adaptation* (eds. I.A. Johnston and A.F. Bennett), 53–78. Cambridge University Press, Cambridge.

Sømme, L. (1982) Supercooling and winter survival in terrestrial arthropods. *Comparative Biochemistry and Physiology A* **73**, 519–543.

Sømme, L. (1995) *Invertebrates in Hot and Cold Arid Environments*. Springer, Berlin.

Sømme, L. (1996) The effect of prolonged exposures at low temperatures in insects. *Cryoletters* **17**, 341–346.

Sømme, L. (1999) The physiology of cold hardiness in terrestrial arthropods. *European Journal of Entomology* **96**, 1–10.

Sømme, L. and Block, W. (1982) Cold hardiness of Collembola at Signy Island, maritime Antarctic. *Oikos* **38**, 168–176.

Sømme, L. and Block, W. (1991) Adaptations to alpine and polar environments in insects and other terrestrial arthropods. In *Insects at Low Temperatures* (eds. R.E. Lee and D.L. Denlinger), 318–359. Chapman and Hall, New York.

Sømme, L. and Zachariassen, K.E. (1981) Adaptations to low temperature in high altitude insects from Mount Kenya. *Ecological Entomology* **6**, 119–204.

Sørensen, J.G., Dahlgaard, J., and Loeschcke, V. (2001) Genetic variation in thermal tolerance among natural populations of *Drosophila buzzatii*: down regulation of Hsp70 expression and variation in heat stress resistance traits. *Functional Ecology* **15**, 289–296.

Sørensen, J.G. and Loeschcke, V. (2001) Larval crowding in *Drosophila melanogaster* induces Hsp70 expression, and leads to increased adult longevity and adult thermal stress resistance. *Journal of Insect Physiology* **47**, 1301–1307.

Sørensen, J.G., Michalak, P., Justesen, J., and Loeschcke, V. (1999) Expression of the heat-shock protein Hsp70 in *Drosophila buzzatii* lines selected for thermal resistance. *Hereditas* **131**, 155–164.

Southwood, T.R.E. (1972) The insect/plant relationship—an evolutionary perspective. In *Symposia of the Royal Entomological Society of London* (ed. H.F. Van Emden), Vol. 6, 3–30. Blackwell Scientific Publications, London.

Sözen, M.A., Armstong, J.D., Yang, M., Kaiser, K., and Dow, J.A.T. (1997) Functional domains are specified to single-cell resolution in *Drosophila* epithelium. *Proceedings of the National Academy of Sciences of the USA* **94**, 5207–5212.

Spaargaren, D.H. (1992) Transport function of branching structures and the 'surface law' for basic metabolic rate. *Journal of Theoretical Biology* **154**, 495–504.

Speight, M.R., Hunter, M.D., and Watt, A.D. (1999) *Ecology of Insects: Concepts and Applications*. Blackwell Science, Oxford.

Spicer, J.I. and Gaston, K.J. (1999) *Physiological Diversity and its Ecological Implications*. Blackwell Science, Oxford.

Spiller, N.J., Koenders, L., and Tjallingii, W.F. (1990) Xylem ingestion by aphids—a strategy for maintaining water balance. *Entomologia Experimentalis et Applicata* **55**, 101–104.

Srygley, R.B. and Chai, P. (1990) Predation and the elevation of thoracic temperature in brightly colored Neotropical butterflies. *American Naturalist* **135**, 766–787.

Stabentheiner, A. (1991) Thermographic monitoring of the thermal behaviour of dancing bees. In *The Behaviour and Physiology of Bees* (eds. L.J. Goodman and R.C. Fisher), 89–101. CAB International, Oxford.

Stabentheiner, A. (2001) Thermoregulation of dancing bees: thoracic temperature of pollen and nectar foragers in relation to profitability of foraging and colony need. *Journal of Insect Physiology* **47**, 385–392.

Stamp, N.E. (1990) Growth versus molting time of caterpillars as a function of temperature, nutrient concentration and the phenolic rutin. *Oecologia* **82**, 107–113.

Stamp, N.E. (1993) A temperate region view of the interaction of temperature, food quality, and predators on caterpillar foraging. In *Caterpillars: Ecological and Evolutionary Constraints on Foraging* (eds. N.E. Stamp and T.M. Casey), 478–508. Chapman and Hall, New York.

Stamp, N.E. and Casey, T.M. (eds.) (1993) *Caterpillars: Ecological and Evolutionary Constraints on Foraging*. Chapman and Hall, New York.

Stanley, S.M., Parsons, P.A., Spence, G.E., and Weber, L. (1980) Resistance of species of the *Drosophila melanogaster* subgroup to environmental extremes. *Australian Journal of Zoology* **28**, 413–421.

Sterner, R.W. and Elser, J.J. (2002) *Ecological Stoichiometry*. Princeton University Press, Princeton.

Stevens, D.J., Hansell, M.H., and Monaghan, P. (2000) Developmental trade-offs and life histories: strategic allocation of resources in caddis flies. *Proceedings of the Royal Society of London B* **267**, 1511–1515.

Stevens, E.D. and Josephson, R.K. (1977) Metabolic rate and body temperature in singing katydids. *Physiological Zoology* **50**, 31–42.

Stevens, G.C. (1989) The latitudinal gradient in geographical range: how so many species coexist in the tropics. *American Naturalist* **133**, 240–256.

Stevenson, R.D. (1985) The relative importance of behavioural and physiological adjustments controlling body temperature in terrestrial ectotherms. *American Naturalist* **126**, 362–386.

Stevenson, R.D. and Josephson, R.K. (1990) Effects of operating frequency and temperature on mechanical power output from moth flight muscle. *Journal of Experimental Biology* **149**, 61–78.

Stockoff, B.A. (1993) Ontogenetic change in dietary selection for protein and lipid by gypsy moth larvae. *Journal of Insect Physiology* **39**, 677–686.

Stoffolano, J.G. (1995) Regulation of a carbohydrate meal in the adult Diptera, Lepidoptera, and Hymenoptera. In *Regulatory Mechanisms in Insect Feeding* (eds. R.F. Chapman and G. de Boer), 210–247. Chapman and Hall, New York.

Stone, G.N. (1993) Endothermy in the solitary bee *Anthophora plumipes*: independent measures of thermoregulatory ability, costs of warm-up and the role of body size. *Journal of Experimental Biology* **174**, 299–320.

Stone, G.N. (1994a) Activity patterns of females of the solitary bee *Anthophora plumipes* in relation to temperature, nectar supplies and body size. *Ecological Entomology* **19**, 177–189.

Stone, G.N. (1994b) Patterns of evolution of warm-up rates and body temperatures in flight in solitary bees of the genus *Anthophora*. *Functional Ecology* **8**, 324–335.

Stone, G.N., Gilbert, F., Willmer, P., Potts, S., Semida, F., and Zalat, S. (1999) Windows of opportunity and the temporal structuring of foraging activity in a desert solitary bee. *Ecological Entomology* **24**, 208–221.

Stone, G.N., Loder, P.M.J., and Blackburn, T.M. (1995) Foraging and courtship behaviour in males of the solitary bee *Anthophora plumipes* (Hymenoptera: Anthophoridae): thermal physiology and the roles of body size. *Ecological Entomology* **20**, 169–183.

Stone, G.N. and Willmer, P.G. (1989*a*) Endothermy and temperature regulation in bees: a critique of 'grab and stab' measurement of body temperature. *Journal of Experimental Biology* **143**, 211–223.

Stone, G.N. and Willmer, P.G. (1989*b*) Warm-up rates and body temperatures in bees: the importance of body size, thermal regime and phylogeny. *Journal of Experimental Biology* **147**, 303–328.

Storey, K.B. (1997) Organic solutes in freezing tolerance. *Comparative Biochemistry and Physiology A* **117**, 319–326.

Storey, K.B. (2002) Life in the slow lane: molecular mechanisms of estivation. *Comparative Biochemistry and Physiology A* **133**, 733–754.

Storey, K.B. and Storey, J.M. (1991) Biochemistry of cryoprotectants. In *Insects at Low Temperatures* (eds. R.E. Lee and D.L. Denlinger), 64–93. Chapman and Hall, New York.

Storey, K.B. and Storey, J.M. (1996) Natural freezing survival in animals. *Annual Review of Ecology and Systematics* **27**, 365–386.

Stutt, A.D. and Willmer, P. (1998) Territorial defence in speckled wood butterflies: do the hottest males always win? *Animal Behaviour* **55**, 1341–1347.

Suarez, R.K., Lighton, J.R.B., Joos, B., Roberts, S.P., and Harrison, J.F. (1996) Energy metabolism, enzymatic flux capacities, and metabolic flux rates in flying honeybees. *Proceedings of the National Academy of Sciences of the USA* **93**, 12616–12620.

Sugumaran, M. (1998) Unified mechanism for sclerotization of insect cuticle. *Advances in Insect Physiology* **27**, 229–334.

Sumner, J.-P., Dow, J.A.T., Earley, F.G.P., Klein, U., Jäger, D., and Wieczorek, H. (1995) Regulation of plasma membrane V-ATPase activity by dissociation of peripheral subunits. *Journal of Biological Chemistry* **270**, 5649–5653.

Sutcliffe, D.W. (1963) The chemical composition of haemolymph in insects and some other arthropods, in relation to their phylogeny. *Comparative Biochemistry and Physiology* **9**, 121–135.

Symonds, M.R.E. and Elgar, M.A. (2002) Phylogeny affects estimation of metabolic scaling in mammals. *Evolution* **56**, 2330–2333.

Tabashnik, B.E. (1982) Responses of pest and non-pest *Colias* butterfly larvae to intraspecific variation in leaf nitrogen and water content. *Oecologia* **55**, 389–394.

Takahashi-Del-Bianco, M., Benedito-Silva, A.A., Hebling, M.J.A., Marques, N., and Marques, M.D. (1992) Circadian oscillatory patterns of oxygen uptake in individual workers of the ant *Camponotus rufipes*. *Physiological Entomology* **17**, 377–383.

Tammariello, S.P., Rinehart, J.P., and Denlinger, D.L. (1999) Desiccation elicits heat shock protein transcription in the flesh fly, *Sarcophaga crassipalpis*, but does not enhance tolerance to high or low temperatures. *Journal of Insect Physiology* **45**, 933–938.

Tanaka, S. (1986) Uptake and loss of water in diapause and non-diapause eggs of crickets. *Physiological Entomology* **11**, 343–351.

Tanaka, S., Wolda, H., and Denlinger, D.L. (1988) Group size affects the metabolic rate of a tropical beetle. *Physiological Entomology* **13**, 239–241.

Tantawy, A.O. and Mallah, G.S. (1961) Studies on natural populations of *Drosophila*. I. Heat resistance and geographical variation in *Drosophila melanogaster* and *D. simulans*. *Evolution* **15**, 1–14.

Tartes, U., Vanatoa, A., and Kuusik, A., (2002) The insect abdomen—a heartbeat manager in insects? *Comparative Biochemistry and Physiology A* **133**, 611–623.

Tatar, M. (1999) Transgenes in the analysis of life span and fitness. *American Naturalist Supplement* **154**, 67–81.

Tauber, M.J., Tauber, C.A., Nyrop, J.P., and Villani, M.G. (1998) Moisture, a vital but neglected factor in the seasonal ecology of insects: hypotheses and tests of mechanisms. *Environmental Entomology* **27**, 523–530.

Taylor, C.R., Schmidt-Nielsen, K., and Raab, J.L. (1970) Scaling of energetic cost of running to body size in mammals. *American Journal of Physiology* **219**, 1104–1107.

Taylor, P. (1977) The respiratory metabolism of tsetse flies, *Glossina* spp., in relation to temperature, blood-meal size and pregnancy cycle. *Physiological Entomology* **2**, 317–322.

Terra, W.R. (1990) Evolution of digestive systems of insects. *Annual Review of Entomology* **35**, 181–200.

Terra, W.R. and Ferreira, C. (1994) Insect digestive enzymes: properties, compartmentalization and function. *Comparative Biochemistry and Physiology B* **109**, 1–62.

Terra, W.R., Ferreira, C., and Baker, J.E. (1996*b*) Compartmentalization of digestion. In *Biology of the Insect Midgut* (eds. M.J. Lehane and P.F. Billingsley), 206–235. Chapman and Hall, London.

Terra, W.R., Ferreira, C., and Bastos, F. (1985) Phylogenetic considerations of insect digestion. Disaccharidases and the spatial organization of digestion in the *Tenebrio molitor* larvae. *Insect Biochemistry* **15**, 443–449.

Terra, W.R., Ferreira, C., and de Bianchi, A.G. (1979) Distribution of digestive enzymes among the endo- and ectoperitrophic spaces and midgut cells of

Rhynchosciara and its physiological significance. *Journal of Insect Physiology* 25, 487–494.

Terra, W.R., Ferreira, C., Jordão, B.P., and Dillon, R.J. (1996a) Digestive enzymes. In *Biology of the Insect Midgut* (eds. M.J. Lehane and P.F. Billingsley), 153–194. Chapman and Hall, London.

Thomas, C.D., Bodsworth, E.J., Wilson, R.J., Simmons, A.D., Davies, Z.G., Musche, M. *et al.* (2001) Ecological and evolutionary processes at expanding range margins. *Nature* 411, 577–581.

Thomas, C.D., Hill, J.K., and Lewis, O.T. (1998) Evolutionary consequences of habitat fragmentation in a localized butterfly. *Journal of Animal Ecology* 67, 485–497.

Thomas, C.D., Singer, M.C., and Boughton, D.A. (1996) Catastrophic extinction of population sources in a butterfly metapopulation. *American Naturalist* 148, 957–975.

Thomson, L.J., Robinson, M., and Hoffmann, A.A. (2001) Field and laboratory evidence for acclimation without costs in an egg parasitoid. *Functional Ecology* 15, 217–221.

Timmins, G.S., Penatti, C.A.A., Bechara, E.J.H., and Swartz, H.M. (1999) Measurement of oxygen partial pressure, its control during hypoxia and hyperoxia, and its effect upon light emission in a bioluminescent elaterid larva. *Journal of Experimental Biology* 202, 2631–2638.

Timmins, W.A., Bellward, K., Stamp, A.J., and Reynolds, S.E. (1988) Food intake, conversion efficiency and feeding behaviour of tobacco hornworm caterpillars given artificial diet of varying nutrient and water content. *Physiological Entomology* 13, 303–314.

Timmins, W.A. and Reynolds, S.E. (1992) Physiological mechanisms underlying the control of meal size in *Manduca sexta* larvae. *Physiological Entomology* 17, 81–89.

Todd, C.M. and Block, W. (1997) Responses to desiccation in four coleopterans from sub-Antarctic South Georgia. *Journal of Insect Physiology* 43, 905–913.

Tokeshi, M. (1992) Dynamics of distribution in animal communities: theory and analysis. *Researches on Population Ecology* 34, 249–273.

Tokeshi, M. (1999) *Species Coexistence. Ecological and Evolutionary Perspectives*. Blackwell Science, Oxford.

Tomkins, J.L., Simmons, L.W., and Alcock, J. (2001) Brood-provisioning strategies in Dawson's burrowing bee, *Amegilla dawsoni* (Hymenoptera: Anthophorini). *Behavioural Ecology and Sociobiology* 50, 81–89.

Toolson, E.C. (1984) Interindividual variation in epicuticular hydrocarbon composition and water loss rates of the cicada *Tibicen dealbatus* (Homoptera: Cicadidae). *Physiological Zoology* 57, 550–556.

Toolson, E.C. (1987) Water profligacy as an adaptation to hot deserts: water loss rates and evaporative cooling in the Sonoran desert cicada, *Diceroprocta apache* (Homoptera: Cicadidae). *Physiological Zoology* 60, 379–385.

Toolson, E.C. (1998) Comparative thermal physiological ecology of syntopic populations of *Cacama valvata* and *Tibicen bifidus* (Homoptera: Cicadidae): modeling fitness consequences of temperature variation. *American Zoologist* 38, 568–582.

Toolson, E.C. and Toolson, E.K. (1991) Evaporative cooling and endothermy in the 13-year periodical cicada, *Magicicada tredecem* (Homoptera: Cicadidae). *Journal of Comparative Physiology B* 161, 109–115.

Treherne, J.E. (1958) The absorption and metabolism of some sugars in the locust, *Schistocerca gregaria* (Forsk.). *Journal of Experimental Biology* 35, 611–625.

Treherne, J.E. and Willmer, P.G. (1975) Hormonal control of integumentary water-loss: evidence for a novel neuroendocrine system in an insect (*Periplaneta americana*). *Journal of Experimental Biology* 63, 143–159.

Trumper, S. and Simpson, S.J. (1993) Regulation of salt intake by nymphs of *Locusta migratoria*. *Journal of Insect Physiology* 39, 857–864.

Trumper, S. and Simpson, S.J. (1994) Mechanisms regulating salt intake in fifth-instar nymphs of *Locusta migratoria*. *Physiological Entomology* 19, 203–215.

Tsuji, J.S., Kingsolver, J.G., and Watt, W.B. (1986) Thermal physiological ecology of *Colias* butterflies in flight. *Oecologia* 69, 161–170.

Tucker, L.E. (1977) Effects of dehydration and rehydration on the water content and Na^+ and K^+ balance in adult male *Periplaneta americana*. *Journal of Experimental Biology* 71, 49–66.

Turner, J.S. and Lombard, A.T. (1990) Body color and body temperature in white and black Namib desert beetles. *Journal of Arid Environments* 19, 303–315.

Turunen, S. (1985) Absorption. In *Comprehensive Insect Biochemistry, Physiology and Pharmacology* (eds. G.A. Kerkut and L.I. Gilbert), Vol. 4, 241–277. Pergamon Press, Oxford.

Turunen, S. and Crailsheim, K. (1996) Lipid and sugar absorption. In *Biology of the Insect Midgut* (eds. M.J. Lehane and P.F. Billingsley), 293–320. Chapman and Hall, London.

Underwood, B.A. (1991) Thermoregulation and energetic decision-making by the honeybees *Apis cerana*, *Apis dorsata* and *Apis laboriosa*. *Journal of Experimental Biology* 157, 19–34.

Unwin, D.M. and Corbet, S.A. (1984) Wingbeat frequency, temperature and body size in bees and flies. *Physiological Entomology* 9, 115–121.

Ushakov, B. (1964) Thermostability of cells and proteins of poikilotherms and its significance in speciation. *Physiological Reviews* 44, 518–560.

van der Have, T.M. (2002) A proximate model for thermal tolerance in ectotherms. *Oikos* **98**, 141–155.

van der Have, T.M. and de Jong, G. (1996) Adult size in ectotherms: temperature effects on growth and differentiation. *Journal of Theoretical Biology* **183**, 329–340.

van der Laak, S. (1982) Physiological adaptations to low temperature in freezing-tolerant *Phyllodecta laticollis* beetles. *Comparative Biochemistry and Physiology A* **73**, 613–620.

van der Merwe, M., Chown, S.L., and Smith, V.R. (1997) Thermal tolerance limits in six weevil species (Coleoptera, Curculionidae) from sub-Antarctic Marion Island. *Polar Biology* **18**, 331–336.

Van Dyck, H. and Matthysen, E. (1998) Thermoregulatory differences between phenotypes in the speckled wood butterfly: hot perchers and cold patrollers? *Oecologia* **114**, 326–334.

Van Nerum, K. and Buelens, H. (1997) Hypoxia-controlled winter metabolism in honeybees (*Apis mellifera*). *Comparative Biochemistry and Physiology A* **117**, 445–455.

van Noordwijk, A.J. and de Jong, G. (1986) Acquisition and allocation of resources: their influence on variation in life history tactics. *American Naturalist* **128**, 137–142.

Van 'T Land, J., Van Putten, P., Zwaan, B., Kamping, A., and Van Delden, W. (1999) Latitudinal variation in wild populations of *Drosophila melanogaster*: heritabilities and reaction norms. *Journal of Evolutionary Biology* **12**, 222–232.

Vannier, G. (1994) The thermobiological limits of some freezing tolerant insects: the supercooling and thermo-stupor points. *Acta Oecologica* **15**, 31–42.

Veenstra, J.A., Lau, G.W., Agricola, H.-J., and Petzel, D.H. (1995) Immunohistological localization of regulatory peptides in the midgut of the female mosquito *Aedes aegypti*. *Histochemistry and Cell Biology* **104**, 337–347.

Vermeij, G.J. and Dudley, R. (2000) Why are there so few evolutionary transitions between aquatic and terrestrial ecosystems? *Biological Journal of the Linnean Society* **70**, 541–554.

Vernon, P. and Vannier, G. (1996) Developmental patterns of supercooling capacity in a subantarctic wingless fly. *Experientia* **52**, 155–158.

Vogt, J.T. and Appel, A.G. (1999) Standard metabolic rate of the fire ant, *Solenopsis invicta* Buren: effects of temperature, mass, and caste. *Journal of Insect Physiology* **45**, 655–666.

Vogt, J.T. and Appel, A.G. (2000) Discontinuous gas exchange in the fire ant, *Solenopsis invicta* Buren: caste differences and temperature effects. *Journal of Insect Physiology* **46**, 403–416.

von Bertalanffy, L. (1957) Quantitative laws in metabolism and growth. *Quarterly Review of Biology* **32**, 217–231.

Waddington, K.D. (1990) Foraging profits and thoracic temperature of honey bees (*Apis mellifera*). *Journal of Comparative Physiology B* **160**, 325–329.

Waide, R.B., Willig, M.R., Steiner, C.F., Mittelbach, G., Gough, L., Dodson, S.I. *et al.* (1999) The relationship between productivity and species richness. *Annual Review of Ecology and Systematics* **30**, 257–300.

Waldbauer, G.P. (1968) The consumption and utilization of food by insects. *Advances in Insect Physiology* **5**, 229–288.

Waldbauer, G.P. and Friedman, S. (1991) Self-selection of optimal diets by insects. *Annual Review of Entomology* **36**, 43–63.

Walker, V.K., Kuiper, M.J., Tyshenko, M.G., Doucet, D., Graether, S.P., Liou, Y.-C. *et al.* (2001) Surviving winter with antifreeze proteins: studies on budworms and beetles. In *Insect Timing: Circadian Rhythmicity to Seasonality* (eds. D.L. Denlinger, J. Giebultowicz, and D.S. Saunders), 199–211. Elsevier, Amsterdam.

Walther, G.-R., Post, E., Convey, P., Menzel, A., Parmesan, C., Beebee, T.J.C. *et al.* (2002) Ecological responses to recent climate change. *Nature* **416**, 389–395.

Ward, D. and Seely, M.K. (1996a) Behavioral thermo-regulation of six Namib Desert tenebrionid beetle species (Coleoptera). *Annals of the Entomological Society of America* **89**, 442–451.

Ward, D. and Seely, M.K. (1996b) Adaptation and con-straint in the evolution of the physiology and behavior of the Namib desert tenebrionid beetle genus *Onymacris*. *Evolution* **50**, 1231–1240.

Wasserthal, L.T. (1975) The rôle of butterfly wings in regulation of body temperature. *Journal of Insect Physiology* **21**, 1921–1930.

Wasserthal, L.T. (1996) Interaction of circulation and tracheal ventilation in holometabolous insects. *Advances in Insect Physiology* **26**, 297–351.

Wasserthal, L.T. (2001) Flight-motor-driven respiratory airflow in the hawkmoth *Manduca sexta*. *Journal of Experimental Biology* **204**, 2209–2220.

Watanabe, H., Noda, H., Tokuda, G., and Lo, N. (1998) A cellulase gene of termite origin. *Nature* **394**, 330–331.

Watson, M.J.O. and Hoffmann, A.A. (1996) Acclimation, cross-generation effects, and the response to selection for increased cold resistance in *Drosophila*. *Evolution* **50**, 1182–1192.

Watson, R.T. (ed.) (2002) *Climate Change 2001: Synthesis Report*. Cambridge University Press, Cambridge.

Watt, A.D., Whittaker, J.B., Docherty, M., Brooks, G., Lindsay, E., and Salt, D.T. (1995) The impact of elevated atmospheric CO_2 on insect herbivores. In *Insects in a Changing Environment* (eds. R. Harrington and N.E. Stork), 197–217. Academic Press, London.

Watt, W.B. (1968) Adaptive significance of pigment polymorphisms in *Colias* butterflies. I. Variation of melanin pigments in relation to thermoregulation. *Evolution* **22**, 437–458.

Watt, W.B. (1977) Adaptation at specific loci. I. Natural selection on phosphoglucose isomerase of *Colias* butterflies: biochemical and population aspects. *Genetics* **87**, 177–194.

Watt, W.B. (1983) Adaptation at specific loci. II. Demographic and biochemical elements in the maintenance of the *Colias* PGI polymorphism. *Genetics* **103**, 691–724.

Watt, W.B. (1991) Biochemistry, physiological ecology, and population genetics—the mechanistic tools of evolutionary biology. *Functional Ecology* **5**, 145–154.

Watt, W.B. (1992) Eggs, enzymes and evolution: natural genetic variants change insect fecundity. *Proceedings of the National Academy of Sciences of the USA* **89**, 10608–10612.

Watt, W.B. (1997) Accuracy, anecdotes, and artifacts in the study of insect thermal ecology. *Oikos* **80**, 399–400.

Watt, W.B., Cassin, R.C., and Swan, M.S. (1983) Adaptation at specific loci. III. Field behavior and survivorship differences among Colias PGI genotypes are predictable from *in vitro* biochemistry. *Genetics* **103**, 725–739.

Weber, R.E. and Vinogradov, S.N. (2001) Nonvertebrate hemoglobins: functions and molecular adaptations. *Physiological Reviews* **81**, 569–628.

Weeks, A.R., McKechnie, S.W., and Hoffmann, A.A. (2002) Dissecting adaptive clinal variation: markers, inversions and size/stress associations in *Drosophila melanogaster* from a central field population. *Ecology Letters* **5**, 756–763.

Wehner, R., Marsh, A.C., and Wehner, S. (1992) Desert ants on a thermal tightrope. *Nature* **357**, 586–587.

Weis-Fogh, T. (1967a) Respiration and tracheal ventilation in locusts and other flying insects. *Journal of Experimental Biology* **47**, 561–587.

Weis-Fogh, T. (1967b) Metabolism and weight economy in migrating animals, particularly birds and insects. In *Insects and Physiology* (eds. J.W.L. Beament and J.E. Treherne), 143–159. Oliver and Boyd, Edinburgh and London.

Weiss, S.L., Lee, E.A., and Diamond, J. (1998) Evolutionary matches of enzyme and transporter capacities to dietary substrate loads in the intestinal brush border. *Proceedings of the National Academy of Sciences of the USA* **95**, 2117–2121.

Welte, M.A., Tetrault, J.M., Dellavalle, R.P., and Lindquist, S.L. (1993) A new method for manipulating transgenes: engineering heat tolerance in a complex, multicellular organism. *Current Biology* **3**, 842–853.

West, G.B., Brown, J.H., and Enquist, B.J. (1997) A general model for the origin of allometric scaling laws in biology. *Science* **276**, 122–126.

West, G.B., Brown, J.H., and Enquist, B.J. (1999) The fourth dimension of life: fractal geometry and allometric scaling of organisms. *Science* **284**, 1677–1679.

West, G.B., Savage, V.M., Gillooly, J., Enquist, B.J., Woodruff, W.H., and Brown, J.H. (2003) Why does metabolic rate scale with size? *Nature* **421**, 713.

West, G.B., Woodruff, W.H., and Brown, J.H. (2002) Allometric scaling of metabolic rate from molecules and mitochondria to cells and mammals. *Proceedings of the National Academy of Sciences of the USA* **99**, 2473–2478.

Westneat, M.W., Betz, O., Blob, R.W., Fezzaa, K., Cooper, W.J., and Lee, W.-K. (2003) Tracheal respiration in insects visualized with synchrotron x-ray imaging. *Science* **299**, 558–560.

Wharton, G.W. (1985) Water balance of insects. In *Comprehensive Insect Physiology, Biochemistry and Pharmacology* (eds. G.A. Kerkut and L.I. Gilbert), Vol. 4, 565–601. Pergamon Press, Oxford.

Wharton, G.W. and Richards, A.G. (1978) Water vapor exchange kinetics in insects and acarines. *Annual Review of Entomology* **23**, 309–328.

White, E.B., DeBach, P., and Garber, M.J. (1970) Artificial selection for genetic adaptation to temperature extremes in *Aphytis lingnanensis* Compere (Hymenoptera: Aphelinidae). *Hilgardia* **40**, 161–192.

White, T.C.R. (1974) A hypothesis to explain outbreaks of looper caterpillars, with special reference to populations of *Selidosema suavis* in a plantation of *Pinus radiata* in New Zealand. *Oecologia* **16**, 279–301.

Whiting, M.F., Bradler, S., and Maxwell, T. (2003) Loss and recovery of wings in stick insects. *Nature* **421**, 264–267.

Whitman, D.W. (1987) Thermoregulation and daily activity patterns in a black desert grasshopper, *Taeniopoda eques*. *Animal Behaviour* **35**, 1814–1826.

Whitmore, A.V. and Bignell, D.E. (1990) Drinking behaviour in the cockroach *Periplaneta americana*: an automated method for the measurement of periodicity and uptake. *Journal of Insect Physiology* **36**, 103–109.

Wieczorek, H., Brown, D., Grinstein, S., Ehrenfeld, J., and Harvey, W.R. (1999) Animal plasma membrane energization by proton-motive V-ATPases. *BioEssays* **21**, 637–648.

Wieczorek, H., Putzenlechner, M., Zeiske, W., and Klein, U. (1991) A vacuolar-type proton pump energizes K^+/H^+ antiport in an animal plasma membrane. *Journal of Biological Chemistry* **266**, 15340–15347.

Wiehart, U.I.M., Nicolson, S.W., Eigenheer, R.A., and Schooley, D.A. (2002) Antagonistic control of fluid secretion by the Malpighian tubules of *Tenebrio molitor*: effects of diuretic and antidiuretic peptides and their second messengers. *Journal of Experimental Biology* **205**, 493–501.

Wieser, W. (1994) Costs of growth in cells and organisms: general rules and comparative aspects. *Biological Reviews* **68**, 1–33.

Wigglesworth, V.B. (1935) The regulation of respiration in the flea, *Xenopsylla cheopsis*, Roths. (Pulicidae). *Proceedings of the Royal Society of London B* **118**, 397–419.

Wightman, J.A. and Rogers, V.M. (1978) Growth, energy and nitrogen budgets and efficiencies of the growing larvae of *Megachile pacifica* (Panzer) (Hymenoptera: Megachilidae). *Oecologia* **36**, 245–257.

Wiklund, C., Nylin, S., and Forsberg, J. (1991) Sex-related variation in growth rate as a result of selection for large size and protandry in a bivoltine butterfly, *Pieris napi*. *Oikos* **60**, 241–250.

Wilkins, M.B. (1960) A temperature-dependent endogenous rhythm in the rate of carbon dioxide output of *Periplaneta americana*. *Nature* **185**, 481–482.

Wilkinson, T.L., Ashford, D.A., Pritchard, J., and Douglas, A.E. (1997) Honeydew sugars and osmoregulation in the pea aphid *Acyrthosiphon pisum*. *Journal of Experimental Biology* **200**, 2137–2143.

Williams, A.E. and Bradley, T.J. (1998) The effect of respiratory pattern on water loss in desiccation-resistant *Drosophila melanogaster*. *Journal of Experimental Biology* **201**, 2953–2959.

Williams, A.E., Rose, M.R., and Bradley, T.J. (1997) CO_2 release patterns in *Drosophila melanogaster*: the effect of selection for desiccation resistance. *Journal of Experimental Biology* **200**, 615–624.

Williams, J.B., Shorthouse, J.D., and Lee, R.E. (2002) Extreme resistance to desiccation and microclimate-related differences in cold-hardiness of gall wasps (Hymenoptera: Cynipidae) overwintering on roses in southern Canada. *Journal of Experimental Biology* **205**, 2115–2124.

Williams, P.H. (1994) Phylogenetic relationships among bumble bees (*Bombus* Latr.): a reappraisal of morphological evidence. *Systematic Entomology* **19**, 327–344.

Williamson, M. and Gaston, K.J. (1999) A simple transformation for sets of range sizes. *Ecography* **22**, 674–680.

Willmer, P. (1986) Microclimatic effects on insects at the plant surface. In *Insects and the Plant Surface* (eds. B. Juniper and T.R.E. Southwood), 65–80. Edward Arnold, London.

Willmer, P. (1991*a*) Thermal biology and mate acquisition in ectotherms. *Trends in Ecology and Evolution* **6**, 396–399.

Willmer, P. and Stone, G. (1997) Temperature and water relations in desert bees. *Journal of Thermal Biology* **22**, 453–465.

Willmer, P., Stone, G., and Johnston, I.A. (2000) *Environmental Physiology of Animals*. Blackwell Science, Oxford.

Willmer, P.G. (1980) The effects of a fluctuating environment on the water relations of larval Lepidoptera. *Ecological Entomology* **5**, 271–292.

Willmer, P.G. (1982) Microclimate and the environmental physiology of insects. *Advances in Insect Physiology* **16**, 1–57.

Willmer, P.G. (1991*b*) Constraints on foraging by solitary bees. In *The Behaviour and Physiology of Bees* (eds. L.J. Goodman and R.C. Fisher), 131–148. CAB International, Oxford.

Willmer, P.G., Hughes, J.P., Woodford, J.A.T., and Gordon, S.C. (1996) The effects of crop microclimate and associated physiological constraints on the seasonal and diurnal distribution patterns of raspberry beetle (*Byturus tomentosus*) on the host plant *Rubus idaeus*. *Ecological Entomology* **21**, 87–97.

Willmer, P.G. and Unwin, D.M. (1981) Field analysis of insect heat budgets: reflectance, size and heating rates. *Oecologia* **50**, 250–255.

Willott, S.J. (1997) Thermoregulation in four species of British grasshoppers (Orthoptera: Acrididae). *Functional Ecology* **11**, 705–713.

Wilson, R.S. and Franklin, C.E. (2002) Testing the beneficial acclimation hypothesis. *Trends in Ecology and Evolution* **17**, 66–70.

Wingrove, J.A. and O'Farrell, P.H. (1999) Nitric oxide contributes to behavioral, cellular, and developmental responses to low oxygen in *Drosophila*. *Cell* **98**, 105–114.

Wittmann, D. and Scholz, E. (1989) Nectar dehydration by male carpenter bees as preparation for mating flights. *Behavioural Ecology and Sociobiology* **25**, 387–391.

Wolda, H. (1988) Insect seasonality: why? *Annual Review of Ecology and Systematics* **19**, 1–18.

Wolf, T.J., Ellington, C.P., Davis, S., and Feltham, M.J. (1996) Validation of the doubly labelled water technique for bumblebees *Bombus terrestris* (L.). *Journal of Experimental Biology* **199**, 959–972.

Wolf, T.J., Schmid-Hempel, P., Ellington, C.P., and Stevenson, R.D. (1989) Physiological correlates of foraging efforts in honey-bees: oxygen consumption and nectar load. *Functional Ecology* **3**, 417–424.

Wolfe, G.R., Hendrix, D.L., and Salvucci, M.E. (1998) A thermoprotective role for sorbitol in the silverleaf whitefly, *Bemisia argentifolii*. *Journal of Insect Physiology* **44**, 597–603.

Wolfersberger, M.G. (2000) Amino acid transport in insects. *Annual Review of Entomology* **45**, 111–120.

Woods, H.A. (1999) Patterns and mechanisms of growth of fifth-instar *Manduca sexta* caterpillars following exposure to low- or high-protein food during early intars. *Physiological and Biochemical Zoology* **72**, 445–454.

Woods, H.A. and Bernays, E.A. (2000) Water homeostasis by wild larvae of *Manduca sexta*. *Physiological Entomology* **25**, 82–87.

Woods, H.A. and Chamberlin, M.E. (1999) Effects of dietary protein concentration on L-proline transport by *Manduca sexta* midgut. *Journal of Insect Physiology* **45**, 735–741.

Woods, H.A., Fagan, W.F., Elser, J.J., and Harrison, J.F. (2004) Allometric and phylogenetic variation in insect phosphorus content. *Functional Ecology* **18**, 103–109.

Woods, H.A. and Harrison, J.F. (2001) The beneficial acclimation hypothesis versus acclimation of specific traits: physiological change in water-stressed *Manduca sexta* caterpillars. *Physiological and Biochemical Zoology* **74**, 32–44.

Woods, H.A. and Harrison, J.F. (2002) Interpreting rejections of the beneficial acclimation hypothesis: when is physiological plasticity adaptive? *Evolution* **56**, 1863–1866.

Woods, H.A. and Kingsolver, J.G. (1999) Feeding rate and the structure of protein digestion and absorption in lepidopteran midguts. *Archives of Insect Biochemistry and Physiology* **42**, 74–87.

Woods, H.A., Makino, W., Cotner, J.B., Hobbie, S.E., Harrison, J.F., Acharya, K. *et al.* (2003) Temperature and the chemical composition of poikilothermic organisms. *Functional Ecology* **17**, 237–245.

Woods, H.A. and Singer, M.S. (2001) Contrasting responses to desiccation and starvation by eggs and neonates of two Lepidoptera. *Physiological and Biochemical Zoology* **74**, 594–606.

Woods, W.A. and Stevenson, R.D. (1996) Time and energy costs of copulation for the sphinx moth, *Manduca sexta*. *Physiological Zoology* **69**, 682–700.

Worland, M.R. and Convey, P. (2001) Rapid cold hardening in Antarctic microarthropods. *Functional Ecology* **15**, 515–524.

Worland, M.R., Grubor-Lajsic, G., and Montiel, P.O. (1998) Partial desiccation induced by sub-zero temperatures as a component of the survival strategy of the Arctic collembolan *Onychiurus arcticus* (Tullberg). *Journal of Insect Physiology* **44**, 211–219.

Worland, R., Block, W., and Rothery, P. (1992) Survival of sub-zero temperatures by two South Georgian beetles (Coleoptera, Perimylopidae). *Polar Biology* **11**, 607–613.

Worthen, W.B. and Haney, D.C. (1999) Temperature tolerance in three mycophagous *Drosophila* species: relationships with community structure. *Oikos* **86**, 113–118.

Worthen, W.B., Jones, M.T., and Jetton, R.M. (1998) Community structure and environmental stress: desiccation promotes nestedness in mycophagous fly communities. *Oikos* **81**, 45–54.

Wright, J.C. and Machin, J. (1993) Atmospheric water absorption and the water budget of terrestrial isopods (Crustacea, Isopoda, Oniscidea). *Biological Bulletin* **184**, 243–253.

Wu, B.S., Lee, J.K., Thompson, K.M., Walker, V.K., Moyes, C.D., and Robertson, R.M. (2002) Anoxia induces thermotolerance in the locust flight system. *Journal of Experimental Biology* **205**, 815–827.

Wyatt, G.R. (1967) The biochemistry of sugars and polysaccharides in insects. *Advances in Insect Physiology* **4**, 287–360.

Yang, Y. and Joern, A. (1994) Gut size changes in relation to variable food quality and body size in grasshoppers. *Functional Ecology* **8**, 36–45.

Yanoviak, S.P. and Kaspari, M. (2000) Community structure and the habitat templet: ants in the tropical forest canopy and litter. *Oikos* **89**, 259–266.

Yarro, J.G. (1985) Effect of host plant on moulting in the African armyworm *Spodoptera exempta* (Walk.) (Lepidoptera: Noctuidae) at constant temperature and humidity conditions. *Insect Science and its Application* **6**, 171–175.

Yocum, G.D. (2001) Differential expression of two *HSP70* transcripts in response to cold shock, thermoperiod, and adult diapause in the Colorado potato beetle. *Journal of Insect Physiology* **47**, 1139–1145.

Yocum, G.D. and Denlinger, D.L. (1992) Prolonged thermotolerance in the flesh fly, *Sarcophaga crassipalpis*, does not require continuous expression or persistence of the 72 kDa heat-shock protein. *Journal of Insect Physiology* **38**, 603–609.

Yocum, G.D. and Denlinger, D.L. (1994) Anoxia blocks thermotolerance and the induction of rapid cold hardening in the flesh fly, *Sarcophaga crassipalpis*. *Physiological Entomology* **19**, 152–158.

Yocum, G.D., Joplin, K.H., and Denlinger, D.L. (1991) Expression of heat shock proteins in response to high and low temperatures in diapausing pharate larvae of the gypsy moth, *Lymantria dispar*. *Archives of Insect Biochemistry and Physiology* **18**, 239–249.

Yocum, G.D., Joplin, K.H., and Denlinger, D.L. (1998) Upregulation of a 23 kDa small heat shock protein transcript during pupal diapause in the flesh fly *Sarcophaga crassipalpis*. *Insect Biochemistry and Molecular Biology* **28**, 677–682.

Yocum, G.D., Ždárek, J., Joplin, K.H., Lee, R.E., Smith, D.C., Manter, K.D. *et al.* (1994) Alteration of the eclosion rhythm and eclosion behavior in the flesh fly,

Sarcophaga crassipalpis, by low and high temperature stress. *Journal of Insect Physiology* **40**, 13–21.

Yoder, J.A. and Denlinger, D.L. (1991) Water balance in flesh fly pupae and water vapor absorption associated with diapause. *Journal of Experimental Biology* **157**, 273–286.

Yoder, J.A. and Denlinger, D.L. (1992) Water vapour uptake by diapausing eggs of a tropical walking stick. *Physiological Entomology* **17**, 97–103.

Yoder, J.A. and Grojean, N.C. (1997) Group influence on water conservation in the giant Madagascar hissing-cockroach, *Gromphadorhina portentosa* (Dictyoptera: Blaberidae). *Physiological Entomology* **22**, 79–82.

Yoder, J.A. and Smith, B.E. (1997) Enhanced water conservation in clusters of convergent lady beetles, *Hippodamia convergens*. *Entomologia Experimentalis et Applicata* **85**, 87–89.

Yu, M.-J. and Beyenbach, K.W. (2001) Leucokinin and the modulation of the shunt pathway in Malpighian tubules. *Journal of Insect Physiology* **47**, 263–276.

Zachariassen, K.E. (1985) Physiology of cold tolerance in insects. *Physiological Reviews* **65**, 799–837.

Zachariassen, K.E. (1991*a*) Routes of transpiratory water loss in a dry-habitat tenebrionid beetle. *Journal of Experimental Biology* **157**, 425–437.

Zachariassen, K.E. (1991*b*) The water relations of over-wintering insects. In *Insects at Low Temperatures* (eds. R.E. Lee and D.L. Denlinger), 47–63. Chapman and Hall, New York.

Zachariassen, K.E. (1996) The water conserving physiological compromise of desert insects. *European Journal of Entomology* **93**, 359–367.

Zachariassen, K.E., Andersen, J., Maloiy, G.M.O., and Kamau, J.M.Z. (1987) Transpiratory water loss and metabolism of beetles from arid areas in East Africa. *Comparative Biochemistry and Physiology A* **86**, 403–408.

Zangerl, A.R., McKenna, D., Wraight, C.L., Carroll, M., Ficarello, P., Warner, R. *et al.* (2001) Effects of exposure to event 176 *Bacillus thuringiensis* corn pollen on monarch and black swallowtail caterpillars under field conditions. *Proceedings of the National Academy of Sciences of the USA* **98**, 11908–11912.

Zanotto, F.P., Gouveia, S.M., Simpson, S.J., Raubenheimer, D. and Calder, P.C. (1997) Nutritional homeostasis in locusts: is there a mechanism for increased energy expenditure during carbohydrate overfeeding? *Journal of Experimental Biology* **200**, 2437–2448.

Zanotto, F.P., Simpson, S.J., and Raubeheimer, D. (1993) The regulation of growth by locusts through post-ingestive compensation for variation in the levels of dietary protein and carbohydrate. *Physiological Entomology* **18**, 425–434.

Zatsepina, O.G., Velikodvorskaia, V.V., Molodtsov, V.B., Garbuz, D., Lerman, D.N., Bettencourt, B.R. *et al.* (2001) A *Drosophila melanogaster* strain from sub-equatorial Africa has exceptional thermotolerance but decreased Hsp70 expression. *Journal of Experimental Biology* **204**, 1869–1881.

Zera, A.J. and Brink, T. (2000) Nutrient absorption and utilization by wing and flight muscle morphs of the cricket *Gryllus firmus*: implications for the trade-off between flight capability and early reproduction. *Journal of Insect Physiology* **46**, 1207–1218.

Zera, A.J. and Denno, R.F. (1997) Physiology and ecology of dispersal polymorphism in insects. *Annual Review of Entomology* **42**, 207–231.

Zera, A.J. and Harshman, L.G. (2001) The physiology of life history trade-offs in animals. *Annual Review of Ecology and Systematics* **32**, 95–126.

Zera, A.J. and Larsen, A. (2001) The metabolic basis of life history variation: genetic and phenotypic differences in lipid reserves among life history morphs of the wing-polymorphic cricket, *Gryllus firmus*. *Journal of Insect Physiology* **47**, 1147–1160.

Zera, A.J., Sall, J., and Grudzinski, K. (1997) Flight-muscle polymorphism in the cricket *Gryllus firmus*: muscle characteristics and their influence on the evolution of flightlessness. *Physiological Zoology* **70**, 519–529.

Zhang, S.-L., Leyssens, A., Van Kerkhove, E., Weltens, R., Van Driessche, W., and Steels, P. (1994) Electrophysiological evidence for the presence of an apical H^+-ATPase in Malpighian tubules of *Formica polyctena*: intracellular and luminal pH measurements. *Pflugers Archives* **426**, 288–295.

Zhou, S., Criddle, R.S., and Mitcham, E.J. (2001) Metabolic response of *Platynota stultana* pupae during and after extended exposure to elevated CO_2 and reduced O_2 atmospheres. *Journal of Insect Physiology* **47**, 401–409.

Ziegler, R. (1985) Metabolic energy expenditure and its hormonal regulation. In *Environmental Physiology and Biochemistry of Insects* (ed. K.H. Hoffmann), 95–118. Springer, Berlin.

Ziman, J. (1978) *Reliable Knowledge. An Exploration of the Grounds for Belief in Science*. Cambridge University Press, Cambridge.

Zudaire, E., Simpson, S.J., and Montuenga, L.M. (1998) Effects of food nutrient content, insect age and stage in the feeding cycle on the FMRFamide immunoreactivity of diffuse endocrine cells in the locust gut. *Journal of Experimental Biology* **201**, 2971–2979.

Index

Printed in the United States
By Bookmasters

Printed in the United States
By Bookmasters